Bio-based Polymers and Nanocomposites

Muhammed Lamin Sanyang ·
Mohammad Jawaid
Editors

Bio-based Polymers and Nanocomposites

Preparation, Processing, Properties & Performance

Springer

Editors
Muhammed Lamin Sanyang
University of The Gambia
Serrekunda, Gambia

Mohammad Jawaid
Laboratory of Biocomposite Technology
Universiti Putra Malaysia
Serdang, Selangor, Malaysia

ISBN 978-3-030-05824-1 ISBN 978-3-030-05825-8 (eBook)
https://doi.org/10.1007/978-3-030-05825-8

Library of Congress Control Number: 2018964241

This Springer imprint is published by the registered company Springer Nature Switzerland AG
The registered company address is: Gewerbestrasse 11, 6330 Cham, Switzerland

Editors are honoured to dedicate this book to Colleagues who motivated and inspired:

Dr. Othman Alothman, Saudi Arabia
Dr. Abou el kacem Qaiss, Morocco
Dr. Anish Khan, Saudi Arabia
Dr. Ir. Mohammed Thariq, Malaysia
Dr. Ali Kilic, Turkey
Prof. Sarat Kumar Swain, India
Prof. Sami Boufi, Tunisia
Dr. Mohamad Midani, Egypt
Dr. Zoheb Karim, Sweden
Dr. Mohammd Mansoob Khan, Brunei
Dr. Aamir Hussain Bhat, Oman
Prof. Rajini Nagarajan, India
Prof. Enamul Hoque, Bangladesh
Prof. Rafaqat Hussain, Pakistan
Prof. Suchart, Thailand
Dr. Rob Elias, United Kingdom

Preface

The ongoing environmental devastation and the rapid depletion of petroleum resources coupled with the constant demand from customers for high-performance lightweight materials triggered the interest to utilize new eco-friendly material from renewable resources. Of late, bio-based materials have emerged in the plastic and composite market and are becoming more commercialized. Hence, this book supplies an overview and extensive knowledge of the latest advanced developments in bio-based polymer and nanocomposites. The book highlights the various types of polymer and nanocomposites from different biorenewable resources. Specifically, it covers the different aspects of bio-based polymer and nanocomposite preparation, processing, properties and performance. This book gives a sound knowledge of modern research application of biopolymers and nanocomposites for diverse applications with standard business values.

The chapters in the book presents an up-to-date information on the subject area of bio-based polymer and nanocomposites by prominent researchers in academia and industry as well as government/private research laboratories across the world. The book can serve as a holistic reference source for suitable for professionals, students and researchers from different disciplines such as material science/engineering, polymer science, composite technology, nanotechnology and biotechnology, working on bio-based polymer and nanocomposites.

Present book will cover the Renewable Resource-based Polymers, Fillers and reinforcements for advanced nanocomposites, Cellulose Nanocrystals and Nanofibrillated cellulose-based nanocomposites, Bacterial Cellulose Nanocomposites, Properties of nanocomposites based on different clays and polyamide 6/acrylonitrile–butadiene–styrene blends, Nanotube/Biopolymer Nanocomposites, Starch-based Nanocomposites: Types and Industrial Applications, Recent Developments in Chitosan-Based Nanocomposites, Structural properties of protein and their role in Polymer Nanocomposites, Polylactic acid (PLA)-Based Nanocomposites and Biopolymers-Based Nanocomposites: Properties and Applications from leading experts from Canada, Sweden, the Netherlands, Malaysia, India, Thailand, Bangladesh, Saudi Arabia, Morocco, Nigeria and United Kingdom.

I am highly grateful to all authors who contributed to this book and made idea it to reality. In addition, I am also thankful to Springer Nature support team, especially Janet Sterritt-Brunner, Anthony Doyle and Maniarasan for technical support and advice during this project.

Serdang, Malaysia Mohammad Jawaid

Contents

About the Editors

Muhammed Lamin Sanyang is currently working as an Environmental Specialist at the Gambia National Petroleum Corporation (GNPC) as well as a Part-time Lecturer at the University of The Gambia (UTG). He completed his postdoctoral fellowship under the Laboratory of Biocomposite Technology at Universiti Putra Malaysia. In 2011, he was awarded the Islamic Development Bank (IDB) Masters scholarship to pursue his master's degree (M.Sc.) in Environmental Engineering at Universiti Putra Malaysia (UPM). Upon completing his M.Sc. programme in 2013, he was again awarded the Commonwealth Scholarship and Fellowship Plan (CSFP) by the Ministry of Higher Education Malaysia to undertake a Ph.D. degree in the field of Green Engineering at the Institute of Advance Technology (ITMA), UPM. He received the "Best Ph.D. student award" for his outstanding researches, publications and completion on time. His main research interests are: Green Engineering, Environmental Engineering, Environmental Impact Assessment (EIA), Environmental Management, and Health Safety and Environment in Oil and Gas (Upstream and Downstream). Dr. Sanyang has published more than 15 journal articles in international journals, authored or co-authored over 8 chapters and 1 edited book. He also served as reviewer for many international journals such as *Polymer Composite, BioResources, Polymers (MPDI), Polymer Bulletin, International Journal of Polymer Science,* and *Current Analytical Chemistry.*

Dr. Mohammad Jawaid is currently working as Fellow Researcher (Associate Professor) at Biocomposite Technology Laboratory, Institute of Tropical Forestry and Forest Products (INTROP), Universiti Putra Malaysia (UPM), Serdang, Selangor, Malaysia, and also Visiting Professor at the Department of Chemical Engineering, College of Engineering, King Saud University, Riyadh, Saudi Arabia, since June 2013. He is also Visiting Scientist to TEMAG Laboratory, Faculty of Textile Technologies and Design at Istanbul Technical University, Turkey. He has more than 14 years of experience in teaching, research and industries. His area of research interests includes hybrid-reinforced/hybrid-filled polymer composites, advance materials: graphene/nanoclay/fire-retardant, lignocellulosic-reinforced/ filled polymer composites, modification and treatment of lignocellulosic fibres and solid wood, biopolymers for packaging applications, nanocomposites, nanocellulose, and polymer blends. He has published 23 books, 52 chapters, and more than 250 peer-reviewed international journal papers and five published review papers under top 25 hot articles in science direct during 2013–2017. He worked as Guest Editor of special issue, *Current Organic Synthesis & Current Analytical Chemistry*, Bentham Publishers, UK; *International Journal of Polymer Science*-Hindawi Publishing; Inderscience Enterprises Ltd.; IOP Conference Proceeding; and is Editorial Board Member of Journal of Polymers, and the Environment, *Journal of Asian Science Technology & Innovation*; and *Recent Innovations in Chemical Engineering Journal*. Besides that, he is also reviewer of several high-impact international peer-reviewed journals of Elsevier, Springer, Wiley, Saga, etc. Presently, he is supervising 20 Ph.D. students (5 Ph.D. as Main Supervisor and 15 as Member of Supervisory Committee) and 4 master students (2 as Main Supervisor and 2 as Member of Supervisory Committee) in the field of hybrid composites, green composites, nanocomposites, natural fibre-reinforced composites, nanocellulose, etc. A total of 15 Ph.D. (3 Ph.D. as Main Supervisor and 12 as Member of Supervisory Committee), and 11 master students (2 as Main Supervisor and 9 as Member of Supervisory Committee) graduated under his supervision in 2014–2018. He has several research grants at university, national and international levels on polymer composites of around RM 3 million (USD 700,000). He also delivered plenary and invited talk in international conference related to composites in India, Egypt, Turkey, Malaysia, Thailand, UK, France, Saudi Arabia and China. Besides that, he is also member of technical committee of several national and international conferences on composites and material science. His H-index, 42 (Google Scholar); 36 (Scopus).

Renewable Resource-Based Polymers

Ahmed Sharif and Md Enamul Hoque

Abstract There are growing concerns that a significant reliance on fossil resources is not sustainable for the production of polymers. The all-out transition toward renewable resources for polymer production is inevitable. Polymers from renewable resources include cellulose, starch, chitosan, lignin, proteins, oils, common commodity polymers (e.g., polyethylene, polyethylene terephthalate), and microbial poly(ester)s. Fundamental research in the production, modification, property enhancement, and new applications of these materials is an important undertaking. In this chapter, existing advances in the exploitation of renewable resources to produce polymers are summarized.

Keywords Renewable · Polymer · Fossil · Cellulose · Starch · Chitosan
Lignin · Proteins

1 Introduction

There are growing concerns over the sustainability of materials and chemicals from petrochemical-based resources for the threat to welfare, health, and the ecosystem (Kanagaratnam et al. 2013; Mac Kinnon et al. 2017; Koutinas et al. 2014; Sahoo et al. 2018). It is an enduring need to substitute polymers from fossil fuel-based resources with the more environmental congenial renewable resources (Gandini and Belgacem 2013). Contemporary progresses in the production and properties of polymers from renewable resources with better characteristics and possibly small

A. Sharif
Department of Materials and Metallurgical Engineering, Bangladesh University of Engineering and Technology (BUET), Dhaka 1000, Bangladesh

M. E. Hoque (✉)
Department of Biomedical Engineering, Military Institute of Science and Technology (MIST), Mirpur Cantonment, Dhaka 1216, Bangladesh
e-mail: enamul1973@gmail.com

M. L. Sanyang and M. Jawaid (eds.), *Bio-based Polymers and Nanocomposites*,
https://doi.org/10.1007/978-3-030-05825-8_1

expenses have stimulated the optimisms of replacing the traditional petrochemical-based polymers (Kumar et al. 2017; Saikia and Karak 2017; Nilsson et al. 2015). With the latest innovations in production technologies and the invention of new exploitable chemical building blocks, investigation in ecological polymers from renewable resources has increased dramatically (Lindblad et al. 2002).

The term "renewable" has become a major attention among the industries and citizens. The used feedstock may be termed as renewable when it is mobilized from resources which are biologically refilled on a human timescale; on the other hand, petrochemical resources take thousands of centuries to be built naturally (Gandini and Lacerda 2015), provided that new harvest balances cultivating bio-based feedstock can be called renewable. For instance, peat is not judged renewable due to sluggish restoration rate, and on the other hand, tropical hardwood can be termed as renewable when appropriately governed. Sustainability is becoming immensely important leading to create an attention in the use of renewable resource in polymer application (Tsanaktsis et al. 2015). Renewable resource-based polymer is a new era, which is entirely based on natural resources, significantly creating a paradigm shift in the properties of polymer-based materials (Wu 2015; Yan and Chen 2015; Das et al. 2015; Han et al. 2016; Zia et al. 2016; Lligadas et al. 2014). The environment is protected due to reduction in emission of harmful gasses by replacement of the fossil fuel and gas by green agricultural resources. Thus, the objective of this chapter is to present the contemporary research achievements by emphasizing the wide possibility of the renewable polymers for a huge number of functions.

Historically, the attainment of fossil fuels depended on cost-effective extensive manufacturing (Bruijnincx et al. 2015). The cheap refining competences offered an economical supply of building block compounds in the petroleum-based industry (Imre and Pukánszky 2015). Similarly, improvements in the development of renewable feedstocks (e.g., cellulosic) for biofuels should facilitate the production of low-cost elementary units for beneficial chemical and polymeric developments (Thakur et al. 2014).

Another aspect is that approximately 1.3 billion tons of waste are being generated all over the world each year and is expected to escalate to 4 billion tons by 2100 (Zhang 2016). Disposing of waste jeopardizes the people life and environment. Incineration of waste like plastic has tremendous impact on the environment, as burning of plastic tends to produce dioxin-like noxious substances (Baeumler et al. 2012). Hence, throwing away of waste resources and materials denotes ruining of huge amount of money (Vestapen et al. 2017). In case of traditional petroleum-based plastics, durability, which makes it a viable candidate, renders it vulnerable after disposal. Having immense resistance to microbial degradation, these materials are not biodegradable and substantially accumulate above the ground creating waste (Botello-Álvarez et al. 2018). The polymers from agro-based resources could conquer the restrictions of the petrochemical resources which are widely used.

2 Renewable Polymers Classifications

A wide spectrum of materials from renewable resources is already potentially available. Most importantly these resources are sustainable to provide a wealthy variety of building blocks and polymers that are presently offered from fossil fuel-based industry. Based on the source and synthesis technology, renewable polymers are classified into four major groups. Figure 1 represents a classification of renewable polymer which is based on synthesis source.

(i) Polymers from vegetable resources, for example, agro-polymers (cellulose or starch)
(ii) Polymers obtained from animal resources, for example, chitin and chitosan
(iii) Polymers obtained from microbial action, for example, the polyhydroxyalkanoates (PHAs)
(iv) Polymers from monomers which is chemically and conventionally synthesized (obtained from agro-resources), for example, polylactic acid (PLA), biopolyethylene (PE), biopolyethylene terephthalate (PET).

3 Some Polymers from Renewable Sources

An assessment of the contemporary polymer field dealing with renewable resources reveals an enormous global activity in both academic and industrial arena. Some renewable resource-based polymers are described on the basis of their preparation, processing, properties, and performance below.

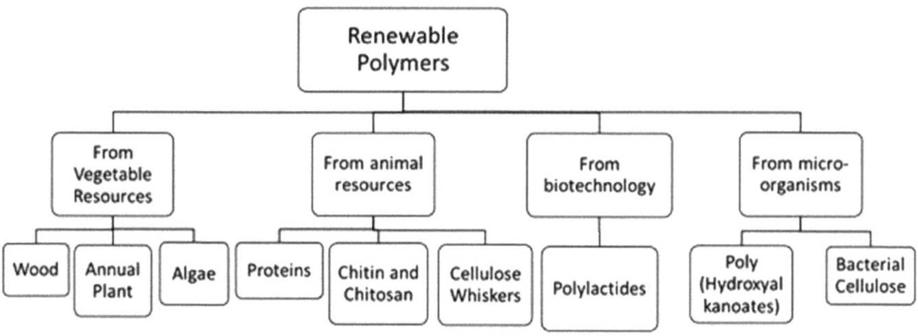

Fig. 1 Classification of renewable polymers

3.1 Cellulose

Cellulose represents, by far, the most appealing candidate in terms of availability. Cellulose is the most abundant natural resource on earth since it represents the chief constituent of plant cell walls (French et al. 1993; Ketabchi et al. 2015, 2016; Hoque et al. 2016). For example, cellulose nanoparticles (CNPs) can abundantly be isolated from Kenaf fiber (Ketabchi et al. 2016). As the most plentiful biopolymer on earth, with some 1012 tons, cellulose is readily renewed. It also possesses unique properties and chemical reactivity. This polymer constituted of D-glucose residues as monomeric units (Fig. 2) (Chiellini et al. 2002a, b). Cellulose molecules are unbranched chains of up to 17,000 1,4 linked ß-D-glucose residues, but shorter chains occur under other circumstances. The two end groups of cellulose are not chemically matching, since one exhibits the "normal" C4–OH group (non-reducing end), while the other has a C1–OH component in equilibrium with the conforming aldehyde function (reducing end) (Gandini 2011a, b).

Cellulose is the dominant integral of woods which are used in an extensive variety of basic materials for buildings and furniture. Clean cellulose is usually branched out from plants exploiting various pulping procedures. Fibers from pulp are exploited in printing papers, industrial papers, household essentials such as toilet papers, paper towels, handkerchiefs, facial tissues, napkins, and so on.

This structure is responsible for the exclusive characteristics of cellulose. Because of the equatorial positions of the hydroxyls on the cellulose chain, they protrude laterally along the extended molecule and are readily available for hydrogen bonding. These hydrogen bonds cause the chains to group together in a highly ordered structure. The interchain hydrogen bonds in the crystalline regions are strong, giving the resultant structure good strength, corresponding high surface energy, high hydrophilic character, and insolubility in most solvents. They also prevent cellulose from melting (nonthermoplastic). Of course, the marked reactivity of cellulose is associated with the three OH groups present in each anhydroglucose units (AGU) (Petersen et al. 1999).

Most celluloses have a high degree of polymerization both as a function of the species and of the isolation and purification procedures. The latter operations often introduce other functionalities, like carbonyl and carboxyl groups, into the macromolecular backbone (Gandini 2011a, b). The glycosidic bonds in cellulose are strong, and this polymer is stable under a wide variety of reaction conditions. The purification step is also known to provide specific features to cellulose, including surface charge and additional chemical reactivity (Petersen et al. 1999).

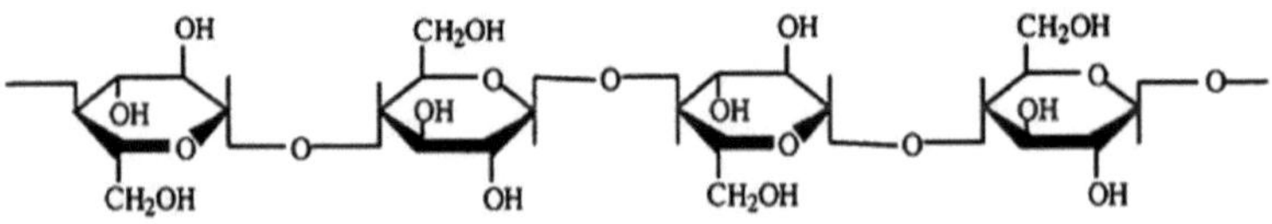

Fig. 2 Chemical structure of cellulose (Chiellini et al. 2002a)

Modified cellulose was the first polymer developed prior to the advancement of modern synthetic plastic industry. Celluloid from nitrocellulose was the first thermoplastic material invented in the nineteenth century (Shi et al. 2011). Technically, worthy chemical modifications of cellulose polymer generally engage reaction (i.e., characteristic of alcohols) with free hydroxyl groups in 2, 3, and 6 positions (Simon et al. 1998). Etherification and esterification of individual hydroxyl groups, of the polysaccharide backbone, are of specific significance for cellulose. The alteration of chemical nature of cellulose enables its synthesis with the same environments exercised for thermoplastic polymers. Several cellulosic derivatives are industrially handy such as cellulose acetate, ethyl cellulose, hydroxyethyl cellulose, and hydroxypropyl cellulose. These derivatives have been widely exploited for the fabrication of membranes and hollow fibers capable of immobilization of enzymes and in the practice of hemodialysis reverse osmosis and chromatographic supports (Lagoa et al. 1999; Hoenich and Stamp 2000). Furthermore, hydroxyalkyl cellulose and carboxymethyl celluloses are being used as a media for drug delivery and as wound dressing (Vert 2001).

The period of nanocellulose can be acknowledged with the inauguration of the third millennium (Klemm et al. 2011; Einchhorn 2011; Einchhorn et al. 2010; Dufresne 2017; Missoum et al. 2013; Habibi 2014). Some noticeable illustrated cellulosic nanoscale substances are enthusiastically investigated because of their exceptional characteristics and innovative functions in materials science: cellulose nanocrystals (CNC) (Habibi et al. 2010), nanofibrillated cellulose (CNF) (Siró and Plackett 2010), and bacterial cellulose (BC) (Pecoraro et al. 2008). For instance, the nanocellulosic fibrils are isolated from cellulose through high-pressure, high-temperature, and high-velocity impact homogenization, grinding, or microfluidization.

Cellulose xanthate in sulfuric acid or sodium sulfate solution is used to produce regenerated cellulose by extrusion or spinning (Shi et al. 2011). Cellophane film and rayon fiber are two important regenerated celluloses. Biodegradable cellophane is a transparent film with a low permeability to air, oil, and grease. Cellophane is still exploited these days in some specialized packaging films such as candy wraps, and industrial films as a base film for tapes. Rayon is an artificial redeveloped cellulose fiber formulated from a range of cellulosic natural resources such as wood, cotton, hemp, and bamboo (Kauffman 1993). Rayon fiber is employed mostly in apparels, bedding products, disposable personal sanitary objects including feminine pads and liners, furnishings, window treatment, medical products, tire cords, lighter fillings, and so on. Cellulose can also produce nanocomposites consisting of carboxymethyl cellulose, chitosan, and hydroxyapatite (HA) that can be used for load-bearing biomedical devices (Michael et al. 2016). Jiang et al. (2008) prepared a three-dimensional scaffold consisting of HA, chitosan, and carboxymethyl cellulose where the mechanical properties were seen to increase due to enhanced dispersion of HA in the polymer matrix.

3.2 Starch

Starch is the chief form of carbohydrate packing in green plants. It is the primary constituent of seeds, tubers, and roots. Starch is manufactured commercially from corn, wheat, rice, tapioca, potato, sago, and other sources. Starch contains six-membered ring glucose units (glucopyranose). There are two major types of starch molecular structure. In one type of structure, starch molecules are highly branched, known as amylopectin. It constitutes the major part of starch (i.e., up to 100% in waxy starches, 72% in normal maize starch, and 80% in potato starch). Amylopectin molecules are composed of chains of glucose units linked by a-1-4 linkages with branches of double helical a-1-6 linkages which gives the crystallinity of starch (Chauhan et al. 2000). The average branch chain length is 20–30 glucosidic units (Zdrahala 1997). The other type of starch structure is arranged by repeating units of 1-4-a-glucose with very few branches and is known as amylose. Both amylopectin and amylose structures are shown in Fig. 3.

Apart from use in food applications, starch is utilized considerably as surface modifying agents in paper and textile industry, as adhesive in corrugated board industry. Other applications of starch include laundry, baby powders, oil production, and production of bioplastics or biodegradable polymers (Bemiller 1997; Hoque et al. 2013a). Relatively cheap polyhydroxy compounds were developed from starch for industrial applications. A suitable mixture of chemical reagents permits for the introduction of ionic charge into the starch molecules (Doane et al. 1992). Starch can be readily transformed to glucose from which a variety of cyclic and acyclic polyols, aldehydes, ketones, acids, esters, and ether can be attained with the introduction of anionic or cationic charge by adding suitable chemical reagents (Tharanathan 2005). Lately, cornstarch was utilized to produce adhesive by cross-linking with polyvinyl alcohol and hexamethoxymethylmelamine together with citric acid as catalyst (Imam et al. 1999).

Aqueous esterification of starch can be generated at low alkalinity, under controlled pH. These reactions can be controlled by regulating the temperature for a range of degree of substitution (DS = $\leq$ 0.2 to $\geq$ 2.4) (Tessler and Billmers 1996; Sagar and Merrill 1995; Mormann and Spitzer 2001). Starch ester can be produced in nonaqueous medium with the use of solvents such as pyridine. Here, pyridine is used to serve the dual purpose of catalyst and solvent (Maim et al. 1951). The reaction of

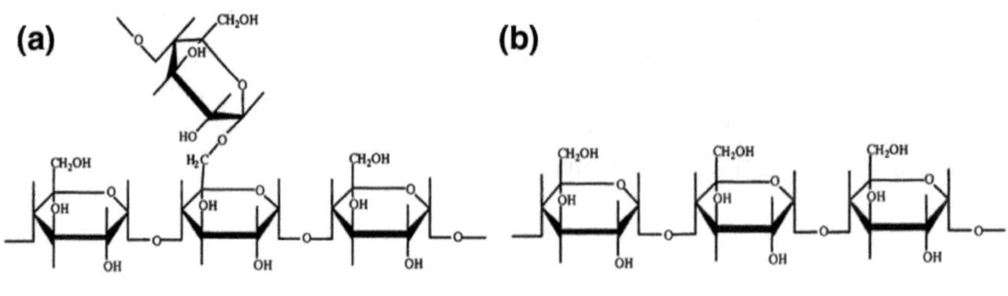

Fig. 3 Schematic representation of **a** amylopectin and **b** amylose structure (Chiellini et al. 2002a)

starches with several oxidizing reagents (such as hypochlorite, permanganate, hydrogen peroxide, persulfate, periodate, and dichromate) reduces molecular weight and also increases the solubility of starch for paper and food applications (Thoma and Stewart 1965; Kruger 1989; Vikso-Nielsen et al. 2009). These oxidized starches have found a large application in adhesive manufacturing.

A high carboxyl and carbony1 containing starch-modified water-soluble products can be developed by a reactive extrusion process with hydrogen peroxide in the presence of ferrous-cupric catalyst (Wing and Willett 1997, 1999). In graft polymerization technique, a free radical is initiated on the starch backbone and then allowed to react with polymerizable vinyl or acrylic monomers with chemical or radioactive induction (Jyothi 2010; Meshram et al. 2009). Such starch graft polymers were suggested for the application as thickeners for aqueous system, flocculants, clarification aids for wastewaters, retention aids in paper making, and many other uses (Labet et al. 2007).

3.3 Lignin

Lignin is the second most plentiful natural polymer surpassed only by cellulose in nature. Naturally, lignin acts as a cement and has a supportive structural function by establishing hydrophobic structure to defend highly hydrophilic cellulose and hemicellulose segments. Free lignin is insoluble even in strong mineral acids and hydrocarbons (Goheen and Royt 1981; Pollak 1952). Lignin in wood can be made soluble in water by changing its polymerization state with physical or chemical treatment (Calvo-Flores and Dobado 2010). Lignin exhibits a complicated three-dimensional aromatic molecule composed of phenyl groups, depending on its origin (Adler 1977). The irregular molecular mass of this biopolymer is an outcome of the random cross-linked polymerization of phenolic moieties by at least ten types of C–C and C–O–C bonds, initiating from radical-coupling reactions between phenolic radicals. It is commonly acknowledged that there are three basic phenol products that are known as monolignols: 10 p-coumaryl alcohol (M1H), coniferyl alcohol (M1G), and sinapyl alcohol (M1S) (Vanholme et al. 2008). Each monolignol produces p-hydroxyphenyl, guaiacyl, and syringyl residues in the polymer (Fig. 4).

The lignin polymer is synthesized by peroxidase-mediated dehydrogenation of monolignol units, giving a heterogeneous structure formed by basic units linked by C–C bonds and aryl–ether linkages with aryl–glycerol and β-aryl ether (dos Santos Abreu et al. 2007). In general, lignin is obtained from biomass with an organic solvent/water/reagents mixture at a high temperature and pressure. The most abundant monolignol in hardwoods is sinapyl alcohol, and this leads to syringyl S-units in the lignin polymer (Lora and Aziz 1985). Lignin is separated by acid precipitation. However, only Alcell and Organocell lignins are commercially available (Piccolo et al. 1997). A partial hydrolysis of lignin can be carried out by treating wood with steam at high temperature and pressure (i.e., about 200 °C),

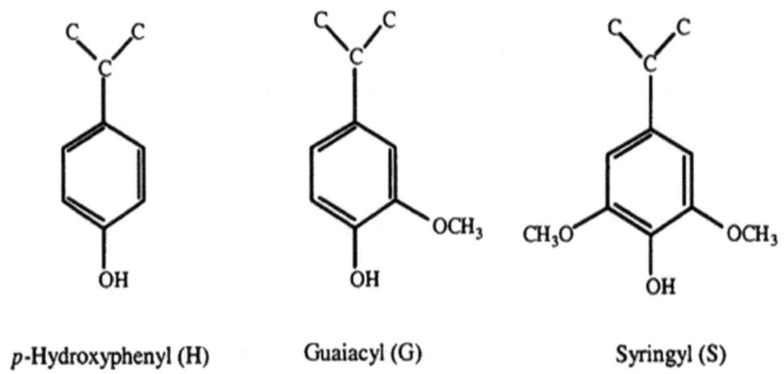

Fig. 4 Representation of typical subunits of lignin

followed by a sudden decompression in the presence of some chemicals (Focher et al. 1991). Under these circumstances, a water-insoluble lignin material with a low level of carbohydrate and wood-extractive impurities develops. Liquefaction of lignocellulosics was earlier performed by several hours of treatment at 300–400 °C in aqueous or organic solvents. Subsequently, liquefaction in organic solvents at temperatures of 240–270 °C without catalyst or at temperatures around 80–150 °C with acidic catalysts was developed with very high yield (90–95%) (Shirashi 1992). Lignin is used as dispersant, binders, humectants, emulsion stabilizers, sequestering agents, fuel source, concrete admixtures, and so on (Northey 1992).

Lignin surface and solubility characteristics can be modified with a wide range of reactions: alkylation, dealkylation, oxyalkylation, amination, carboxylation, acylation, halogenation, nitration, hydrogenolysis, methylolation, oxidation and reduction, sulfomethylation, sulfonation, silylation, phosphorylation, and nitroxide formation (Wang et al. 1992). The glass transition temperature, T_g, of lignin varies widely depending on the method of isolation and molecular weight. Chemical modification of lignin has mainly been used to dissolve lignin into organic solvents and to determine the functional groups of lignin in the synthesis of copolymers with modified lignins (Thielemans and Wool 2005).

The sulfite-pulping procedure is used to employed to produce ligninsulfonates from waste liquid from softwood with salts of sulfurous acid (sulfites or bisulfites) (Lindsey and Tollens 1892). Ligninsulfonates can be used as binders, dispersant agents for pesticides, emulsifiers, and heavy-metal sequestrants (Mathiasson and Kubat 1994; Browning 1955; Wetzel et al. 2006). Low-cost binders from ligninsulfonates are commonly used in coal briquettes or ceramics, briquetting of mineral dust (fines, shavings, turnings), and wood-related material such as plywood or particle boards (Lora and Glasser 2002). In craft pulping of wood, the lignin is made water soluble through the introduction of alkaline solution. Kraft lignin is a good precursor for the preparation of char and active carbon (Carrott and Carrott 2007). Furthermore, lignin is an economical basic material for carbon fiber production. Several synthetic methods exist for preparing carbon fiber from lignin (Kadla et al. 2002).

3.4 Chitin and Chitosan

Chitin, an important structural polysaccharide, is well known to consist of 2-acetamido-2-deoxy-β-D-glucose through a β (1 → 4) linkage (Rinaudo 2006) (Fig. 5). In fact, chitin exists in three different polymorphic forms (α, β, and γ) (Blackwell 1982; Rudall and Kenchington 1973; Tong and Yao 1997). Recent studies have confirmed that the γ form is a variant of α family (Atkins 1985). This naturally abundant mucopolysaccharide can be found in the cell walls of lower plants and fungi, insect cuticles and in the exoskeleton of arthropods and mollusks. An estimated billion tons of chitins are synthesized every year in nature (Singh and Ray 2000). It is highly hydrophobic and is insoluble in water and most organic solvents. It has low chemical reactivity. For its inertness, it gained less attention with respect to cellulose, remaining an almost unutilized resource. It is soluble in hexafluoro isopropanol, hexafluoroacetone, chloroalcohols in conjugation with aqueous solutions of mineral acids (French et al. 1993) and dimethylacetamide containing 5% lithium chloride.

Chitosan is the fully or partially *N*-deacetylated derivative of chitin (Fig. 6) (Sahoo and Nayak 2011). Chitosan is soluble in acetic acid and other organic solvents. It is more favorable in many applications than chitin. Chitin and chitosan are of commercial interest due to their high percentage of nitrogen (6.89%) compared to synthetically substituted cellulose (1.25%). Because of the presence of the amino functionality, chitin and chitosan could be suitably modified to impart desired properties and distinctive biological functions including solubility (Tharanathan and Kittur 2003; Kurita 2001; Gorochovceva and Makuška 2004).

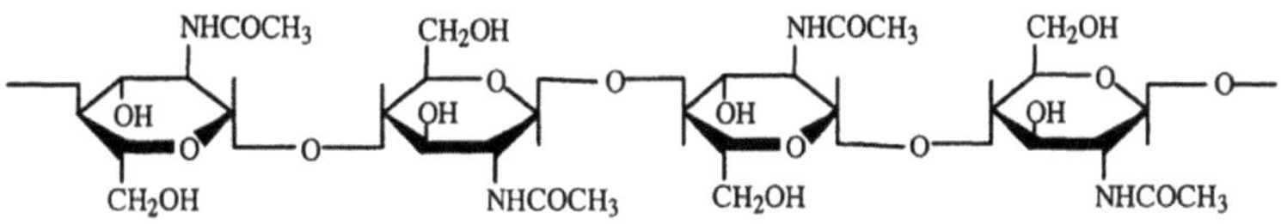

Fig. 5 Representation of typical chitin structure

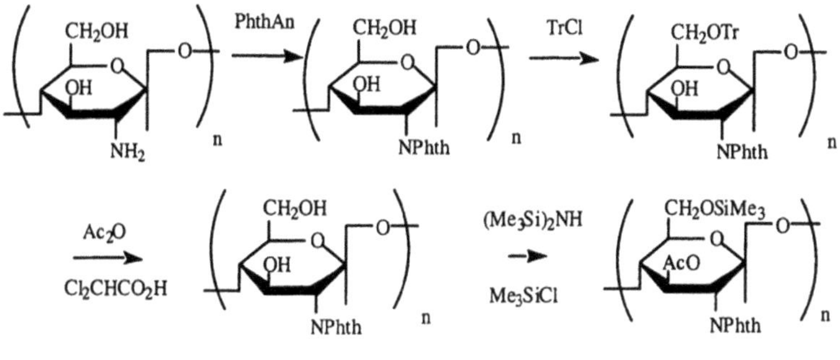

Fig. 6 Representation of typical chitosan structure

Water-resistant adhesive proteins used by marine animals (mussels) to adhere to wet or submerged surface have stimulated the studies to confer water-resistant adhesive properties to semidilute solutions of chitosan. High viscosities and water-resistant adhesive strength of chitosan semidilute solutions have been obtained by tyrosinase catalyzed reaction with 3,4-dihydroxyphenethylamine (dopamine) or by reaction with glutaraldehyde (Nakajima et al. 1984). The deacetylation process is rarely complete, and most commercial and laboratory products tend to be a copolymer of *N*-acetylglucosamine and *N*-glucosamine repeat units. The ratio of the two repeating units depends on the source and preparation conditions of chitosan, but in many cases, the glucosamine units predominate.

Fibers made of chitin and chitosan are useful as absorbable sutures and wound dressing materials (Nakajima et al. 1984). Novel methods have been recently devised for the preparation of chitin threads for the fabrication of absorbable suture materials, dressings, and biodegradable substrates for the growth of human skin cells (keratinocytes and fibroblasts) (Tamura 2004). It has also been claimed that wound dressings made of chitin and chitosan fibers have applications in wastewater treatment. Here, the removal of heavy metal ions by chitosan through chelation has received much attention (Muzzarelli 1997). Yannas et al. proposed a design for artificial skin, applicable to long-term chronic use, focusing on a nonantigenic membrane, which performs as a biodegradable template for synthesis of neodermal tissue (Yannas et al. 1982).

Artificial kidney membranes prepared from modified and alburnin-blended chitosan membranes were found to be potential candidates for dialysis applications (Reinhart and Peppas 1984). Chitosan can provide ionic conductivity when dissolved in acetic acid. The transport of protons which gives conductivity in chitosan is thought to occur through many microvoids in the polymer. The choice of a more suitable electrode material may produce a better battery system (Arof et al. 1995).

3.5 Proteins

Proteins are built up of chain of amino acid residues joined by amide linkages with the presence of reactive groups such as amino, carboxyl, amido, hydroxyl, thiol, and imidazole groups. Proteins have a wide variety of applications in fermentation processes, coatings, encapsulant in the pharmaceutical and food industries, adhesives, surfactants, and plastics (De Graaf and Kolster 1998). Proteins can be used for making of slowly degrading packaging or mulching films (Chiellini et al. 2002a, b). Examples of industrial proteins are plant proteins, such as wheat and corn gluten, soy, pea, and potato proteins; and animal proteins, such as casein, whey, keratin, collagen, and gelatin. Recently, gelatin gets much technological appreciation for its solubility in hot water, polyampholyte character, availability in a wide range of viscosity, and thermally reversible gel formation. Gelatin is currently used in a variety of applications comprising manufacturing of pharmaceutical products, X-ray and photographic films development and food processing (De Graaf and Kolster 1998; Feughelman 2002).

Collagen is used as the basic material for the manufacturing of leather products (Yannas 1972). Proteins are also used for the fabrication of materials such as edible films, films and coating, adhesives, thermoplastics, and surfactants (Cuq et al. 1998; Kester 1986). Chemical modifications of proteins improve its economic consideration as thermoplastic or thermosetting materials (De Graaf et al. 1998). Esterification of protein carboxyl and amide groups by fatty alcohol that lead to a protein derivative with improved functional properties can result into a lower water sensitivity. In this case esterified protein becomes less soluble at basic pH compared to native proteins, indicating a hydrophobation effect (Ayhllon-Meixueiro et al. 2001).

3.6 Polylactic Acid (PLA)

Polylactic acid (PLA) is a biodegradable aliphatic polyester. Most importantly, PLA can be synthesized from bio-derived monomers. Lactide monomer is produced from lactic acid, which is produced by fermentation of a renewable agricultural source corn (Auras et al. 2004). Lactic acid is a chiral molecule available in the L and D stereoisomer forms (Fig. 7). Lactide is a cyclic dimer of lactic acid that has three possible stereoisomers: (i) L-lactide (LLA), which is composed of two L-lactic acids, (ii) D-lactide (DLA), which is composed of two D-lactic acids, and (iii) meso-lactide (MLA), which is composed of an L-lactic acid and a D-lactic acid (Kulkarni et al. 1971). Lactic acid in the L-form is actually produced by fermentation from nearly any renewable resources.

To produce lactic acid (2-hydroxy propionic acid) which is the single monomer of PLA, fermentation or chemical synthesis is used. Its two optically active configurations, the L(+) and D(−) stereoisomers, are produced by bacterial (heterofermentative and homofermentative) fermentation of carbohydrates. PLA can be made through a polymerization process from the lactic acid by polycondensation or ring-opening polymerization or can be made by direct methods such as azeotropic dehydration or enzymatic polymerization (Fig. 7) (Lasprilla et al. 2012). The direct polymerization and ring-opening polymerization methods are commonly exploited. Polycondensation reaction produces low molecular weight poly(lactic acid). Lactide ring-opening polymerization provides a direct and easy access to the corresponding high molecular weight polylactide. Industrially important high molecular weight PLA with a controlled optical and crystal property is formed with vacuum distillation of L-lactide, by ring-opening polymerization. Stannous octoate (bis 2-ethyl hexanoate, SnOct2) for the L-lactide ring-opening polymerization is used as active initiator, which in turn causes a low degree of racemization at high temperature (Pang et al. 2010).

Some unique properties (such as high mechanical strength, good appearance, good barrier, and low toxicity) of PLA have broadened its application. PLA exhibits rapid loss of molecular weight at processing temperature while undergoing thermal treatment. PLA is thermally unstable, and during thermal processing or under hydrolytic conditions, the ester linkages of PLA tend to degrade (Cheng et al.

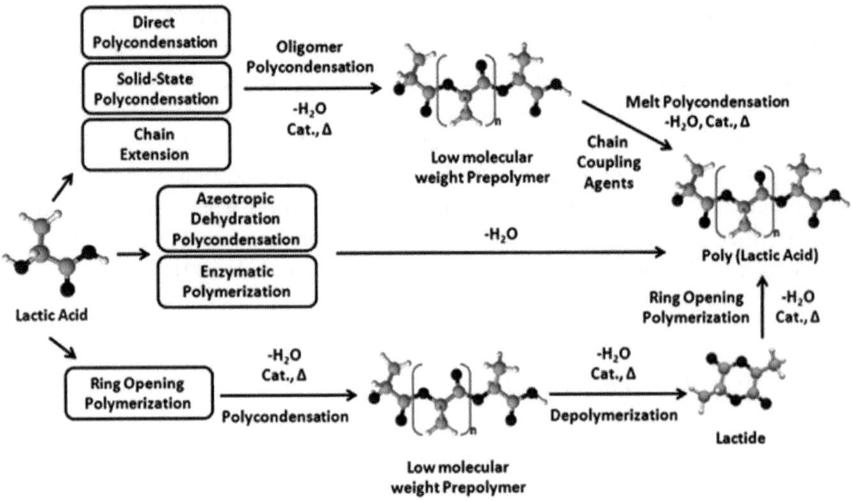

Fig. 7 Synthesis of PLA from L- and D-lactic acids. Adapted from Auras et al. (2004)

2009). The degradation rate of PLA rapidly increases above the melting point, although it undergoes thermal degradation at temperatures lower than the melting point of the polymer. PLA films have good tensile strength, lower than polyethylene terephthalate (PET) but higher than polystyrene (PS). The permeability coefficients of O_2 and CO_2 are lower than PS and comparable to those of PET, in the context of barrier properties of PLA. PLA has the highest value in comparison to high-density polyethylene (HDPE), polypropylene (PP) and PS for tensile modulus and flexural modulus (Akbari et al. 2015).

Crystallinity influences many polymer properties including modulus, stiffness, tensile strength, hardness, melting point, and crease point. The crystallization rate of PLA at temperatures between 100 and 118 °C is very high (Drumright et al. 2000). It was also concluded that the high crystallization rate of PLA below 120 °C has to be ascribed to the high rate of radial growth of the spherulites (spherical semicrystalline regions inside nonbranched linear polymers) after measuring crystallization rates of PLA over a wide temperature range (Müller et al. 2015).

PLA applications include a wide variety of biomedical products (Hutmacher et al. 2008; Yousefi et al. 2015; Hoque et al. 2018), food packaging for meat and soft drinks, films for agro-industry, and nonwoven materials in hygiene products. PLA is approved by the US Food and Drug Administration (FDA) for the biomedical applications in the orthopedic area that include fixation screws, suture anchors, meniscal darts, suture reinforcements, and skin replacement materials (Jagur-Grodzinski 1999; Nair and Laurencin 2007). Tormala et al. proposed fully resorbable composites by reinforcing matrices with resorbable PLLA fibers and calcium phosphate-based glass fibers (Törmälä et al. 1991). One of the first commercially available fiber-formed bioresorbable medical products is based on copolymers of glycolic acid (GA) in combination with L-lactide (Vicryl)

(Albertsson and Varma 2003). Porous PLA scaffolds have been found to be potential reconstruction matrices for damaged tissues and organs (Södergård and Stolt 2002; Avinc and Khoddami 2010; Nainar et al. 2014; Hoque et al. 2018). The capability of PLA to produce hybrid paper plastic packaging that is compostable and to recycle back to lactic acid by alcoholysis or hydrolysis has made it suitable for various applications. As a "green" food packaging polymer, PLA is now considered as the growing alternative. It has a wide range of application in the field of fresh products as thermoformed PLA containers are being used in retail markets as containers (Lunt 1998). Besides, PLA/natural fiber composites have been known for its potential applications in the areas of automotive parts, construction materials, food packaging, and medical devices (Asha et al. 2017).

3.7 *Vegetable Oil-Derived Polymer*

Vegetable oils, such as soybean oil, palm oil, and rapeseed oil, are extracted primarily from the seeds of oilseed plants and have a wide variety of applications: as foods, fuels (biofuels), lubricants, paints, cosmetics, pharmaceuticals, plasticizers, and construction materials (Hoque and Gee 2013). They are also attractive monomers for polymer chemistry due to their natural abundance and reactive functionality (Wang and Schuman 2013). From seeds of certain plants, vegetable oils are extracted, the mainstream of which are oil palm (Elaeis), soybean (Glycine max), sunflower (Helianthus), and oilseed rape (Brassica napus). These seeds have a wide range of applications and are produced on a very large scale (i.e., 156 megatons in 2012) (Zhu et al. 2016). Majority of them are used as food, biofuels, and chemical feedstocks. The key chemical building blocks of vegetable oils are triglycerides which is the triesters of glycerin and fatty acids. Of the five commonly occurring fatty acids, two are saturated (palmitic and stearic) and three are unsaturated (oleic, linoleic, and linolenic).

Triglyceride structure consisting of glycerol esterified by three long-chain aliphatic acids with variable number of carbon atoms (Fig. 8). During esterification, the number of C=C unsaturation in the chain makes the main difference for different types of fatty acid structures including the hydroxyl moieties (Pagliaro et al. 2007). Fatty acids account for 95% of the total weight of triglycerides and their content varies depending on the plant, the crop, the season, and the growing conditions (Fig. 8i). The structures of some frequently studied fatty acids are depicted in Fig. 8 (ii). By transesterification reactions, they are commonly processed to produce glycerol and fatty esters. In terms of polymer production, glycerol is used as a precursor for the production of monomers such as epichlorohydrin and lactic acid. It is a widely used building block in polymer science finding application in the synthesis of polyurethanes, polyesters, or telomeres (Behr et al. 2008). On the other hand, fatty acids have been used for a long period for the development of polymeric structures, both directly and as building blocks for the synthesis of more sophisticated monomers (Behr et al. 2008).

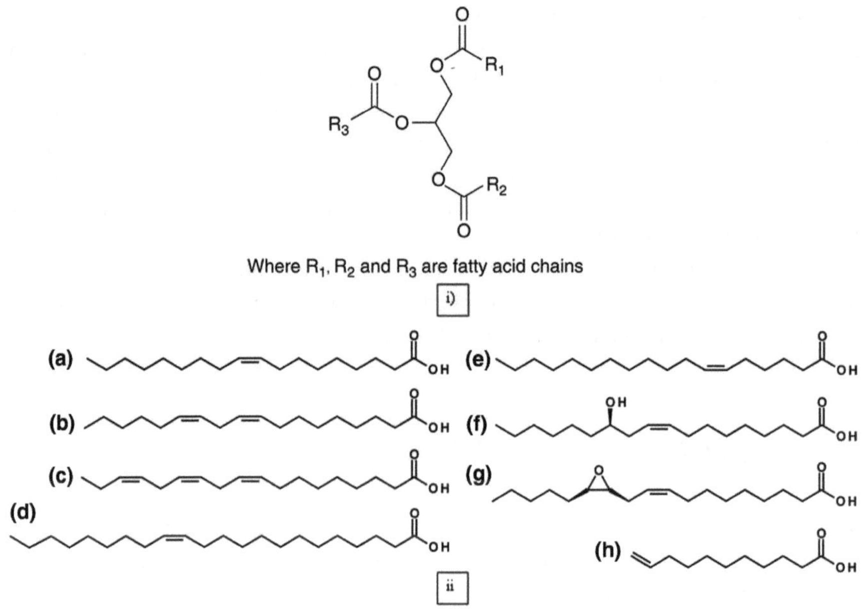

Fig. 8 (i) General triglyceride structure and (ii) fatty acids commonly used in polymer chemistry: **a** oleic acid, **b** linoleic acid, **c** linolenic acid, **d** erucic acid, **e** petroselinic acid, **f** ricinoleic acid, **g** vernolic acid, **h** 10-undecenoic acid

The triglyceride oils need to be functionalized and separated to make well-defined polymers and monomers. By reactions at the olefin functional groups for unsaturated fatty acids, functionalization is achieved. Commonly used olefin functionalization reactions include epoxidation, isomerization, hydroformylation, reduction, and metathesis reactions to produce polyol (Williams and Hillmyer 2008). The polyurethanes are prepared by the reaction between a vegetable oil-derived polyol and various isocyanates. These polyurethane materials are characterized with particular attention being paid to their degradability and biocompatibility. The epoxy resin formation with epoxidized plant oils and fatty acids was the most vividly studied topic (Maisonneuve et al. 2013). The epoxidation of unsaturated fatty acids or triglycerides can be achieved in a straightforward reaction with, e.g., molecular oxygen, hydrogen peroxide as well as by chemo-enzymatic reactions. Additionally, acrylated epoxidized soybean oil as well as maleinized soy oil monoglyceride and maleinized hydroxylated oil were used to prepare composite materials with glass fibers as well as natural flax and hemp fibers (Maisonneuve et al. 2013).

Vegetable oils symbolize an excellent foundation for obtaining aliphatic diacids or diesters and diamines for polyamides production. Partially fatty acid-based polyamides from azelaic acid (PA-6,9 and PA-6,6,9), sebacic acid (PA-6,10), dimer fatty acids, and brassylic acid have been synthesized from different researchers (Meier et al. 2007). Polyamide synthesis from plant oils already found industrial application in the preparation of Nylon-11, since the necessary monomer,

11-amino-undecanoic acid, can be obtained from castor oil by further chemical modification of 10-undecenoic acid (Maisonneuve et al. 2013).

The radical polymerization of acrylated or maleinized plant oil derivatives was investigated to produce plant oil-based thermosets. Another way is the direct cross-linking of soybean oil with polybutadiene or p-dinitrosobenzene and of linseed oil with phenolic resins. In this regard recently, the synthesis of polyurethane foams and the development of biocompatible adhesives were also carried out from plant oil (Meier et al. 2007; Yilmaz and Kusefoğlu 2005; Mutlu and Kusefoglu 2009). Moreover, the great potential of fatty acids as building blocks for olefin metathesis polymerization (ADMET) is now materialized commercially (de Espinosa and Meier 2011). On the other hand, the production of monomers and polymers by thiol–ene coupling reactions with fatty acid derivatives is a newly growing area (Black and Rawlins 2009). These multifunctional thiols and enes were synthesized by ring opening of epoxidized soybean oil with polythiols or allyl alcohol, respectively. These products with high functionality provided good coating film properties by UV curing in the presence of petrochemical-based enes or thiols.

3.8 *Terpenes*

Terpenes and terpenoids are derived from plants. Both terpenes and terpenoids have a common isoprene unit (2-methyl-1,4-butadiene) as a common carbon skeleton building block in their chemical structures. The best-known polyterpene is natural rubber. Other types of terpenes are being investigated as monomers for polymer production, although on a much smaller scale. One of them is turpentine, which is extracted from pine trees (*Pinus* spp.) and is composed mainly of α-pinene (45–97%) and β-pinene (0.5–28%), and limonene, which is extracted from the peel of citrus fruits (De Carvalho and da Fonseca 2006). The general formula of monoterpene is $C_{10}H_{16}$ (Erman 1985). Among the huge variety, α-pinene, β-pinene, limonene, and myrcene are the most common types of terpenes which can be readily isolated in viable amounts from turpentine.

Among terpenes, β-pinene is by far the most sustainable monomer because of the unhindered availability of its alkenyl unsaturation. Its cationic polymerization has been extensively studied, first with classical Lewis-acid initiators and then using quasi-living systems (Silvestre and Gandini 2008). Myrcene was polymerized through a double mechanism consisting of a cyclization step through ring-closing metathesis followed by its cationic activation to give a high molecular weight polymer (Kobayashi et al. 2009). The synthesis of polyamides and polyurethanes was also done from limonene-based diamines and diamides prepared by a similar click strategy involving thiolamine-hydrochlorides (Firdaus and Meier 2013). Monomer limonene oxide was copolymerized with CO_2 using β-diiminate zinc complexes to yield polycarbonate structures (Byrne et al. 2004). A polyester network, with ultra-small pores exploited for gas separation application, was synthesized from triterpene as monomer by the isolation of betulin (which bear two OH groups) from birch bark and its polycondensation with 1,3,5-benzenetricarbonyl trichloride (Jeromenok et al. 2011).

3.9 Furans

Furan derivatives are present in nature in a wide category of structures. Furan is one of the major representatives of the five-member unsaturated heterocycle family. Furan displays the lowest aromatic and the highest dienic character. In furan chemistry alkylation, halogenation, sulfonation, and nitration occur regioselectively at C2 and/or C5. A typical consequence in macromolecular synthesis is the ease with which furan, and some of its derivatives undergo the Diels–Alder (DA) reaction as dienes (Gandini and Belgacem 1997; Moreau et al. 2004; Gandini 2010, 2011a, b; Manfredi and Rivero 2012; Hou et al. 1998; Sutton et al. 2013).

Some of the novel furan monomers displayed characteristics entirely compatible with their conventional petrochemical alternatives. 2-Furyl oxirane monomer can take part in anionic polymerization with the initiation by a very weak nucleophiles like amines and alcohols, for the synthesis of block, star, and grafted copolymers (Gandini and Belgacem 1997; Moreau et al. 2004; Gandini 2010, 2011a, b; Manfredi and Rivero 2012; Hou et al. 1998; Sutton et al. 2013).

Polyamides were made by "direct condensation" through the polycondensation of both 2,5- and 3,4-dicarboxylic acids with a number of aromatic diamines (Gandini and Belgacem 1997; Moreau et al. 2004; Gandini 2010, 2011a, b; Manfredi and Rivero 2012; Hou et al. 1998; Sutton et al. 2013). These polyamides showed regular structures, high molecular weights and interesting crystallization and thermal properties. Furans resulting from furfural (F) can be condensed with aldehydes and ketones to produce difuran monomers bearing different terminal functionalities. Polyesters, polyamides, and polyurethanes were prepared with these monomers by step-growth polymerizations. When hydroxymethylfurfural (HMF) was used to prepare novel furan monomers, a new family of polycondensation structures arose and their macromolecules displayed remarkable properties (Gandini and Belgacem 1997; Moreau et al. 2004; Gandini 2010, 2011a, b; Manfredi and Rivero 2012; Hou et al. 1998; Sutton et al. 2013). Other materials based on the polymerization of furan monomers include photosensitive macromolecules with applications in the printing industry, conjugated semiconducting polymers with luminescent features, polymer electrolytes for solid-state batteries based on the combination of furan and chitosan chemistries, and insulating foams (Gandini and Belgacem 1997; Moreau et al. 2004; Gandini 2010, 2011a, b; Manfredi and Rivero 2012; Hou et al. 1998; Sutton et al. 2013).

3.10 Polyhydroxyalkanoates

Biodegradable polyhydroxyalkanoates (PHAs) plastics are produced by bacteria in several grades with the difference in composition and molecular weight (Braunegg et al. 1998). The bacteria produces the PHAs from a carbon source that is when overfed, the PHA polymer is formed andon the contrary the polymer is consumed when the bacteria are short of a feeding source. PHAs have properties similar to those of some polyolefins. They are fully and rapidly biodegraded under appropriate conditions. The simplest PHA is polyhydroxybutyrate (PHB) polymer and the only type of PHAs relevant for practical applications (Braunegg 2002). The first step of bacterial synthesis of PHB consists in the conversion of a selected carbon source to acetate. Then, an enzyme cofactor is attached via the formation of a thioester bond. The enzyme, called coenzyme A (CoA), is a universal carrier of acyl groups in biosynthesis, and acetyl-CoA is a basic metabolic molecule found in all PHA-producing organisms by the sequential action of three enzymes, 3-ketothiolase, acetoacetyl-CoA reductase, and PHA synthase (Choi and Lee 1999). A dimer acetoacetyl-CoA is formed via reversible condensation and subsequently reduced to a monomer unit (R)-3-hydroxybutyryl-CoA. PHB is formed via the polymerization of the latter, maintaining the asymmetric center (Macrae and Wilkinson 1958).

Appropriate exchanged propiolactones can be used to produce PHAs chemically using aluminum or zinc alkyl catalysts with water as the cocatalyst. In terms of stereoregularity, synthetic PHAs can be made almost identical to the corresponding biopolymers (Chodak 2008). This resemblance results even in the excellent biodegradability of the material (Araki and Hayase 1979). However, high lactone monomer costs are the major limitation for industrially application of the synthetic PHAs (Philip et al. 2007).

The structure of PHAs is based on polyester macromolecules bearing optically active carbon atoms. The crystalline structure of PHB is orthorhombic with dimensions of the basic crystalline cell a = 5.76 Å, b = 13.20 Å, c = 5.96 Å. A partially planar zigzag structure can be formed as a result of the mechanical uniaxial load from the amorphous phase between lamellae (Yamane et al. 2001; Orts et al. 1990). PHB is a completely biodegradable, highly hydrophobic thermoplastics, containing almost 80% crystallinity, with high melting temperature, resistance to organic solvents and possessing excellent mechanical strength and modulus resembling that of polypropylene (Pollet and Avérous 2011). Noticeable brittleness, very low deformability, high propensity to a rapid thermal degradation, difficult processing by conventional thermoplastic technologies (mainly due to fast thermal degradation), and rather high price compared to other high-volume plastics may be named as the major factors hindering a wider application of PHB (Follain et al. 2014).

In order to improve properties of PHB, random copolymers have been prepared by replacing the methyl group on the PHB main chain by ethyl or longer substituents with the help of changing the substrate on which the bacteria grow (Doi et al. 1987). Bacterial synthesis using sodium octanoate as a sole carbon source was reported to result in a formation of copolymer with the majority of polyhydroxy-octanoate segments (Gagnon et al. 1992). An application of epoxidized soybean oil has been reported recently for plasticizing blends of PHBV and cellulose acetate butyrate for packaging applications. The material with longer side chains is claimed to be suitable for using as degradability enhancers for poly(olefins) (Marschessault et al. 1990). In another study, it is claimed that the ductility of PHB can be improved significantly by hot rolling treatment (Barham and Keller 1986). In case of chemical modification, poly(hydroxy ether) prepared from Bisphenol A was reported to react with PHB by transesterification during annealing at 180 °C (Yuan and Ruckenstein 1998). In another case, grafting of acrylic acid onto PHB and PHBV was initiated by gamma irradiation (Mitomo et al. 1996).

Injection blow-molded bottles for packaging of biodegradable containers for hair shampoo or motor oil and disposable razor handles were produced with PHB (Ramsay and Ramsay 1990; Dalev et al. 1998). Excellent barrier properties against gas permeation of isotropic PHB foils may be considered for application in food-stuffs packaging (Basta 1984). PHB-coated paper has been shown to be completely biodegradable and also easier to recycle than conventionally coated paper (Chodak 2002). A direct electrostatic coating technique is possible to use for depositing PHB on a low dielectric substrate such as paper (Marchessault et al. 1993). PHB fibers were considered to be mainly used for the production of scaffolds (Van de Velde and Kiekens 2002). PHB composites with apatite can be used as biodegradable bone fracture fixations or even as bone repair materials (Ni and Wang 2002).

3.11 Polycarbonates

As carbon dioxide (CO_2) is abundant and inexpensive, it is an interesting synthetic feedstock for the production of polymers. CO_2 fixation through reaction with the highly reactive three-membered epoxide ring to afford cyclic or polymeric carbonates is an extensive field of research (Kember et al. 2011; Miao et al. 2008). The copolymerization involves the alternating insertion of carbon dioxide and epoxide in the growing polymer chain. The use of a catalyst is required in order to achieve the selective synthesis of polycarbonates with high yield and under relatively mild conditions of temperature and CO_2 pressure. In the copolymerization cycle, a Lewis acidic metal species binds and ring opens an epoxide, generating a metal alkoxide species. CO_2 introduces into the metal alkoxide bond giving a metal carbonate, and this group can combine and ring open a further molecule of epoxide; duplication of the cycle ultimately directs to the production of high molecular weight polycarbonate (Coates and Moore 2004). A variety of homogeneous and heterogeneous Lewis acidic metal complexes initiate the polymerization, including Zn(II), Co(II), and Cr(III) alkoxides, carboxylates, and carbonates (Cohen et al. 2005). It was also found that the addition of phenylphosphonic acid [$PhP(O)(OH)_2$] or phosphoric acid [H_3PO_4] as chain transfer agent in CO_2–PO copolymerization catalyzed by a cobalt salen complex produced flame-retarding poly(propylene carbonate)s (Cohen et al. 2005).

The mechanical properties of polycarbonates can differ significantly as a function of the nature of the epoxide used in the copolymerization with CO_2. Poly (cyclohexene carbonate) is a brittle material at room temperature, whereas poly (ethylene carbonate) displays rubber-like properties (Taherimehr and Pescarmona 2014). The low thermal degradation temperature of polycarbonates is problematic for processing by common thermoplastic techniques such as molding, mixing, or extrusion. The use of co-monomers and end groups has improved the thermal stability (Hwang et al. 2003).

Polycarbonate polyols of low molecular weight may be suitable to prepare foams, coatings, and adhesives, whereas high molecular weight polycarbonates may be used as rigid plastics or elastomers (Zhu et al. 2016). Recently, it was shown in a discovery that poly(propylene carbonate) (PPC) could be used in the application as organic filler in packaging material and containers as with some modification it exhibited good adhesion, smooth processability, and better thermal degradability (Allen 2014). PPC has potential application in medical implants for its production of nontoxic consequences by enzymatic degradation in the body (Kim et al. 2008). PPC can also be used in combination with other polymers (such as starch) to make biodegradable plastics (Luinstra 2008).

3.12 Biopolyethylene (PE) and Biopolyethylene Terephthalate (PET)

Polyethylene (PE) is an important engineering polymer traditionally produced from petroleum-based resources. PE is now being manufactured from dehydration of ethanol produced by microbial fermentation of plant-based renewable resources. Bioethanol is developed from sugar cane, sugar beet, maize, wood, wheat, corn, and other plant wastes through microbial strain and biological fermentation process (Babu et al. 2013). At the end of the fermentation process, ethanol is distilled in order to remove water and to yield azeotropic mixture of hydrous ethanol. Ethanol is then dehydrated at high temperatures over a solid catalyst to produce ethylene and, subsequently, polyethylene (Chen et al. 2007). Polyethylene from natural resources has exactly the same chemical, physical, and mechanical properties as petrochemical polyethylene. These bio-based PE grades are mainly targeted toward food packing, cosmetics, personal care, automotive parts, and toys.

For partially bio-based polyethylene terephthalate (PET) (30 or 70%) production, one of the two fossil precursors {e.g., ethylene glycol (EG) and purified terephthalic acid (PTA)} is substituted with a bio-based material. Distinctively for a fully bio-based PET, both precursors are produced from biomass (Tuck et al. 2012). PTA accounts for approximately 70% of the weight of PET resin, while EG contributes to the rest. In the case of bio-based PET, para-xylene is produced from bioisobutanol by fermentation of cellulosic materials. PTA is currently produced from the oxidation of aromatic para-xylene. Then, PTA is condensed with ethylene glycol to make PET. Bio-based versions of PTA and EG can be produced from a variety of biomass materials (Mülhaupt 2013). Currently, PET is being used for clothing and beverage bottles on a huge scale.

4 Conclusions and Future Perspectives

Undoubtedly, renewability of polymer material is an increasing new point to maintain the sustainability of ecosystem. As the foreseeable limit of fossil feedstocks and the increasing the environmental concerns, it is necessary to develop novel bio-renewable resource-based polymer materials on a large scale. Therefore, readjusting in carbon cycle is taken into serious consideration by the researchers. In this process, renewability of polymers is an escalating new fact to uphold the sustainability of the ecosystem. Polymers from renewable resources will perform a continual growing task in the common plastic goods as well as in medical products. Apart from economic and environmental advantages, the new property outcomes are evitable in the renewable base polymers. Essential research attempts in this field are set to persist. It is obligatory at this stage in the development of this renewable research field to assemble an abundant database in preparation for the gradual declining of fossil resources. This will eventually enable us to pick wise decisions and choices for industrial implementation.

References

Adler E (1977) Lignin chemistry—past, present and future. Wood Sci Technol 11(3):169–218

Akbari A, Majumder M, Tehrani A (2015) Polylactic acid (PLA) carbon nanotube nanocomposites. In: Handbook of polymer nanocomposites. Processing, performance and application. Springer, Berlin, pp 283–297

Albertsson AC, Varma IK (2003) Recent developments in ring opening polymerization of lactones for biomedical applications. Biomacromolecules 4(6):1466–1486

Allen SD (2014) U.S. Patent No. 8,748,555. U.S. Patent and Trademark Office, Washington, DC

Araki T, Hayase S (1979) Biodegradability of synthetic β-substituted poly-β-esters. J Polym Sci Polym Chem Ed 17(6):1877–1881

Arof L, Subban RHY, Radhakrishna S (1995) In: Prasad PN (ed) Polymer and other advanced materials: emerging technologies and business. p 539

Asha AB, Sharif A, Hoque ME (2017) Interface interaction of jute fiber reinforced PLA biocomposites for potential applications. In: Jawaid M, Salit MS, Alothman OY (eds) Green biocomposites: design and applications. Springer, Switzerland

Atkins E (1985) Conformations in polysaccharides and complex carbohydrates. J Biosci 8(1–2):375–387

Auras R, Harte B, Selke S (2004) An overview of polylactides as packaging materials. Macromol Biosci 4(9):835–864

Avinc O, Khoddami A (2010) Overview of poly(lactic acid)(PLA) fibre. Fibre Chem 42(1):68–78

Ayhllon-Meixueiro F, Tropini V, Silvestre F (2001) Fatty esterification of plant proteins. In: Biorelated polymers. Springer, Boston, pp 231–236

Babu RP, O'connor K, Seeram R (2013) Current progress on bio-based polymers and their future trends. Prog Biomater 2(1):8
Baeumler A, Ijjasz-Vasquez E, Mehndiratta S (2012) Sustainable low-carbon cities in China: why it matters and what can be done. In: Sustainable low-carbon city development in China, pp 39–67
Barham PJ, Keller A (1986) The relationship between microstructure and mode of fracture in polyhydroxybutyrate. J Polym Sci Part B Polym Phys 24(1):69–77
Basta N (1984) Biopolymers challenge petrochemicals. High Technol 4(2):66–70
Behr A, Eilting J, Irawadi K, Leschinski J, Lindner F (2008) Improved utilisation of renewable resources: new important derivatives of glycerol. Green Chem 10(1):13–30
Bemiller JN (1997) Starch modification: challenges and prospects. Starch-Stärke 49(4):127–131
Black M, Rawlins JW (2009) Thiolene UV-curable coatings using vegetable oil macromonomers. Eur Polym J 45(5):1433–1441
Blackwell J (1982) The macromolecular organization of cellulose and chitin. In: Cellulose and other natural polymer systems. Springer, Boston, pp 403–428
Botello-Álvarez JE, Rivas-García P, Fausto-Castro L, Estrada-Baltazar A, Gomez-Gonzalez R (2018) Informal collection, recycling and export of valuable waste as transcendent factor in the municipal solid waste management: a Latin-American reality. J Clean Prod 182:485–495
Braunegg G (2002) Sustainable poly(hydroxyalkanoate)(PHA) production. In: Degradable polymers. Springer, Dordrecht, pp 235–293
Braunegg G, Lefebvre G, Genser KF (1998) Polyhydroxyalkanoates, biopolyesters from renewable resources: physiological and engineering aspects. J Biotechnol 65(2–3):127–161
Browning WC (1955) Lignosulfonate stabilized emulsions in oil well drilling fluids. J Petrol Technol 7(06):9–15
Bruijnincx PC, Rinaldi R, Weckhuysen BM (2015) Unlocking the potential of a sleeping giant: lignins as sustainable raw materials for renewable fuels, chemicals and materials. Green Chem 17(11):4860–4861
Byrne CM, Allen SD, Lobkovsky EB, Coates GW (2004) Alternating copolymerization of limonene oxide and carbon dioxide. J Am Chem Soc 126(37):11404–11405
Calvo-Flores FG, Dobado JA (2010) Lignin as renewable raw material. Chemsuschem 3 (11):1227–1235
Carrott PJM, Carrott MR (2007) Lignin–from natural adsorbent to activated carbon: a review. Biores Technol 98(12):2301–2312
Chauhan GS, Mahajan S, Guleria LK (2000) Polymers from renewable resources: sorption of Cu^{2+} ions by cellulose graft copolymers. Desalination 130(1):85–88
Chen G, Li S, Jiao F, Yuan Q (2007) Catalytic dehydration of bioethanol to ethylene over TiO_2/γ-Al_2O_3 catalysts in microchannel reactors. Catal Today 125(1–2):111–119
Cheng Y, Deng S, Chen P, Ruan R (2009) Polylactic acid (PLA) synthesis and modifications: a review. Front Chem China 4(3):259–264
Chiellini E, Chiellini F, Cinelli P (2002a) Polymers from renewable resources. In: Degradable polymers. Springer, Dordrecht, pp 163–233
Chiellini E, Cinelli P, Antone SD, Ilieva VI (2002b) Environmentally degradable polymeric materials (EDPM) in agricultural applications—an overview. Polimery 47(7–8):538–544
Chodak I (2002) Polyhydroxyalkanoates: properties and modification for high volume applications. In: Degradable polymers. Springer, Dordrecht, pp 295–319
Chodak I (2008) Polyhydroxyalkanoates: origin, properties and applications. In: Monomers, polymers and composites from renewable resources. pp 451–477
Choi J, Lee SY (1999) Factors affecting the economics of polyhydroxyalkanoate production by bacterial fermentation. Appl Microbiol Biotechnol 51(1):13–21
Coates GW, Moore DR (2004) Discrete metal-based catalysts for the copolymerization of CO_2 and epoxides: discovery, reactivity, optimization, and mechanism. Angew Chem Int Ed 43 (48):6618–6639

Cohen CT, Chu T, Coates GW (2005) Cobalt catalysts for the alternating copolymerization of propylene oxide and carbon dioxide: combining high activity and selectivity. J Am Chem Soc 127(31):10869–10878

Cuq B, Gontard N, Guilbert S (1998) Proteins as agricultural polymers for packaging production. Cereal Chem 75(1):1–9

Dalev P, Vassileva E, Mark JE, Fakirov S (1998) Enzymatic degradation of formaldehyde-crosslinked gelatin. Biotechnol Tech 12(12):889–892

Das O, Sarmah AK, Bhattacharyya D (2015) A sustainable and resilient approach through biochar addition in wood polymer composites. Sci Total Environ 512:326–336

De Carvalho CC, da Fonseca MMR (2006) Biotransformation of terpenes. Biotechnol Adv 24(2): 134–142

de Espinosa LM, Meier MA (2011) Plant oils: the perfect renewable resource for polymer science?! Eur Polym J 47(5):837–852

De Graaf LA, Kolster P (1998) Industrial proteins as a green alternative for 'petro' polymers: potentials and limitations. In: Macromolecular symposia, vol 127, no 1. Hüthig & Wepf Verlag, Basel, pp 51–58

De Graaf LA, Kolster P, Vereijken JM (1998) Modification of wheat gluten for non-food applications. Plant proteins from European crops. Springer, Berlin, pp 335–339

Doane WM, Swanson CL, Fanta GF (1992) Emerging polymeric materials based on starch. In: Rowell RM, Scultz TP, Narayan R (eds) Emerging technologies for materials and chemicals from biomass, ACS symposium series (USA) 476, pp 197–230

Doi Y, Kunioka M, Nakamura Y, Soga K (1987) Biosynthesis of copolyesters in Alcaligenes eutrophus H16 from carbon-13 labeled acetate and propionate. Macromolecules 20(12):2988–2991

dos Santos Abreu H, do Nascimento AM, Maria MA (2007) Lignin structure and wood properties. Wood Fiber Sci 31(4):426–433

Drumright RE, Gruber PR, Henton DE (2000) Polylactic acid technology. Adv Mater 12(23): 1841–1846

Dufresne A (2017) Nanocellulose: from nature to high performance tailored materials. Walter de Gruyter GmbH & Co KG, Berlin

Einchhorn SJ (2011) Cellulose nanowhiskers: promising materials for advanced applications. Soft Matter 7(2):303–315

Einchhorn SJ, Dufresne A, Aranguren MM, Capadona JR, Rowan SJ, Weder C et al (2010) Review: current international research into cellulose nanofibres and composites. J Mater Sci 45:1–33

Erman WF (1985) In: Gassman PG (ed) Chemistry of the monoterpenes, Part A. Studies in organic chemistry. Marcel Dekker, New York, pp 929–1033

Feughelman M (2002) Natural protein fibers. J Appl Polym Sci 83(3):489–507

Firdaus M, Meier MA (2013) Renewable polyamides and polyurethanes derived from limonene. Green Chem 15(2):370–380

Focher B, Marzetti A, Crescenzi V (eds) (1991) Steam explosion techniques: fundamentals and industrial applications. In: Proceedings of the international workshop on steam explosion techniques: fundamentals and industrial applications, Milan, Italy, 20–21 Oct 1988. CRC Press, Boca Raton

Follain N, Chappey C, Dargent E, Chivrac F, Crétois R, Marais S (2014) Structure and barrier properties of biodegradable polyhydroxyalkanoate films. J Phys Chem C 118(12):6165–6177

French AD, Bertoniere NR, Battista OA, Cuculo JA, Gray DG (1993) Cellulose, Chapter in: Kirk-Othmer encyclopedia of chemical technology, 4th edn., vol. 5. Wiley, New York, pp 476–496

Gagnon KD, Lenz RW, Farris RJ, Fuller RC (1992) The mechanical properties of a thermoplastic elastomer produced by the bacterium *Pseudomonas oleovorans*. Rubber Chem Technol 65 (4):761–777

Gandini A (2010) Furans as offspring of sugars and polysaccharides and progenitors of a family of remarkable polymers: a review of recent progress. Polym Chem 1(3):245–251

Gandini A (2011a) Furan monomers and their polymers: synthesis, properties and applications. In: Biopolymers—new materials for sustainable films and coatings, pp 179–209
Gandini A (2011b) The irruption of polymers from renewable resources on the scene of macromolecular science and technology. Green Chem 13(5):1061–1083
Gandini A, Belgacem MN (1997) Furans in polymer chemistry. Prog Polym Sci 22(6):1203–1379
Gandini A, Belgacem MN (2013) The state of the art of polymers from renewable resources. In: Handbook of biopolymers and biodegradable plastics, pp 71–85
Gandini A, Lacerda TM (2015) From monomers to polymers from renewable resources: recent advances. Prog Polym Sci 48:1–39
Goheen DH, Royt CR (1981) Lignin. Kirk-othmer encyclopedia of polymer science and technology, vol 14, 3rd edn. Wiley, New York, pp 294–312
Gorochovceva N, Makuška R (2004) Synthesis and study of water-soluble chitosan-*O*-poly (ethylene glycol) graft copolymers. Eur Polym J 40(4):685–691
Habibi Y (2014) Key advances in the chemical modification of nanocelluloses. Chem Soc Rev 43 (5):1519–1542
Habibi Y, Lucia LA, Rojas OJ (2010) Cellulose nanocrystals: chemistry, self-assembly, and applications. Chem Rev 110(6):3479–3500
Han Y, Yuan L, Li G, Huang L, Qin T, Chu F, Tang C (2016) Renewable polymers from lignin via copper-free thermal click chemistry. Polymer 83:92–100
Hoenich NA, Stamp S (2000) Clinical investigation of the role of membrane structure on blood contact and solute transport characteristics of a cellulose membrane. Biomaterials 21(3): 317–324
Hoque ME, Gee LP (2013) Biodiesel from plant resources—sustainable solution to ever increasing fuel oil demands. J Sustain Bioenergy Syst 3:163–170
Hoque ME, Ye TJ, Yong LC, Dahlan KM (2013) Sago starch-mixed low-density polyethylene (LDPE) biodegradable polymer: synthesis and characterization. J Mater Article ID 365380, 7 pages. https://doi.org/10.1155/2013/365380
Hoque ME, Khan AM, Islam MS, Asim M, Jawaid M, Othman OA (2016) The effect of natural degradation on the mechanical and morphological properties of tropical woods. Cellul Chem Technol 50(7–8):723–730
Hoque ME, Daei JMG, Khalid M (2018) Next generation biomimetic bone tissue engineering matrix from poly(L-lactic acid) PLA/calcium carbonate composites doped with silver nanoparticles. Curr Anal Chem 14(3):268–277
Hou XL, Cheung HY, Hon TY, Kwan PL, Lo TH, Tong SY, Wong HN (1998) Regioselective syntheses of substituted furans. Tetrahedron 54(10):1955–2020
Hutmacher DW, Hoque ME, Wong YS (2008) Design, fabrication and physical characterization of scaffolds made from biodegradable synthetic polymers in combination with RP systems based on melt extrusion. In: Bidanda B, Bartolo P (eds) Virtual prototyping & bio manufacturing in medical applications. Springer, USA
Hwang Y, Jung J, Ree M, Kim H (2003) Terpolymerization of CO_2 with propylene oxide and ε-caprolactone using zinc glutarate catalyst. Macromolecules 36(22):8210–8212
Imam SH, Mao L, Chen L, Greene RV (1999) Wood adhesive from crosslinked poly(vinyl alcohol) and partially gelatinized starch: preparation and properties. Starch-Stärke 51(6):225–229
Imre B, Pukánszky B (2015) From natural resources to functional polymeric biomaterials. Eur Polym J 68:481–487
Jagur-Grodzinski J (1999) Biomedical application of functional polymers. React Funct Polym 39(2):99–138
Jeromenok J, Böhlmann W, Antonietti M, Weber J (2011) Intrinsically microporous polyesters from betulin-toward renewable materials for gas separation made from birch bark. Macromol Rapid Commun 32(22):1846–1851
Jiang L, Li Y, Wang X, Zhang L, Wen J, Gong M (2008) Preparation and properties of nano-hydroxyapatite/chitosan/carboxymethyl cellulose composite scaffold. Carbohydr Polym 74(3):680–684

Jyothi AN (2010) Starch graft copolymers: novel applications in industry. Compos Interfaces 17(2–3):165–174
Kadla JF, Kubo S, Gilbert RD, Venditti RA (2002) Lignin-based carbon fibers. In: Chemical modification, properties, and usage of lignin. Springer, Boston, MA, pp 121–137
Kanagaratnam S, Hoque ME, Sahri MM, Spowage A (2013) Investigating the effect of deforming temperature on the oil-binding capacity of palm oil based shortening. J Food Eng 118(1):90–99
Kauffman GB (1993) Rayon: the first semi-synthetic fiber product. J Chem Educ 70(11):887
Kember MR, Buchard A, Williams CK (2011) Catalysts for CO_2/epoxide copolymerisation. Chem Commun 47(1):141–163
Kester J (1986) Edible films and coatings: a review. Food Technol 40(12):47–59
Ketabchi MR, Khalid M, Hoque ME, Ratnam CT, Walvekar R (2015) Eco-friendly and cost-effective isolation of cellulose microfibres and nano-crystals from kenaf fibres. In: Proceedings, 13th international conference on environment, ecosystems and development (EED 2015), 23–25 Apr 2015, Kuala Lumpur, Malaysia
Ketabchi MR, Khalid M, Ratnam CT, Manicham S, Walvekar R, Hoque ME (2016) Sonosynthesis of cellulose nanoparticles (CNP) from kenaf fiber: effects of processing parameters. Fibers Polym 17(9):1352–1358
Kim G, Ree M, Kim H, Kim IJ, Kim JR, Im Lee J (2008) Biological affinity and biodegradability of poly(propylene carbonate) prepared from copolymerization of carbon dioxide with propylene oxide. Macromol Res 16(5):473–480
Klemm D, Kramer F, Moritz S, Lindström T, Ankerfors M, Gray D, Dorris A (2011) Nanocelluloses: a new family of nature-based materials. Angew Chem Int Ed 50(24):5438–5466
Kobayashi S, Lu C, Hoye TR, Hillmyer MA (2009) Controlled polymerization of a cyclic diene prepared from the ring-closing metathesis of a naturally occurring monoterpene. J Am Chem Soc 131(23):7960–7961
Koutinas AA, Chatzifragkou A, Kopsahelis N, Papanikolaou S, Kookos IK (2014) Design and techno-economic evaluation of microbial oil production as a renewable resource for biodiesel and oleochemical production. Fuel 116:566–577
Kruger LH (1989) U.S. Patent No. 4,838,944. U.S. Patent and Trademark Office, Washington, DC
Kulkarni RK, Moore EG, Hegyeli AF, Leonard F (1971) Biodegradable poly(lactic acid) polymers. J Biomed Mater Res 5(3):169–181
Kumar S, Krishnan S, Samal SK, Mohanty S, Nayak SK (2017) Itaconic acid used as a versatile building block for the synthesis of renewable resource based resins and polyesters for future prospective: a review. Polym Int 66(10):1349–1363
Kurita K (2001) Controlled functionalization of the polysaccharide chitin. Prog Polym Sci 26(9): 1921–1971
Labet M, Thielemans W, Dufresne A (2007) Polymer grafting onto starch nanocrystals. Biomacromolecules 8(9):2916–2927
Lagoa R, Murtinho D, Gil MH (1999) Membranes of cellulose derivatives as supports for immobilization of enzymes. In: Imam SH, Greene RV, Zaidi BR (eds) Biopolymers—utilizing nature's, advanced, materials. Acs symposium series 723, ACS, Washington DC, pp 228–234
Lasprilla AJ, Martinez GA, Lunelli BH, Jardini AL, Maciel Filho R (2012) Poly-lactic acid synthesis for application in biomedical devices—a review. Biotechnol Adv 30(1):321–328
Lindblad MS, Liu Y, Albertsson AC, Ranucci E, Karlsson S (2002) Polymers from renewable resources. In: Degradable aliphatic polyesters. Springer, Berlin, pp 139–161
Lindsey JB, Tollens B (1892) Ueber Holz-Sulfitflüssigkeit und Lignin. Justus Liebigs Annalen der Chemie 267(2–3):341–366
Lligadas G, Tüzün A, Ronda JC, Galià M, Cádiz V (2014) Polybenzoxazines: new players in the bio-based polymer arena. Polym Chem 5(23):6636–6644
Lora JH, Aziz S (1985) Organosolv pulping: a versatile approach to wood refining. Tappi (United States) 68(8)
Lora JH, Glasser WG (2002) Recent industrial applications of lignin: a sustainable alternative to nonrenewable materials. J Polym Environ 10(1–2):39–48

Luinstra GA (2008) Poly(propylene carbonate), old copolymers of propylene oxide and carbon dioxide with new interests: catalysis and material properties. Polym Rev 48(1):192–219
Lunt J (1998) Large-scale production, properties and commercial applications of polylactic acid polymers. Polym Degrad Stab 59(1–3):145–152
Mac Kinnon M, Brouwer J, Samuelsen S (2018) The role of natural gas and its infrastructure in mitigating greenhouse gas emissions, improving regional air quality, and renewable resource integration. Prog Energy Combust Sci 64:62–92. https://doi.org/10.1016/j.pecs.2017.10.002
Macrae RM, Wilkinson JF (1958) The influence of culture conditions on polyhydroxybutyrate synthesis by *Bacillus megaterium*. In: Proceedings of the royal society of Edinburgh, vol 27. pp 73–78
Maim CJ, Mench JW, Kendall DL, Hiatt GD (1951) Aliphatic acid esters of cellulose. Preparation by acid-chloride-pyridine procedure. Ind Eng Chem 43(3):684–688
Maisonneuve L, Lebarbé T, Grau E, Cramail H (2013) Structure–properties relationship of fatty acid-based thermoplastics as synthetic polymer mimics. Polym Chem 4(22):5472–5517
Manfredi LB, Rivero G (2012) Furan resins: synthesis, characterization and applications. Chem Phys Res J 5(1/2):45
Marchessault RH, Rioux P, Saracovan I (1993) Direct electrostatic coating of paper. Nordic Pulp Paper Res J (Sweden) 8(1):211–216
Marschessault RH, Monasterios CJ, Morin FG, Sundararajan PR (1990) Chiral poly(β-hydroxyalkanoates): an adaptable helix influenced by the alkane side-chain. Int J Biol Macromol 12 (2):158–165
Mathiasson A, Kubat DG (1994) Lignin as binder in particle boards using high frequency heating. Holz als Roh-und Werkstoff 52(1):9
Meier MA, Metzger JO, Schubert US (2007) Plant oil renewable resources as green alternatives in polymer science. Chem Soc Rev 36(11):1788–1802
Meshram MW, Patil VV, Mhaske ST, Thorat BN (2009) Graft copolymers of starch and its application in textiles. Carbohydr Polym 75(1):71–78
Miao CX, Wang JQ, He LN (2008) Catalytic processes for chemical conversion of carbon dioxide into cyclic carbonates and polycarbonates. Open Org Chem J 2(1)
Michael FM, Khalid M, Walvekar R, Ratnam CT, Ramarad S, Siddiqui H, Hoque ME (2016) Effect of nanofillers on the physico-mechanical properties of load bearing bone implants. Mater Sci Eng C 67:792–806
Missoum K, Belgacem MN, Bras J (2013) Nanofibrillated cellulose surface modification: a review. Materials 6(5):1745–1766
Mitomo H, Sasaoka T, Yoshii F, Makuuchi K, Saito T (1996) Radiation-induced graft polymerization of acrylic acid onto poly(3-hydroxybutyrate) and its copolymer. Sen'i Gakkaishi 52(11):623–626
Moreau C, Belgacem MN, Gandini A (2004) Recent catalytic advances in the chemistry of substituted furans from carbohydrates and in the ensuing polymers. Top Catal 27(1–4):11–30
Mormann W, Spitzer D (2001) Silylation of OH-polymers by reactive extrusion. In: Macromolecular symposia, vol 176, no 1. WILEY-VCH Verlag GmbH, Weinheim, pp 279–288
Mülhaupt R (2013) Green polymer chemistry and bio-based plastics: dreams and reality. Macromol Chem Phys 214(2):159–174
Müller AJ, Ávila M, Saenz G, Salazar J (2015) Crystallization of PLA-based materials. Poly(lactic acid) science and technology: processing, properties, additives and applications. Royal Society of Chemistry, London, pp 68–98
Mutlu H, Kusefoglu SH (2009) Synthesis and characterization of polymers from soybean oil and p-dinitrosobenzene. J Appl Polym Sci 113(3):1925–1934
Muzzarelli RAA (1997) Human enzymatic activities related to the therapeutic administration of chitin derivatives. Cell Mol Life Sci CMLS 53(2):131–140
Nainar SMM, Begum S, Ansari MNM, Hoque ME, Aini SS, Ng MH, Ruszymah BHI (2014) Effect of compatibilizers on in vitro biocompatibility of PLA-HA bioscaffold. Bioinspired Biomimetic Nanobiomater 3(4):208–216

Nair LS, Laurencin CT (2007) Biodegradable polymers as biomaterials. Prog Polym Sci 32(8–9):762–798

Nakajima M, Atsumi K, Kifune K (1984) Development of absorbable sutures from chitin. In: Chitin, chitosan, and related enzymes. Academic Press, New York, pp 407–410

Ni J, Wang M (2002) In vitro evaluation of hydroxyapatite reinforced polyhydroxybutyrate composite. Mater Sci Eng C 20(1–2):101–109

Nilsson TY, Wagner M, Inganäs O (2015) Lignin modification for biopolymer/conjugated polymer hybrids as renewable energy storage materials. Chemsuschem 8(23):4081–4085

Northey RA (1992) In: Rowell RM, Scultz TP, Narayan R (eds) Low cost uses of lignin, in emerging technologies for materials and chemicals from biomass, ACS symposium series 476. ACS, Washington, D.C., pp 146–175

Orts WJ, Marchessault RH, Bluhm TL, Hamer GK (1990) Observation of strain-induced β form in poly(β-hydroxyalkanoates). Macromolecules 23(26):5368–5370

Pagliaro M, Ciriminna R, Kimura H, Rossi M, Della Pina C (2007) From glycerol to value-added products. Angew Chem Int Ed 46(24):4434–4440

Pang X, Zhuang X, Tang Z, Chen X (2010) Polylactic acid (PLA): research, development and industrialization. Biotechnol J 5(11):1125–1136

Pecoraro É, Manzani D, Messaddeq Y, Ribeiro SJ (2008) Bacterial cellulose from Glucanacetobacter xylinus: preparation, properties and applications. In: Monomers, polymers and composites from renewable resources, pp 369–383

Petersen K, Nielsen PV, Bertelsen G, Lawther M, Olsen MB, Nilsson NH, Mortensen G (1999) Potential of biobased materials for food packaging. Trends Food Sci Technol 10(2):52–68

Philip S, Keshavarz T, Roy I (2007) Polyhydroxyalkanoates: biodegradable polymers with a range of applications. J Chem Technol Biotechnol Int Res Process Environ Clean Technol 82(3): 233–247

Piccolo RS, Santos F, Frollini E (1997) Sugar cane bagasse lignin in resol-type resin: alternative application for ligninphenol-formaldehyde resins. J Macromol Sci Part A 34(1):153–164

Pollak A (1952) Lignin. Kirk-Othmer encyclopedia of chemical technology, 1st edn., vol 8. Wiley, New York, pp 327–338

Pollet E, Avérous L (2011) Production, chemistry and properties of polyhydroxyalkanoates. In: Biopolymers—new materials for sustainable films and coatings. pp 65–86

Ramsay JA, Ramsay BA (1990) Poly-beta-hydroxyalkanoic acids(PHAs): unique, microbially produced thermoplastics. Appl Phycol Forum 7(3):1–5

Reinhart CT, Peppas NA (1984) Solute diffusion in swollen membranes. Part II. Influence of crosslinking on diffusive properties. J Membr Sci 18:227–239

Rinaudo M (2006) Chitin and chitosan: properties and applications. Prog Polym Sci 31(7): 603–632

Rudall KM, Kenchington W (1973) The chitin system. Biol Rev 48(4):597–633

Sagar AD, Merrill EW (1995) Properties of fatty-acid esters of starch. J Appl Polym Sci 58 (9):1647–1656

Sahoo D, Nayak PL (2011) Chitosan: the most valuable derivative of chitin. Biopolym Biomed Environ Appl 129–166

Sahoo S, Mohanty S, Nayak SK (2018) Biobased polyurethane adhesive over petroleum based adhesive: use of renewable resource. J Macromol Sci Part A 55(1):36–48

Saikia A, Karak N (2017) Renewable resource based thermostable tough hyperbranched epoxy thermosets as sustainable materials. Polym Degrad Stab 135:8–17

Shi B, Topolkaraev V, Wang J (2011) Biopolymers, processing, and biodegradation. In: Renewable and sustainable polymers. American Chemical Society, Washington, pp 117–132

Shirashi N (1992) Liquefaction of lignocellulosics in organic solvents and its applications. In: Rowell RM, Scultz TP, Narayan R (eds) Emerging technologies for materials and chemicals from biomass, ACS symposium series 476, pp 136–145

Silvestre AJ, Gandini A (2008) Terpenes: major sources, properties and applications. In: Monomers, polymers and composites from renewable resources, pp 17–38

Simon J, Müller HP, Koch R, Müller V (1998) Thermoplastic and biodegradable polymers of cellulose. Polym Degrad Stab 59(1–3):107–115

Singh DK, Ray AR (2000) Biomedical applications of chitin, chitosan, and their derivatives. J Macromol Sci Part C Polym Rev 40(1):69–83

Siró I, Plackett D (2010) Microfibrillated cellulose and new nanocomposite materials: a review. Cellulose 17(3):459–494

Södergård A, Stolt M (2002) Properties of lactic acid based polymers and their correlation with composition. Prog Polym Sci 27(6):1123–1163

Sutton AD, Waldie FD, Wu R, Schlaf M, Louis A III, Gordon JC (2013) The hydrodeoxygenation of bioderived furans into alkanes. Nat Chem 5(5):428

Taherimehr M, Pescarmona PP (2014) Green polycarbonates prepared by the copolymerization of CO_2 with epoxides. J Appl Polym Sci 131(21)

Tamura H (2004) Destruction of rigid crystalline structure to prepare chitin solution. Adv Chitin Sci 7:84–87

Tessler MM, Billmers RL (1996) Preparation of starch esters. J Environ Polym Degrad 4(2):85–89

Thakur VK, Thakur MK, Raghavan P, Kessler MR (2014) Progress in green polymer composites from lignin for multifunctional applications: a review. ACS Sustain Chem Eng 2(5):1072–1092

Tharanathan RN (2005) Starch—value addition by modification. Crit Rev Food Sci Nutr 45(5): 371–384

Tharanathan RN, Kittur FS (2003) Chitin—the undisputed biomolecule of great potential. Crit Rev Food Sci Nutr 43(1):61–87

Thielemans W, Wool RP (2005) Lignin esters for use in unsaturated thermosets: lignin modification and solubility modeling. Biomacromolecules 6(4):1895–1905

Thoma JA, Stewart L (1965) In: Whistler RL, Pashall EF (eds) Starch: chemistry and technology, vol 1. Academic Press, New York, pp 209–249

Tong H, Yao SN (1997) Study of the microstructure and the microtopography for shell of crab and lobster. J Anal Sci 13(3):206–209

Törmälä P, Vasenius J, Vainionpää S, Laiho J, Pohjonen T, Rokkanen P (1991) Ultra-high-strength absorbable self-reinforced polyglycolide (SR-PGA) composite rods for internal fixation of bone fractures: in vitro and in vivo study. J Biomed Mater Res 25(1):1–22

Tsanaktsis V, Terzopoulou Z, Exarhopoulos S, Bikiaris DN, Achilias DS, Papageorgiou DG, Papageorgiou GZ (2015) Sustainable, eco-friendly polyesters synthesized from renewable resources: preparation and thermal characteristics of poly(dimethyl-propylene furanoate). Polym Chem 6(48):8284–8296

Tuck CO, Pérez E, Horváth IT, Sheldon RA, Poliakoff M (2012) Valorization of biomass: deriving more value from waste. Science 337(6095):695–699

Van de Velde K, Kiekens P (2002) Biopolymers: overview of several properties and consequences on their applications. Polym Testing 21(4):433–442

Vanholme R, Morreel K, Ralph J, Boerjan W (2008) Lignin engineering. Curr Opin Plant Biol 11(3):278–285

Vert M (2001) Biopolymers and artificial biopolymers in biomedical applications, an overview. In: Biorelated polymers. Springer, Boston, MA, pp 63–79

Vestapen B, Pawa F, Dortman B, Bagastyo A, Pratono AH, Rahmani P, Zurbrugg C (2017) Municipal solid waste management: market-driven upcycling of urban solid waste in Indonesia. In: Mahadwartha PA, Pratono AH, Verstappen BM, Zurbrügg C (eds) Generating value from organic waste (pp. 1–5), Master of Management Program, Faculty of Business and Economics, Universitas Surabaya, Surabaya, Indonesia

Vikso-Nielsen A, Andersen C, Pedersen S, Hjort C (2009) U.S. Patent No. 7,618,795. U.S. Patent and Trademark Office, Washington, DC

Wang R, Schuman TP (2013) Vegetable oil-derived epoxy monomers and polymer blends: a comparative study with review. Express Polym Lett 7(3):272–292

Wang J, Manley RSJ, Feldman D (1992) Synthetic polymer-lignin copolymers and blends. Prog Polym Sci 17(4):611–646

Wetzel S, Duchesne LC, Laporte MF (2006) Bioproducts from Canada's forests: new partnerships in the bioeconomy. Springer Science & Business Media, Berlin

Williams CK, Hillmyer MA (2008) Polymers from renewable resources: a perspective for a special issue of polymer reviews. Polym Rev 48(1):1–10

Wing RE, Willett JL (1997) Water soluble oxidized starches by peroxide reactive extrusion. Ind Crops Prod 7(1):45–52

Wing RE, Willett JL (1999) In Imam SH, Greene RY, Zaidi BR (eds) Thermochemical processes for derivatization of starches with different amylose content in biopolymers—utilizing nature's, advanced, materials. ACS symposium series 723, ACS, Washington DC, pp 55–64

Wu CS (2015) Renewable resource-based green composites of surface-treated spent coffee grounds and polylactide: characterisation and biodegradability. Polym Degrad Stab 121:51–59

Yamane H, Terao K, Hiki S, Kimura Y (2001) Mechanical properties and higher order structure of bacterial homo poly(3-hydroxybutyrate) melt spun fibers. Polymer 42(7):3241–3248

Yan N, Chen X (2015) Don't waste seafood waste: turning cast-off shells into nitrogen-rich chemicals would benefit economies and the environment. Nature 524(7564):155–158

Yannas IV (1972) Collagen and gelatin in the solid state. J Macromol Sci Rev Macromol Chem 7(1):49–106

Yannas IV, Burke JF, Orgill DP, Skrabut EM (1982) Wound tissue can utilize a polymeric template to synthesize a functional extension of skin. Science 215(4529):174–176

Yilmaz M, Kusefoğlu SH (2005) Gelation of soybean oil with polybutadiene. J Appl Polym Sci 96(6):2240–2246

Yousefi AM, Hoque ME, Prasad RGSV, Uth N (2015) Current strategies in multiphasic scaffold design for osteochondral tissue engineering: a review. J Biomed Mater Res Part A 103(7): 2460–2481

Yuan Y, Ruckenstein E (1998) Miscibility and transesterification of phenoxy with biodegradable poly(3-hydroxybutyrate). Polymer 39(10):1893–1897

Zdrahala RJ (1997) Thermoplastic starch revisited. Structure/property relationship for "dialed-in" biodegradability. In: Macromolecular symposia, vol 123, no 1. Hüthig & Wepf Verlag, Basel, pp 113–121

Zhang MSY (2016) Low-carbon indicator system–sino: evaluating low-carbon city development level in China, Doctoral dissertation, University of Duisburg-Essen

Zhu Y, Romain C, Williams CK (2016) Sustainable polymers from renewable resources. Nature 540(7633):354

Zia KM, Noreen A, Zuber M, Tabasum S, Mujahid M (2016) Recent developments and future prospects on bio-based polyesters derived from renewable resources: a review. Int J Biol Macromol 82:1028–1040

Fillers and Reinforcements for Advanced Nanocomposites

Nilesh Kumar Shrivastava, Muhammad Akmal Ahmad Saidi, Norhayani Othman, Mohamad Zurina and Azman Hassan

Abstract The performance and properties of nanocomposites largely depend on its nanofiller and reinforcement. This chapter presents an overview of the different types of nanofiller and reinforcement in biopolymer nanocomposite. It mainly focuses on the preparation, processing, properties, and the application of the bio-nanocomposite. Bio-nanocomposite based on biopolymer such as poly(lactic acid) (PLA), poly(e-caprolactone) (PCL), poly(vinyl alcohol) (PVA), poly(hydroxybutyrate) (PHB), poly(hydroxybutyrate-co-hydroxyvalerate) (PHBV), and chitosan, and nanofiller such as graphene, carbon nanotube, layered silicate reinforcement, sepiolite, and halloysite nanofiller has become a topic of discussion, and their performance and future application become a focus of interest. Enhanced mechanical and thermal properties imparted by the nanofiller reinforcement on the bio-nanocomposite and its optimum loading had been reviewed. It can be concluded that the properties of bio-nanocomposite could be further enhanced by the utilization of compatibilizer, coupling agent, and nanofiller treatment in bio-nanocomposite, which play an important role in enhancing the compatibility between the component of bio-nanocomposite and state of nanofiller dispersion and distribution in bio-nanocomposite.

Keywords Nanofiller · Reinforcement · Mechanical properties · Thermal properties · Graphene · Carbon nanotubes · Layered silicates · Sepiolite Halloysite

N. K. Shrivastava · M. A. A. Saidi · N. Othman (✉) · M. Zurina · A. Hassan
Faculty of Engineering, Department of Bioprocess and Polymer Engineering,
School of Chemical and Energy Engineering, Universiti Teknologi Malaysia,
81310 Skudai, Johor, Malaysia
e-mail: norhayani@utm.my

A. Hassan
Centre for Advanced Composite Materials, Universiti Teknologi Malaysia,
81310 Skudai, Johor, Malaysia

M. L. Sanyang and M. Jawaid (eds.), *Bio-based Polymers and Nanocomposites*,
https://doi.org/10.1007/978-3-030-05825-8_2

1 Introduction

Micrometer-size range fillers, namely carbon black, calcium carbonate, glass fibers, and talc, have been used extensively as reinforcement in plastic industry. Fillers have been added into polymers to provide an improvement of the final product properties which also known as reinforcement fillers. However, these fillers only provide an improvement in composite properties at high concentration, which leads to an increase in viscosity of polymer mixtures and in turn makes the flow difficult (Chang et al. 2003; Vaisman et al. 2006). Recently, it has been noticed that the introduction of a small quantity of nano-sized fillers has a positive impact on the properties of virgin polymers, without limiting the processability (Ray and Bousmina 2005; Pramanik et al. 2016; Pavlidou and Papaspyrides 2008). Nanoparticles are defined as particulate dispersions or solid particles with a size in the range of 10–1000 nm. To promote the mechanical properties of the nanocomposites, it is necessary to achieve a stronger interfacial interaction between the fillers and matrix as well as enhance the external load transfer efficiency (Nikolic et al. 2017; Nima et al. 2015; Kalappa et al. 2013).

Advance nanocomposites refer to the cutting edge, state of the art on the effective use of nanoscaled fillers and reinforcements to enhance the performance of advanced nanocomposites, both in industrial and manufacturing applications. The development of advanced nanocomposite based on various types of nanofillers, particularly nanocelluloses, nanotubes, nanoplatelets, and nanoparticles, has been extended toward a wide range of applications (Rhim et al. 2013; Gouvêa et al. 2018). Many works have been reported in an attempt to have a better understanding of the role of fillers and reinforcements in advanced nanocomposites fabrication and synthesis process, as well as performance optimization of new materials in terms of mechanical, thermal, structural, and multi-functional properties (Rhim et al. 2013; Salam et al. 2011; Gouvêa et al. 2018). The advance nanocomposite has been introduced in potential applications, such as high-performance composites for aircraft/automobile industries and biomaterials.

In this chapter, nature, and properties of fillers, feasible approaches for processing of bio-nanocomposites, as well as the effect of nanofillers on the mechanical, electrical, and thermal properties as well as biodegradability of bio-nanocomposites will be discussed briefly. Incorporation of nanofillers such as graphene, carbon nanotubes, nanoclay, sepiolite, and halloysite into biopolymers leads to improvement of biodegradability and properties of biopolymers. So far, the most widely studied bio-nanocomposites are poly(lactic acid) (PLA), poly(hydroxybutyrate) (PHB), starch, poly(vinyl alcohol) (PVA), and cellulose.

2 Preparation, Processing, Properties, and Application Performance of Nanocomposites

2.1 Graphene Reinforced Bio-nanocomposites

Graphene is a nanofiller with a sheet of a monolayer of carbon atoms; each carbon is sp^2 hybridized as illustrated in Fig. 1. Graphene is the stiffest and strongest known material and has Young's modulus and ultimate strength of up to 1 TPa and 130 GPa, respectively (Lee et al. 2008). Graphene oxide (GO) is an oxidized form of graphene, which contains epoxide, carbonyl, and hydroxyl groups moieties on the surface that allow the formation of hydrogen bonds. Reduced GO can be obtained chemically or thermally, which has lower oxygen content than GO. The hydrogen bonding between rGO and the polymer matrix, as well as between multiple sheets of rGO, is the main factor contributing to poor distribution of rGO in polymer matrix (Barrett et al. 2014). Incorporating other chemical moieties can

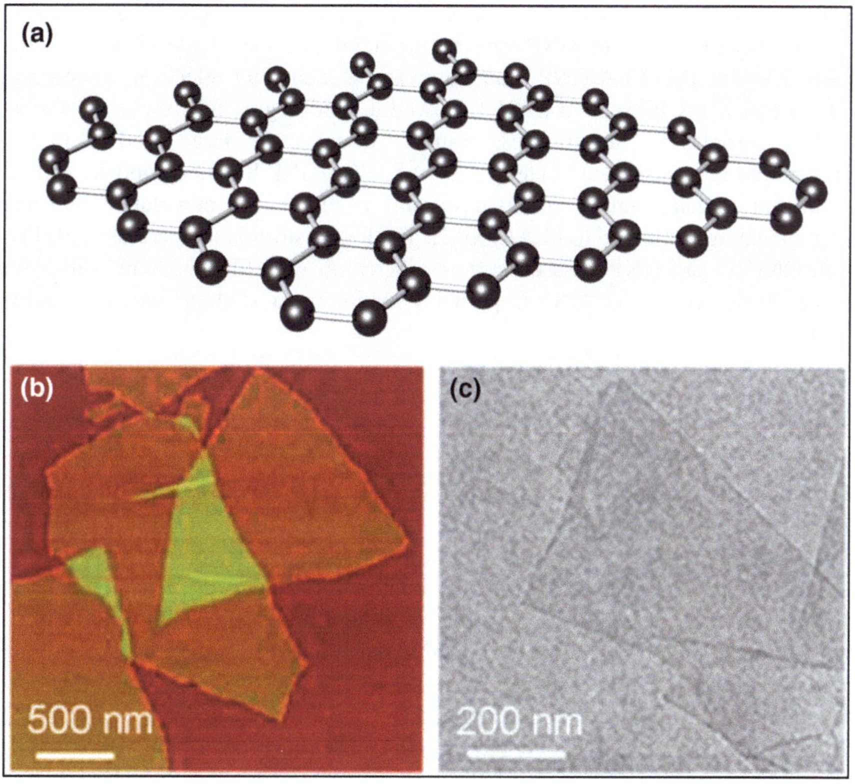

Fig. 1 **a** Crystalline structure of graphene film, **b** a tapping mode AFM image of GO sheets on mica surface, **c** TEM image of the GO (Zhang et al. 2010)

also be done to maximize the interfacial interaction between rGO and polymer matrix (Kim et al. 2010).

Graphite nanoplatelets (GNPs) which are also called graphite nanosheets (GNS) are another form of graphene consisting stacked 2D graphene sheets. It possesses many outstanding properties including electrical properties (room-temperature electron mobility of 2.5 × 105 $cm^2\ V^{-1}\ s^{-1}$), thermal properties (thermal conductivity above 3000 $Wm^{-1}\ K^{-1}$), and mechanical properties (Young's modulus of 1 TPa and strength of 130 GPa) (Kim et al. 2010). Therefore, graphene, including GNP, is a promising reinforcement for high-performance composites (Sadasivuni et al. 2014). A number of PLA/GNP-based nanocomposites have been successfully prepared (Xu et al. 2010; Kim et al. 2010; Sun and He 2012; Nuona et al. 2015), and the main areas which were investigated are mechanical properties, thermal and electric conductivity, thermal degradation behavior, and biodegradation behavior.

Studies on the influence of filler size on the properties of PLA/GNP nanocomposites have been reported (Pinto et al. 2016; Gao et al. 2017). Pinto et al. (2016) added two types of GNPs at 0.25 wt%, GNP-C and GNP-M, in PLA. They found that incorporation of GNP-C and GNP-M in PLA increased their Young's modulus by 14%. Also, tensile strength increased by 20% with GNP-C and by 6% with GNP-M. In a recent study, Gao et al. (2017) reported an investigation on the influence of two types of GNP varying in lateral size—small (GNP-S) and large (GNP-L)—on the PLA/GNP nanocomposites. Scanning electron microscopy (SEM) and X-ray diffraction (XRD) showed a good degree of dispersion of small GNP-S particles up to 10 wt% content, whereas the large GNP-L particle agglomeration was seen at content $\geq$7 wt%. PLA/GNP-S nanocomposite showed significant enhancement in conductivity and a reduction in percolation threshold after annealing compared to PLA/GNP-L. The heat distortion temperature (HDT) of both GNP-S- and GNP-L-based systems increased with filler content, with composites based on larger GNP-L particles exhibiting a slightly superior thermal stability.

The biocompatibility and degradation behavior of the PLA/GNP nanocomposites have been studied by several researchers (Pinto et al. 2013, 2016; Li et al. 2016; Botta et al. 2018). Pinto et al. (2016) reported that addition of low content (0.25 wt %) GNPs (GNP-C and GNP-M) into PLA by melt mixing was found not to interrupt the polymer degradation process, as significant decrease in molecular weight can be observed. Comparing with neat PLA, PLA/GNP-M and PLA/GNP-C composites present similar human cell (human foreskin fibroblasts) adhesion and growth at the surface before degradation, and after 6 months degradation in physiological conditions, they do not release toxic products toward human fibroblasts. Recently, a study on degradation behavior of PLA/GNP nanocomposites after reprocessing has been reported by Botta et al. (2018). The results indicated that the re-extruded nanocomposite samples, even that reprocessed one time, enabled a higher degree of particle dispersion and less aggregate structure. The presence of GNP decreased the degradation rate as a function of the reprocessing cycles that, on the contrary, showed a more drastic reduction of the molecular weight. The stabilizing effect of

GNP could facilitate the reprocessability of PLA without sacrificing other outstanding material properties of the matrix.

Numerous researchers have investigated the effect of incorporation of graphene (rGO, GNS) on the properties of PHA/graphene nanocomposites (Wang et al. 2013; Barrett et al. 2014; Ambrosio-Martin et al. 2015a, b; Pramanik et al. 2016; Montagna et al. 2017b; Zine and Sinha 2017; Gouvêa et al. 2018). Wang et al. (2013) investigated the effect of GNS on thermal and mechanical properties PHBV/GNS nanocomposites. The thermal stability of PHBV was significantly improved with the incorporation of GNS from thermogravimetric analysis. The presence of GNS also does not change the crystal structure of PHBV. Dynamic mechanical analysis showed that the storage modulus of the PHBV/GNS nanocomposites was greatly improved. A study on the effect of rGO nanoparticles with PHAs was reported by Barrett et al. (2014). The study used three different medium chain PHAs; poly (hydroxyoctanoate) (PHO), poly(hydroxyoctenoate) (PHOe), and poly(hydroxydecanoate) (PHD). Depending on matrix type, Young's modulus increased by 180–590%, tensile strength increased by 2–16%, and the elongation at break decreased by 41–89% at the 2.5% rGO content. DSC results show that the addition of up to 2.5% rGO content increases the melting and the glass transition temperature of those PHAs. Recently, Montagna et al. (2017b) reported the influence of GNS on the properties of PHBV nanocomposites. A good dispersion of GNS in PHBV matrix can be anticipated from XRD and Raman results. The addition of GNS did not result in any changes in the degree of crystallinity of the nanocomposites. However, double melting peaks were observed on the second heating run, evidencing the GNS acted as a nucleation agent for crystallization. Besides that, the addition the GNS has no significant change on the thermal stability of the PHBV.

The potential use of PHA/graphene as a biodegradable nanocomposite has also been explored (Sridhar et al. 2013; Montagna et al. 2017a, 2018). Sridhar et al. (2013) had reported biodegradable PHBV/graphene nanocomposites prepared by solution casting. The biodegradation test was carried out by means of soil degradation and found that graphene does not interfere with the biodegradation of PHBV. Another study on PHBV/GNS nanocomposites was reported by Montagna et al. (2017a). The nanocomposites were prepared using solution casting, and the biodegradation study was carried out by using filamentous fungi (*Penicillium funiculosum*) in a solid medium. They concluded that PHBV/GNS nanocomposites can be totally degraded in the presence of *Penicillium funiculosum*. On the other hand, in the most recent investigation, Montagna et al. (2018) reported that GNS inhibits the oxidative degradation of PHBV/GNS nanocomposites. The investigation was carried out by preparing PHBV/GNS films by solution casting at different GNS contents (0.25, 0.50 and 1.00 wt%), and the oxidative degradation was performed by exposing the films to artificial ultraviolet (UV) radiation for 52 h. An increase in oxidation indicates the occurrence of photodegradation. From the FTIR spectrum, neat PHBV film and PHBV/0.25 wt% GNS nanocomposite showed an increase in oxidation after UV exposure. However, the other PHBV/GNS nanocomposites showed the final oxidation levels similar or equal to those of PHBV/GNS nanocomposites without exposure.

2.2 Carbon Nanotubes-Reinforced Bio-nanocomposites

Since the discovery of carbon nanotubes (CNTs) by Iijima in 1991, it allured the academicians and industrialist because of its outstanding electrical, mechanical, and thermal properties (Iijima 1991). CNTs are tougher than steel, weightless when compared with aluminum, and far more electrically conductive than copper (Moniruzzaman and Winey 2006). Besides being flexible, CNTs also have low density, high strength, and larger surface area. Conceptualization of CNTs structure is like wrapping of graphene into a long barrel as presented in Fig. 2. CNTs are categorized as single-walled carbon nanotubes (SWCNTs) and multi-walled carbon nanotubes (MWCNTs). The major difference between them apart from the number of walls is SWCNT displays better electrical properties, while MWCNT exhibits better chemical resistance (Prajapati et al. 2011). These excellent properties and surface chemistry of CNTs inspired academicians and industrialist to prepare CNT-based composites. The aforementioned properties make CNTs as efficient filler compared to conventional fillers for high-end applications (Qi et al. 2003). Ajayan et al. (1994) are the first to report the utilization of CNT wherein he used aligned CNT as a reinforcing filler in an epoxy matrix to study its rheology on nanometer scales.

However, like other fillers, CNTs also have few limitations which limit their applications to a certain extent. The major issue with CNTs is their dispersion in polymer matrices. This is because they are grown as entangled bundles which are quite difficult to disperse. Researchers have tried many different techniques to disperse CNTs uniformly into the polymer matrix. Some of the techniques which were commonly used for dispersion of CNTs are the chemical modification of CNTs, using different mixing techniques such as solution mixing, melt blending, and in situ polymerization (Chrissafis 2010). The ideal CNTs/polymer

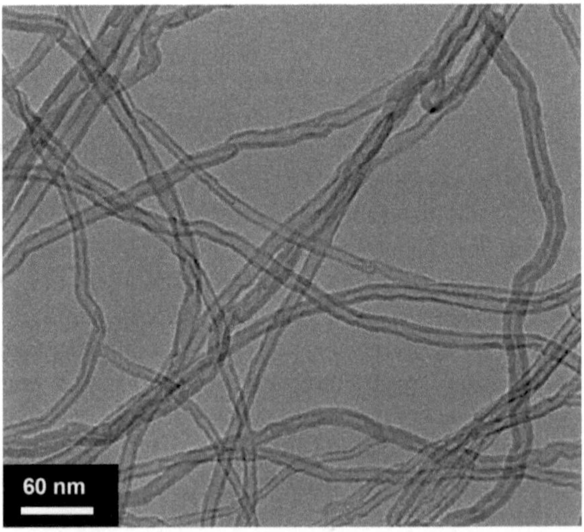

Fig. 2 TEM image of pristine MWCNT (Zhang et al. 2016)

nanocomposites depend not on just one parameter but on several parameters like the CNTs/polymer interfacial adhesion, dispersion, and alignment of CNTs (Vaisman et al. 2006). Thus, techniques like ultrasonication, use of surface-active agent, and chemical treatment of CNTs surface (Ochoa et al. 2017) are projected to enhance not the only dispersion of CNTs but also interfacial adhesion. Amongst biopolymers, PLA has attracted tremendous attention among researchers.

The PLA/CNTs nanocomposites are compatible because of the unique structure and properties of CNTs. Villmow et al. (2008) reported on the effect of melt mixing technique to disperse MWCNT in PLA matrix. The intention of the study was to develop a guideline for plastic manufacturers to achieve an acceptable dispersion and distribution of MWCNT within the PLA matrix to ensure high mechanical performance at low percolation threshold. It was reported that the electrical percolation threshold was achieved at 0.5 wt% of MWNT content. Another researcher Wu et al. (2015) reported the relationship between the aspect ratio of CNTs and the percolation threshold in PLA/CNT nanocomposites, in order to optimize the rheological, electrical, and mechanical properties. It was also reported that the CNT with high aspect ratio could effectively promote the modulus of PLA composites. Furthermore, using "in situ" approach, it was found that good dispersion of CNTs in PLA matrix can be done by polycondensation of lactic acid in xylene.

In a different study, morphological, thermal, mechanical, and electrical characteristics of PLA/CNT-g-PLA nanocomposites were investigated as a function of the CNTs content by Yoon et al. (2010). PLA/CNT-g-PLA nanocomposites exhibited notably enhanced tensile properties and better crystallization ability. However, change in CNT-g-PLA content has an influence on conductivity properties of PLA/CNT-g-PLA nanocomposites; this is due to the high coverage of PLA chains around CNTs.

Apart from percolation threshold and content of CNTs, solubility is another factor, which plays a vital role in the dispersion of CNTs. The solubility of CNTs within the polymer matrix is very important for achieving better dispersion. Usually, surface functionalization of CNTs by carboxyl group (–COOH) or hydroxyl group (–OH) can be done to enhance the solubility of CNTs within polymer matrix via hydrogen bonding between functional moieties and ester linkage of polymer chains. This surface functionalization is an oxidation process that occurs in the presence of oxygen, air, concentrated sulfuric acid, nitric acid, aqueous hydrogen peroxide, and acid mixture (Wang et al. 2005a). Wu et al. (2010) evaluated the effect of various functionalized MWCNTs on the rheology and thermal stability of melt-compounded PLA-based nanocomposites by using –COOH and –OH-functionalized MWCNTs as well as purified MWCNTs as models. The use of –COOH-functionalized MWCNTs in the nanocomposites caused the best possible interaction between PLA matrix and MWCNTs as highlighted by rheological measurements and TEM analyses. However, it still resulted in some CNT bundles, which limited the thermal stability of PLA matrix.

Many other attempts have been made by researchers to enhance the dispersion extent of carbon nanotubes in the PLA matrix. One simple attempt is surface grafting of –COOH-functionalized CNTs onto PLA of different molecular weights reported by Wu et al. (2010). This treatment involved reaction between –COOH-functionalized CNTs and thionyl chloride to form acyl-chloride-functionalized CNTs, which eventually interacted with PLA during melt mixing. However, this technique resulted in disappointing results as reported by Kuan et al. (2008). They investigated the surface functionalization process of CNTs which involves reaction of maleic anhydride with CNTs surface and followed by the coupling reaction with OH-functionalized PLA. The results from this study suggested the presence of CNT agglomerates within PLA-based nanocomposites indicating inefficiency of grafting reaction.

Moving toward application of CNTs, some researchers focused on utilizing the high electrical conductivity of CNTs by incorporating it with biopolymers-based melt blends (Wu et al. 2011) and fibers (Pötschke et al. 2010). In binary melt blending, CNTs were also used as third component to promote interaction and compatibility of immiscible polymer blends and in turn improve the properties of the blends (Wu et al. 2009). In this work, Wu et al. (2009) investigated the effect of COOH-functionalized MWCNT on the morphological and performance measures of PCL/PLA blends. It was noticed that the functionalized MWCNTs selectively dispersed in the PCL matrix and on the interface between two polymer phases. Thus enhanced the interfacial adhesion of resulting ternary PCL/PLA/carboxylic MWCNTs systems in contrast of the blank PCL/PLA blends. As a result, this ternary system could improve rheological, conductive as well as mechanical properties, compared with binary blends and blank PCL/PLA blends.

Apart from PLA, other biopolymers such as chitosan (Salam et al. 2011; Wang et al 2005a) and PCL (Meng et al. 2010) were also used for making CNT/biopolymer composite. Wang et al. (2005a) prepared chitosan/MWCNT nanocomposites by solution-evaporation method. The MWCNT were found to be homogeneously dispersed throughout the chitosan matrix and greatly improved the tensile strength and modulus by 93 and 99% of the nanocomposite. Lee et al. (2011) realized production of homogeneous CNT/PCL systems without agglomeration of MWCNT can be achieved by dispersing various and controllable concentrations of ionically modified MWCNT in tetrahydrofuran (THF), followed by dissolving PCL in homogeneous MWCNT/THF solution. The resultant nanocomposites showed high mechanical properties compared to pure PCL.

2.3 *Layered Silicate Reinforced Bio-nanocomposites*

The nanoclay is one of the earliest nanomaterials used in the preparation of polymer nanocomposites to enhance its properties. In last decade or so, polymer/clay nanocomposites have been researchers' area of choice owing to excellent tensile properties, high gas barrier, thermal stability, flame retardancy, high aspect ratio, and low cost as it is abundantly available on earth cluster (Gorrasi et al. 2002; Miranda-Trevino and Coles 2003). The high aspect ratio (length/thickness) of nanoclay plays an important role in the enhancement of mechanical and physical properties of the polymers. Different types of tetrahedral–octahedral sheet arrangement nanoclays that have been used by researchers for the preparation of polymer nanocomposites are saponite, hectorite, sauconite, stevensite, swinefordite, montmorillonite, beidellite, nontronite, volkonskoite, etc. Among these, montmorillonite (MMT) is the most commonly used nanoclay due to its abundance, environmentally friendliness, large surface area and aspect ratio, and well-studied chemistry. The structure of MMT consists of layers made up of two tetrahedrally coordinated silicon atoms fused to an edge-shared octahedral sheet of either aluminum or magnesium hydroxide as illustrated in Fig. 3. The thickness of single-layer MMT is about 1 nm, and its length varies from tens of nanometers to more than one micron, depending on the layered silicate. These stacked layers have a regular van der Waals gap in between the nanoclay platelets, known as interlayer gallery.

MMT has certain limitations which hinders their application as filler in the polymer. The main limitation of MMT is its hydrophilic surface as we know that most polymers are organophilic compounds because of this mismatch surface characteristic MMT tends to agglomerate in the nanocomposite. The high hydrophilic characteristics of MMT are due by counterbalancing the negatively charged platelets by alkali and alkali earth cations (Na^+, K^+, Ca^{2+}, etc.) located in the galleries. One way to overcome this is by exchanging the inorganic cations by ammonium or phosphonium cations carrying at least one long alkyl chain. Thus, the resulting clays are organomodified layered silicates (commonly known as organo clay) which are compatible to process with most of the polymers. The preferred choice of polymers that have been extensively studied by the researcher for the preparation of nanoclay-based bio-nanocomposites are PLA, PHAs, PCL, poly (butylene succinate/adipate) (PBS/A), and poly(butylene adipate-co-terephthalate)

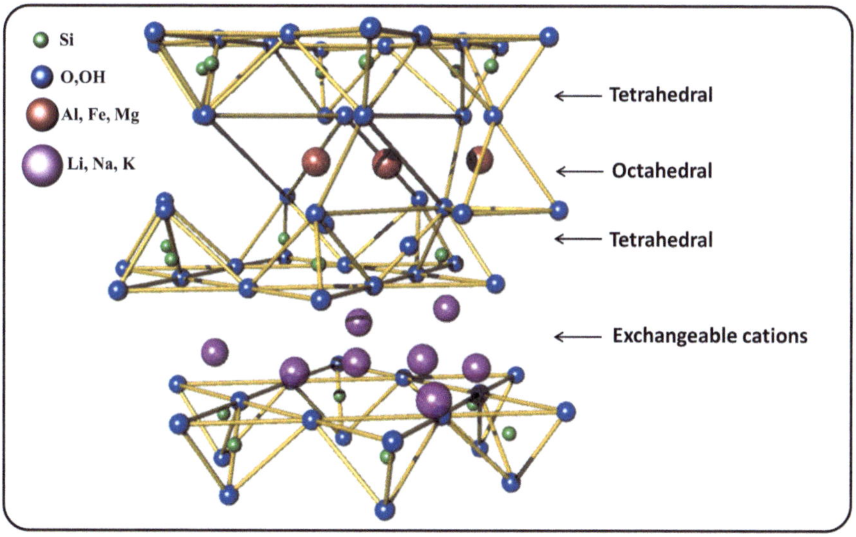

Fig. 3 Structure of montmorillonite nanoclay

(PBA/T) (Bordes et al. 2009). Addition of nanoclay improved the thermal stability, gas permeability, flame retardancy, and mechanical properties of aforesaid biopolymers. The addition of nanoclays to polymers, i.e., preparation of polymer/layered silicate nanocomposites, can be performed using one the following methods: (i) solvent intercalation, (ii) melt intercalation, and (iii) in situ intercalation (Bordes et al. 2009). This section will focus and presents the recent advancements in nanoclay-based bio-nanocomposites.

The preferred polymer of choice for the preparation of nanoclay-based bio-nanocomposite is PLA. PLA is the most sorts after biopolymer because of its excellent mechanical, thermal properties, and biocompatibility. However, its impact resistance, gas permeability, and heat distortion temperature are on the lower side which limits its applications. To overcome these limitations, it needs to be reinforced and one way of doing this is by using organoclays with different organomodifiers (Nazir et al. 2016). Ogata et al. (1997) was the first to attempt this and prepared PLA-based nanocomposites by solvent intercalation technique. Though the dispersion of layered silicates in the nanocomposite was not uniform as they formed immiscible tactoids of stacked silicate monolayers, the results showed increase in Young's modulus with the increase in clay content. Later, Chang et al. (2003) prepared the organomodified MMT clay and examined the influence of the aspect ratio of silicate layer and clay content on the PLA matrix. X-ray diffraction (XRD) and transmission electron microscopy (TEM) were shown the intercalation of clay in the PLA, and in this case, 2 wt% of organomodified MMT clay lead to improve strength about 65%. However, above these clay contents, properties decreased due to agglomeration.

Another preparation technique that was adequately tried by researchers is melt intercalation of PLA into silicate layers (Vaia and Giannelis 1997). In this technique, the polymer chains leave the bulk and penetrate into the interlayer space between the silicate layers, guiding to various nanocomposites with intercalated structures to exfoliated structures (Ray and Okamoto 2003). Arjmandi et al. (2017) prepared organomodified MMT/cellulose nanowhisker (CNW)-reinforced PLA nanocomposite by solution casting. They found $\sim$10–90% increased ductility with the addition of 5 phr MMT and 1 phr CNW in hybrid nanocomposites. Balakrishnan et al. (2010) prepared PLA/MMT nanocomposite and found that the impact strength decreased with increasing of MMT content. To overcome this problem, they blend LLDPE with PLA and increase the toughness by 22% by adding 10% of LDPE in PLA/MMT nanocomposite. Compared to solvent intercalation technique, melt intercalation technique improves the specificity for the intercalation of polymers by eliminating the competing host solvent (Pavlidou and Papaspyrides 2008). However, polymerization of PLA is complicated; hence, nanoclay/PLA nanocomposite prepared by in-situ polymerization has not been reported yet.

Until now, no fully exfoliated structure was reported by any researcher for layered silicate/PLA nanocomposites. The best-obtained structure is intercalated which was achieved by melt-processing technique. Also, some enhancement in mechanical, heat distortion temperature, gas permeability, and rate of crystallization with improved biodegradability was reported for PLA-based nanocomposites. Ray and Okamoto (2003) prepared PLA/MMT nanocomposites with trimethyl octadecylammonium cations as organomodified layered silicates. The prepared nanocomposites had intercalated silicate layers well distributed in PLA matrix. From rheological measurements, he explained that this improvement is because of the formation of mesoscopic structure of randomly oriented intercalated silicate layers in the PLA matrix. Another study reported that the addition of MMT to PLA under compost increases its biodegradability. The reason for the increase is biodegradability was correlated to the uniform dispersion of layered silicates into the PLA matrix; higher the dispersion, higher the degradation rate (Ray and Bousmina 2005). It has been proved that layered silicates have an influence on photooxidation processes on PLA backbone, in the field of durable outdoor applications (Bocchini et al. 2010).

Apart from PLA, PHAs is another important commercial bio-derived and biodegradable plastics due to properties such as water-insoluble and relatively resistant to hydrolytic degradation, non-toxic and good ultraviolet resistance. Maiti and Batt (2003) reported the preparation of organomodified MMT (OMMT)-reinforced PHB nanocomposites using melt extrusion technique. The morphological analysis showed well-distributed intercalated nanocomposites with decreasing d-spacing with an increase in clay content.

In a similar study by Lim et al. (2003) on PHB based nanocomposites prepared by solvent intercalation technique, the content of clay used was varied from 3, 6, and 9 wt%. XRD results showed the formation of intercalated structure with interlayer distance of 35 Å, independent of clay content. FTIR results supported the observation made from XRD results, as FTIR confirmed the presence of two distinct phases. Further, thermal analysis of these nanocomposites by TGA showed an increase in the onset temperature and decrease in the degradation rate with nanocomposite containing 3 wt% of C25A. This enhancement in properties was rendered by the nanoscale OMMT layers dispersion, which decreased the diffusion of volatile decomposition products. However, upon increasing the clay content to >6 wt%, the onset of the thermal degradation did not increase, but the degradation rate decreased. The reason for no change in the onset of thermal degradation is attributed to organomodifier's thermal sensitivity, and the decrease in degradation rate is due to the restricted thermal motion of the polymer chains in the OMMT interlayer.

After the successful preparation of PHB-based nanocomposites, researchers focused on the preparation of PHBV-based nanocomposites, since PHBV exhibits better properties and ease of processability than PHB. Choi et al. (2003) prepared PHBV/C30B nanocomposites with low clay content by melt intercalation using a Brabender mixer and reported its microstructure, thermal, and mechanical properties. The morphological results confirmed the formation of intercalated nanostructures and demonstrated that the nanodispersed organoclay acted as a nucleating agent, thus enhancing the temperature and rate of PHBV crystallization. Thermal analysis revealed a reduction in PHBV melting temperature which led to reduced crystallite size. Also, it was found by TGA that 3% of filler (C30B) enhanced the temperature at 3% weight loss by 10 °C. This enhancement in temperature was due to nanodispersion of the silicate layers into the matrix thus creating an effective barrier to the permeation of oxygen and combustion gas. Lastly, mechanical analysis evidenced that the effective reinforcement of clays to PHBV can be achieved as the results obtained show a good Young's modulus improvement from 480 to more than 790 MPa due to strong hydrogen bonding between PHBV and C30B.

Another study involving PHBV along with OMMT was carried out by Chen et al. (2004) and found that OMMT acted as a nucleating agent in PHBV matrix resulting in increased nucleation and crystallization rate, forming perfect PHBV crystals. Wang et al. (2005b) postulated that the nanoscaled OMMT layers affect the crystallization in two opposite ways—(1) a small part of OMMT can increase the crystalline nuclei thus causing a more rapid crystallization rate, and (2) owing to the interaction of OMMT layers with PHBV chains, most of the OMMT layers restrict the motion of the PHBV chains. Therefore, the crystallization rate increased, whereas the relative degree of crystallinity decreased with increasing amount of clay in the PHBV/OMMT nanocomposites.

2.4 *Sepiolite Reinforced Bio-nanocomposites*

Sepiolite is a fibrous magnesium silicates type of clay that organized in bundles. A single sepiolite fiber is approximately 0.2–4 μm long, depending on the origin, 10–30 nm wide and 5–10 nm thick. In comparison to the layered silicates, the specific surface area of the sepiolite is lower for the same aspect ratio; therefore, the sepiolite can be dispersed easier in a polymer matrix. Sepiolite features a large specific surface area in the region of 140–320 m^2/g (Galàn and Singer 2011).

The structure–property relationships of montmorillonite (MMT) and sepiolite nano-biocomposites based on poly(3hydroxybutyrate-co-4hydroxybutyrate) (P(3HB-co-4HB)) were reported by Khandal et al. (2016). The nanocomposites were prepared by melt mixing of various contents of MMT and sepiolite with and without organomodification. The presence of sepiolite increased the tensile strength of P(3HB-co-4HB) more than MMT. Furthermore, the addition of the organomodified sepiolite further improved the modulus of P(3HB-co-4HB) nanocomposites. This is due to good interactions through silanol groups along the length of the fiber with P(3HB-co-4HB). In another study, it is found that addition of 6 wt% sepiolite improved the tensile strength, modulus, and energy at break of porous poly(vinyl alcohol) (PVA) by up to 104, 331 and 22% (Killeen et al. 2012). The Young's modulus and compressive collapse strength of the gelatin foam with a cell size of 159 mm and a porosity of 98% were increased by 288 and 308%, respectively, in the presence of 9.1 wt% sepiolite (Frydrych et al. 2011).

The existence of strong interactions between sepiolite and polymer matrix not only improved the mechanical properties but also improved other properties such as water resistance and reduction of water absorption (Alcântara et al. 2014). It is found to increase the mean pore size and porosity of the PVA scaffold (Killeen et al.

Table 1 Tensile properties of sepiolite biopolymer nanocomposites

Polymer matrix	Sepiolite content (wt%)	Young's modulus (MPa)	Tensile strength (MPa)	References
PLA	0.5–3	1.9–2.5 (unmodified) 2.4–4.0 (silane treated)	17–14[a] (unmodified) 18–14.5[b]	Nima et al. (2015)
PHBV	3,5	544–658 (unmodified) 596–654 (organomodified)	15.5–15.6 (unmodified) 17.2–17.0 (organomodified)	Khandal et al. (2016)
Gelatin	0.5–9.1	1.847–6.031	–	Frydrych et al. (2011)
Starch Chitosan Alginate	5–50	1250–2250 3000–3500 4000–5000	–	Alcântara et al. (2014)

[a]Maximum value of 18.5 MPa at 1.5wt%
[b]Maximum value of 22 MPa at 1.5wt%

2012). The possible interaction mechanism between the biopolymers and the clay occurs mainly through the interaction of the OH groups in the biopolymers backbone and the silanol groups on the silicate surface. In the case of the presence of other functionalities, such as carboxylate and amino groups, respectively, it may be also implicated. The strong interactions between hydrophilic polymers and sepiolite can be indicated by the results from Fourier transform infrared spectroscopy and differential scanning calorimetry which show substantial shifts in the characteristic absorption peaks and the glass transition temperature in addition to the reinforcement effects of sepiolite on polymeric materials (Frydrych et al. 2011).

The tensile properties of selected sepiolite biocomposites are listed in Table 1. The degree of improvement in the properties is found to be not only a function of interfacial polymer–clay interactions and the morphology/shape of the nanoparticles but also the dispersion of the particles in the polymer matrix. In order to achieve a good dispersion of sepiolite in polymer matrix, few attempts on surface modification of sepiolite have been studied (Nikolic et al. 2017; Nima et al. 2015; Liu et al. 2012; Khandal et al. 2016). One of the modifications made was covalent grafting with short (3-mercaptopropyl)trimethoxysilane (SEP-SI) or long hexadecylamine (SEP-HDA) by using ionic exchange. Organomodified sepiolites dispersed better than native ones. The sepiolite modified with HDA showed the best mechanical results, as a result of better interaction between longer alkyl hydrophobic part of the organic modification and the hydrophobic PCL matrix (Nikolic et al. 2017). The bundles of sepiolite have been dispersed into small aggregates containing several nanorods without destroying the crystal structure after purification and organic modification (Liu et al. 2012) Sepiolite nanofibers were found to well dispersed in the PLA matrix, exhibiting a random orientation. The silane surface treatment of sepiolite increased the adhesion of the PLA matrix to the sepiolite nanoclay and resulted in better mechanical properties of the films as compared to pure PLA (Nima et al. 2015). Young's modulus of PLA films increased linearly with increasing sepiolite loadings. All the nanocomposite films exhibited about 40% increase in modulus compared to the pure PLA films. The improvement in modulus was more pronounced with the addition of silane-modified sepiolite as a result of better dispersion and adhesion of sepiolite particles within PLA. The silane treatment of sepiolite provided larger interfacial region; thus, stress can be transferred effectively from PLA matrix to sepiolite nanoclay. Optimum content of 1.5% wt sepiolite increased the tensile strength before decreased with the addition of 3 wt% sepiolite, which is attributed to the aggregation of sepiolite particles.

2.5 *Halloysite-Reinforced Bio-nanocomposites*

Halloysite nanotubes (HNTs) are novel 1D natural nanomaterials with predominantly hollow tubular nanostructures and high aspect ratios. Due to their high mechanical strength, thermal stability, biocompatibility, and abundance, HNTs have a number of exciting potential applications in polymer bio-nanocomposites. Biopolymer such as PLA, PHBV, and PCL are an important biodegradable, biocompatible, and thermoplasticity polyester, which has been widely considered for environmentally friendly packaging, medical applications, and pharmaceutical controlled release systems. There are some drawbacks to their future application. Most of the biopolymer is stiff and brittle, which results in very poor mechanical properties. Other classes of biopolymer such as regenerated cellulose, starch base biopolymer, pectin, polyethylene glycol also suffer the poor mechanical properties.

Biopolymer nanocomposites reinforced with HNTs possess highly increased tensile and flexural strength (Torres et al. 2017), elastic moduli, and improved toughness. The functionalization of HNTs with the various methods further enhances the mechanical properties of biopolymer/HNT nanocomposite (Prakash et al. 2017; Marius et al. 2012; Kalappa et al. 2013). Table 2 recorded the mechanical properties enhancement of a few bio-nanocomposites at their optimum HNT loading. Prakash et al. 2017 reported that the tensile strength of PLA/HNT bio-nanocomposite prepared by melt mixing method increased to 62.6 MPa with the addition of 4wt% of APTES-modified HNT which is 26.5% higher than pure PLA and 15% higher than unmodified (4 wt%) PLA/HNT nanocomposite. The impact strength of 4 wt% APTES-modified HNT was 29.8 MPa, which is 20% higher than the unmodified PLA/HNT nanocomposites and 40% higher than pure PLA. HNT is also capable of enhancing the mechanical properties of biopolymer thin film, especially for packaging application. Nurbaiti et al. (2014) stated that both tensile strength and Young's modulus of regenerated cellulose/HNT film were enhanced with the addition of 5 wt% HNT loading. The enhancement of mechanical properties of potato starch/HNT nanoclay nanocomposite films was also reported by Hassan and Nafchi (2014).

Table 2 Mechanical properties of a few HNT biopolymer nanocomposites at their optimum HNT loading

Polymer matrix	Optimum HNT loading	Tensile strength (MPa)	Young's modulus (MPa)	Impact strength	References
PLA/HNT	12%	60.0	2770	–	Sandrine et al. (2017)
PLA/ silane-modified HNT	4 wt%	62.6	1198	29.8 J/m	Prakash et al. (2017)
PLA/HNT	4 wt%	54.8	1152	25.9 J/m	Prakash et. al. (2017)
PLA/silane-modified HNT	6 wt%	70.0	2800	2.8 kJ/m^2	Marius et al. (2012)
PLA/HNT	40%	74.1	–	24.9 J/m	Mingxian et al. (2013)
PLA/HNT	6 wt%	62.3	3378	3.71 kJ/m^2	Kalappa et al. (2013)
PLA/treated HNT	6 wt%	70.3	3644	3.88 kJ.m^2	Kalappa et al. (2013)
Potato starch/HNT	5%	9.82	376	–	Hassan and Nafchi (2014)
Pectin/ polyethylene glycol blend/HNT	30 wt%	27	4200	–	Giuseppe et al. (2013)
PCL	10 wt%	84.7	919	–	Kang and Young (2013)
PHBV	3 wt%	39.4	36,000	20 J/m	Larissa et al. (2011)
Regenerated cellulose	5%	45.0	35,000	–	Nurbaiti et al. (2014)

HNT biopolymer nanocomposites are introduced afterward. The differences of reinforcing effect for biopolymer were due to the structural characteristics of nanoparticles, dispersion state, and interfacial interactions in these systems. Comparing with other nanoparticles, HNT exhibited better reinforcing ability for biopolymer in term of the mechanical properties. The excellent reinforcing effect of HNT on biopolymer could be attributed to the uniformly distributed rigid nanotubes in biopolymer matrix and the hydrogen bonding interactions between HNT and biopolymer (Prakash et al. 2017; Mingxian et al. 2013; Kalappa et al. 2013).

HNT biopolymer nanocomposites also exhibit elevated thermal resistance and unique crystallization behavior. The thermal stability of PLA was enhanced by the addition of HNT in PLA/HNT bio-nanocomposite (Prakash et al. 2017; Mingxian et al. 2013). A similar observation was also reported in regenerated cellulose/HNT nanocomposite films (Nurbaiti et al. 2014) and PCL/HNT nanocomposite (Torres et al. 2017). Marius et al. 2012 recommended the use of PLA/HNT nanocomposite in high-temperature application due to the thermal stability of the nanocomposites.

The bio-nanocomposite-filled HNT has the potential applications in environmental protection (packaging), biomaterials, and biomedical since HNT polymer nanocomposites have demonstrated good drug encapsulation and sustained release abilities, gaining them extensive use as tissue engineering scaffolds and drug carriers.

3 Conclusions

The potential of nanofillers becomes distinct by their unique combination of properties. Incorporation of nanofillers in polymer composites not only improves the mechanical properties but can also result in enhancements in thermal stability, flame retardancy, and gas permeability. Additionally, some nanofillers such as graphene and carbon nanotube also provide improved conductive properties of biocomposites as they possess excellent electrical and thermal conductivities. Surface modification is a promising approach to enhance the interaction between matrix and nanofillers. Regardless of the decrease in thermal degradation rate of the bio-nanocomposites due to high thermal stability of nanoclay, there is no significant influence by nanoclay on biodegradability of bio-nanocomposites. Factors affecting the interaction and compatibility between matrix and nanofillers are complex; however, increasing understanding of computational materials design will lead to more effective targeting of interventions of nanofillers in composites.

References

Ajayan P, Stephan O, Colliex C et al (1994) Aligned carbon nanotube arrays formed by cutting a polymer resin-nanotube composite. Science 265(5176):1212–1214

Alcântara ACS, Darder M, Aranda P et al (2014) Polysaccharide–fibrous clay bionanocomposites. Appl Clay Sci 96:2–8

Ambrosio-Martin J, Gorrasi G, Lopez-Rubio A et al (2015a) On the use of ball milling to develop PHBV—graphene nanocomposites (I)—morphology, thermal properties, and thermal stability. J Appl Polym Sci. https://doi.org/10.1002/app.42101

Ambrosio-Martin J, Gorrasi G, Lopez-Rubio A et al (2015b) On the use of ball milling to develop poly(3-hydroxybutyrate-co-3-hydroxyvalerate)-graphene nanocomposites (II)—mechanical, barrier, and electrical properties. J Appl Polym Sci. https://doi.org/10.1002/app.42217

Arjmandi R, Hassan A, Haafiz MM et al (2017) Hybrid montmorillonite/cellulose nanowhiskers reinforced polylactic acid nanocomposites: a comparative study based on formulation design. In: Cellulose-reinforced nanofibre composites. Elsevier, Netherlands, pp 25–44

Balakrishnan H, Hassan A, Wahit MU et al (2010) Novel toughened polylactic acid nanocomposite: mechanical, thermal and morphological properties. Mater Des 31(7):3289–3298

Barrett JS, Abdala AA, Srienc F (2014) Poly(hydroxyalkanoate) elastomers and their graphene nanocomposites. Macromolecules 47(12):3926–3941

Bocchini S, Fukushima K, Blasio AD et al (2010) Polylactic acid and polylactic acid-based nanocomposite photooxidation. Biomacromolecules 11(11):2919–2926

Bordes P, Pollet E, Avérous L (2009) Nano-biocomposites: biodegradable polyester/nanoclay systems. Prog Polym Sci 34(2):125–155
Botta L, Scaffaro R, Sutera F et al (2018) Reprocessing of PLA/graphene nanoplatelets nanocomposites. Polymers. https://doi.org/10.3390/polym10010018
Chang JH, An YU, Cho D et al (2003) Poly(lactic acid) nanocomposites: comparison of their properties with montmorillonite and synthetic mica (II). Polymer 44(13):3715–3720
Chen G, Hao G, Guo T et al (2004) Crystallization kinetics of poly(3-hydroxybutyrate-co-3-hydroxyvalerate)/clay nanocomposites. J Appl Polym Sci 93(2):655–661
Choi WM, Kim TW, Park OO et al (2003) Preparation and characterization of poly (hydroxybutyrate-co-hydroxyvalerate)–organoclay nanocomposites. J Appl Polym Sci 90 (2):525–529
Chrissafis K (2010) Detail kinetic analysis of the thermal decomposition of PLA with oxidized multi-walled carbon nanotubes. Thermochim Acta 511(1–2):163–167
Frydrych M, Wan C, Stengler R et al (2011) Structure and mechanical properties of gelatin/ sepiolite nanocomposite foams. J Mater Chem 21:9103–9111
Galàn E, Singer A (eds) (2011) Developments in clay science. Elsevier, Netherlands
Gao Y, Picot OT, Bilotti E et al (2017) Influence of filler size on the properties of poly(lactic acid) (PLA)/graphene nanoplatelet (GNP) nanocomposites. European Polymer Journal 86:117–131
Giuseppe C, Giuseppe L, Stefana M (2013) Sustainable nanocomposites based on halloysite nanotubes and pectin/polyethylene glycol blend. Polym Degrad Stab 98:2529–2536
Gorrasi G, Tortora M, Vittoria V et al (2002) Transport and mechanical properties of blends of poly(ε-caprolactone) and a modified montmorillonite poly(ε-caprolactone) nanocomposite. J Polym Sci Part B Polym Phys 40(11):1118–1124
Gouvêa RF, Del Aguila EM, Paschoalin VMF et al (2018) Extruded hybrids based on poly (3-hydroxybutyrate-co-3-hydroxyvalerate) and reduced graphene oxide composite for active food packaging. Food Pack Shelf Life 16:77–85
Hassan FS, Nafchi AM (2014) Preparation and characterization of bionanocomposite films based on potato starch/halloysite nanoclay. Int J Biol Macromol 67:458–462
Iijima S (1991) Helical microtubules of graphitic carbon. Nature 354(6348):56–58
Kalappa P, Benoıt L, Michel S, Marie FL, Patricia K (2013) Poly(lactic acid)/halloysite nanotubes nanocomposites: structure, thermal, and mechanical properties as a function of halloysite treatment. J Appl Polym Sci 128:1895–1903
Kang SL, Young WC (2013) Thermal, mechanical, and rheological properties of poly (ε-caprolactone)/halloysite nanotube nanocomposites. J Appl Polym Sci 128(5):2807–2816
Khandal D, Pollet E, Avérous L (2016) Elaboration and behavior of poly (3-hydroxybutyrate-co-4-hydroxybutyrate)-nano-biocomposites based on montmorillonite or sepiolite nanoclays. Eur Polym J 81:64–76
Killeen D, Frydrych M, Chen B (2012) Porous poly(vinyl alcohol)/sepiolite bone scaffolds: preparation, structure and mechanical properties. Mater Sci Eng C 32:749–757
Kim H, Abdala AA, Macosko CW (2010) Graphene/polymer nanocomposites. Macromolecules 43(16):6515–6530
Kuan C-F, Kuan HC, Ma CCM et al (2008) Mechanical and electrical properties of multi-wall carbon nanotube/poly(lactic acid) composites. J Phys Chem Solids 69(5–6):1395–1398
Larissa NC, Janaina SC, Raquel SM (2011) PHBV nanocomposites based on organomodified montmorillonite and halloysite: the effect of clay type on the morphology and thermal and mechanical properties. Compos Part A 42:1601–1608
Lee C, Wei X, Kysar JW et al (2008) Measurement of the elastic properties and intrinsic strength of monolayer graphene. Science 321(5887):385–388
Lee H-H, Shin US, Jin GZ et al (2011) Highly homogeneous carbon nanotube-polycaprolactone composites with various and controllable concentrations of ionically-modified-MWCNTs. Bull Korean Chem Soc 32(1):157–161
Li MX, Kim SH, Choi SW et al (2016) Effect of reinforcing particles on hydrolytic degradation behavior of poly(lactic acid) composites. Compos Part B 96:248–254

Lim ST, Hyun YH, Lee CH et al (2003) Preparation and characterization of microbial biodegradable poly(3-hydroxybutyrate)/organoclay nanocomposite. J Mater Sci Lett 22 (4):299–302
Liu M, Pu M, Ma H (2012) Preparation, structure and thermal properties of polylactide/sepiolite nanocomposites with and without organic modifiers. Composites Science and Technology 72 (13):1508–1514
Maiti P, Batt C (2003) Renewable plastics: synthesis and properties of PHB nanocomposites. Polym Mater Sci Eng 88:58–59
Marius M, Anne LD, Yoann P et al (2012) Polylactide (PLA)—halloysite nanocomposites: production, morphology and key-properties. J Polym Environ 20:932–943
Meng Z, Zheng W, Li L et al (2010) Fabrication and characterization of three-dimensional nanofiber membrane of PCL–MWCNTs by electrospinning. Mater Sci Eng C 30(7):1014–1021
Mingxian L, Yun Z, Changren Z (2013) Nanocomposites of halloysite and polylactide. Appl Clay Sci 75–76:52–59
Miranda-Trevino JC, Coles CA (2003) Kaolinite properties, structure and influence of metal retention on pH. Appl Clay Sci 23(1–4):133–139
Moniruzzaman M, Winey KI (2006) Polymer nanocomposites containing carbon nanotubes. Macromolecules 39(16):5194–5205
Montagna LS, Montanheiro TLDA, Borges AC et al (2017a) Biodegradation of PHBV/GNS nanocomposites by *Penicillium funiculosum*. J Appl Polym Sci. https://doi.org/10.1002/app.44234
Montagna LS, Montanheiro TLDA, Machado JPB et al (2017b) Effect of graphite nanosheets on properties of poly(3-hydroxybutyrate-co-3-hydroxyvalerate). Int J Polym Sci. https://doi.org/10.1155/2017/9316761
Montagna LS, Montanheiro TLDA, Passador FR et al (2018) The Influence of artificial photodegradation on properties of poly(3-hydroxybutyrate-co-3-hydroxyvalerate) (PHBV)/graphite nanosheets (GNS) nanocomposites. J Polym Environ 26(4):1511–1519
Nazir MS, Kassim MHM, Mohapatra L et al (2016) Characteristic properties of nanoclays and characterization of nanoparticulates and nanocomposites. In: Nanoclay reinforced polymer composites. Springer, Berlin, pp 35–55
Nikolic MS, Petrovic R, Veljovic D et al (2017) Effect of sepiolite organomodification on the performance of PCL/sepiolite nanocomposites. Eur Polym J 97:198–209
Nima M, Zurina M, Nazila D (2015) Study of silane treatment on poly-lactic acid(PLA)/sepiolite nanocomposite thin films. J Appl Polym Sci 132(6)
Nuona A, Li X, Zhu X et al (2015) Starch/polylactide sustainable composites: interface tailoring with graphene oxide. Compos Part A Appl Sci Manuf 69:247–254
Nurbaiti AH, Mat UW, Qipeng G et al (2014) Development of regenerated cellulose/halloysites nanocomposites via ionic liquids. Carbohydr Polym 99:91–97
Ochoa M, Collazos N, Le T et al (2017) Nanocellulose-PE-b-PEG copolymer nanohybrid shish-kebab structure via interfacial crystallization. Carbohydr Polym 159:116–124
Ogata N, Jimenez G, Kawai H et al (1997) Structure and thermal/mechanical properties of poly(L-lactide)-clay blend. J Polym Sci Part B Polym Phys 35(2):389–396
Pavlidou S, Papaspyrides C (2008) A review on polymer–layered silicate nanocomposites. Prog Polym Sci 33(12):1119–1198
Pinto AM, Moreira S, Gonçalves IC et al (2013) Biocompatibility of poly(lactic acid) with incorporated graphene-based materials. Colloids Surf B 104:229–238
Pinto AM, Gonçalves C, Gonçalves IC et al (2016) Effect of biodegradation on thermo-mechanical properties and biocompatibility of poly(lactic acid)/graphene nanoplatelets composites. Eur Polym J 85:431–444
Pötschke P, Andres T, Villmow T et al (2010) Liquid sensing properties of fibres prepared by melt spinning from poly(lactic acid) containing multi-walled carbon nanotubes. Compos Sci Technol 70(2):343–349

Prajapati V, Sharma P, Banik A (2011) Carbon nanotubes and its applications. Int J Pharm Sci Res 2(5):1099–1107
Prakash K, Ratnam CT, Sivakumar M (2017) Development of silane grafted halloysite nanotube reinforced polylactide nanocomposites for the enhancement of mechanical, thermal and dynamic-mechanical properties. Appl Clay Sci 135:583–595
Pramanik N, De J, Basu RK et al (2016) Fabrication of magnetite nanoparticle doped reduced graphene oxide grafted polyhydroxyalkanoate nanocomposites for tissue engineering application. RSC Adv 6(52):46116–46133
Qi P, Vermesh O, Grecu M et al (2003) Toward large arrays of multiplex functionalized carbon nanotube sensors for highly sensitive and selective molecular detection. Nano Lett 3(3):347–351
Ray SS, Bousmina M (2005) Biodegradable polymers and their layered silicate nanocomposites: in greening the 21st century materials world. Prog Mater Sci 50(8):962–1079
Ray SS, Okamoto M (2003) Polymer/layered silicate nanocomposites: a review from preparation to processing. Prog Polym Sci 28(11):1539–1641
Rhim JW, Park HM, Ha CS (2013) Bio-nanocomposites for food packaging applications. Prog Polym Sci 38(10):1629–1652
Sadasivuni KK, Ponnamma D, Thomas S et al (2014) Evolution from graphite to graphene elastomer composites. Prog Polym Sci 39(4):749–780
Salam MA, Makki MS, Abdelaal MY (2011) Preparation and characterization of multi-walled carbon nanotubes/chitosan nanocomposite and its application for the removal of heavy metals from aqueous solution. J Alloys Compd 509(5):2582–2587
Sandrine T, Marius M, Philippe D (2017) Bionanocomposites based on PLA and halloysite nanotubes: from key properties to photooxidative degradation. Polym Degrad Stab 145:60–69
Sridhar V, Lee I, Chun HH et al (2013) Graphene reinforced biodegradable poly (3-hydroxybutyrate-co-4-hydroxybutyrate) nano-composites. Express Polym Lett 7(4):320–328
Sun Y, He C (2012) Synthesis and stereocomplex crystallization of poly(lactide)–graphene oxide nanocomposites. ACS Macro Lett 1(6):709–713
Torres E, Fombuena V, Vallés-Lluch A et al (2017) Improvement of mechanical and biological properties of holycaprolactone loaded with hydroxyapatite and halloysite nanotubes. Mater Sci Eng C 75:418–424
Vaia RA, Giannelis EP (1997) Polymer melt intercalation in organically-modified layered silicates: model predictions and experiment. Macromolecules 30(25):8000–8009
Vaisman L, Wagner HD, Marom G (2006) The role of surfactants in dispersion of carbon nanotubes. Adv Colloid Interface Sci 128:37–46
Villmow T, Pötschke P, Pegel S et al (2008) Influence of twin-screw extrusion conditions on the dispersion of multi-walled carbon nanotubes in a poly(lactic acid) matrix. Polymer 49 (16):3500–3509
Wang SF, Shen L, Zhang WD et al (2005a) Preparation and mechanical properties of chitosan/ carbon nanotubes composites. Biomacromolecules 6(6):3067–3072
Wang S, Song C, Chen G et al (2005b) Characteristics and biodegradation properties of poly (3-hydroxybutyrate-co-3-hydroxyvalerate)/organophilic montmorillonite (PHBV/OMMT) nanocomposite. Polym Degrad Stab 87(1):69–76
Wang BJ, Zhang YJ, Zhang JQ et al (2013) Crystallization behavior, thermal and mechanical properties of PHBV/graphene nanosheet composites. Chin J Polym Sci 31(4):670–678
Wu D, Zhang Y, Zhang M et al (2009) Selective localization of multiwalled carbon nanotubes in poly(ε-caprolactone)/polylactide blend. Biomacromolecules 10(2):417–424
Wu D, Wu L, Zhou W et al (2010) Relations between the aspect ratio of carbon nanotubes and the formation of percolation networks in biodegradable polylactide/carbon nanotube composites. J Polym Sci Part B Polym Phys 48(4):479–489
Wu D, Lin D, Zhang J et al (2011) Selective localization of nanofillers: effect on morphology and crystallization of PLA/PCL blends. Macromol Chem Phys 212(6):613–626

Wu D, Lv Q, Feng S et al (2015) Polylactide composite foams containing carbon nanotubes and carbon black: synergistic effect of filler on electrical conductivity. Carbon 95:380–387

Xu JZ, Chen T, Yang CL et al (2010) Isothermal crystallization of poly(L-lactide) induced by graphene nanosheets and carbon nanotubes: a comparative study. Macromolecules 43 (11):5000–5008

Yoon JT, Lee SC, Jeong YG (2010) Effects of grafted chain length on mechanical and electrical properties of nanocomposites containing polylactide-grafted carbon nanotubes. Compos Sci Technol 70(5):776–782

Zhang J, Yang H, Shen G et al (2010) Reduction of graphene oxide via L-ascorbic acid. Chem Comm 46(7):1112–1114

Zhang D, Liu X, Wu G (2016) Forming CNT-guided stereocomplex networks in polylactide-based nanocomposites. Compos Sci Technol 128:8–16

Zine R, Sinha M (2017) Nanofibrous poly(3-hydroxybutyrate-co-3-hydroxyvalerate)/collagen/graphene oxide scaffolds for wound coverage. Mater Sci Eng C 80:129–134

Cellulose Nanocrystals-Based Nanocomposites

Malladi Nagalakshmaiah, Malladi Rajinipriya, Sadaf Afrin, Mohd Ayub Ansari, Mohammad Asad and Zoheb Karim

Abstract In this chapter, an effort has been made to summarize the outstanding research and development related to cellulose nanocrystal-reinforced nanocomposites. A detailed study showed the isolation of crystalline part of cellulose fibers using various chemicals is reported. Furthermore, different functional groups emerged since used chemicals during isolation steps are discussed and their interference during composites production is reported (effect on dispersion, distribution, mechanical properties, etc). Various processing routes are also reported for the production of dimensional nanocomposites. Authors have tried to show a comparative study of various processing routes and impact on final properties. The opted processing routes somehow affect the properties, which someway indicate the possible application in the future. In the last, two emerging applications of cellulose nanocrystal-based nanocomposites have been discussed in short. Water purification and fabrication of scaffold for regeneration of bone are new and budding fields, required biodegradable and biocompatible dimensional structures for green future. Therefore, fabricated cellulose nanocrystal-based composites might be a possible solution of these hurdles, which not only make the process green but also directly convert the waste materials into valuable products.

M. Nagalakshmaiah · M. Rajinipriya
Center for Innovation in Technological Ecodesign (CITE), University of Sherbrooke, Sherbrooke, QC, Canada

S. Afrin · Z. Karim (✉)
MoRe Research Örnsköldsvik AB, Box 70, 891 22 Örnsköldsvik, Sweden
e-mail: zoheb.karim@more.se

M. A. Ansari
Department of Chemistry, Bipin Bihari College, Jhnasi 284001, India

M. Asad
Chemistry Department, Faculty of Science, King Abdulaziz University, Jeddah 21589, Saudi Arabia

M. Asad
Faculty of Science, Center of Excellence for Advanced Materials Research (CEAMR), King Abdulaziz University, Jeddah 21589, Saudi Arabia

M. L. Sanyang and M. Jawaid (eds.), *Bio-based Polymers and Nanocomposites*,
https://doi.org/10.1007/978-3-030-05825-8_3

Keywords Cellulose nanocrystals · Functional properties · Polymer nanocomposites · Affinity membranes · Porous structure

1 Introduction

In recent times, lignocellulosic biomass has vast attention and used for the production of novel functional and structural dimensional nanocomposites. These nanocomposites are sustainable and used in a large number of applications like packaging, automotive parts, biomedical and electronic sensors. Cellulose is the major component present in the wood and abundantly available natural polymer on the earth. Cellulose is composed with the cellobiase which is the dimer of β-1, 4-anhydro-D-glucopyranose unit (Klemm et al. 2005; Ramires and Dufresne 2011).

Cellulose fibers are arranged in bundles of the microfibers in the plant cell wall. These microfibers are arranged with alternative amorphous and crystalline regions. Lately, terminology "nano" has a marked interest among scientists, researchers and industries since decades. The highlight of "nano" dimension is the high surface area to volume ratios. In this aspect, nanocellulose emerged as novel materials and has great attention for researchers and industries due to their incredible physical and mechanical properties. Furthermore, surface functionalization of cellulose nanoparticles makes them ideal for the preparation of nanocomposites (Karim and Afrin 2015; Nagalakshmaiah et al. 2016a; Karim et al. 2017a; Afrin and Karim 2017).

Nanocellulose can be broken down into nanofibrillated cellulose (NFC), also referred as cellulose nanofibrils (CNF) and cellulose nanocrystals (CNC) also called whiskers or nanocrystalline cellulose (NCC) as shown in Fig. 1. The aqueous suspensions of cellulose nanofibrils can be prepared by mechanical treatment, enzymatic treatment and chemical treatment to yield high surface area. Further acid hydrolysis results the distributed colloidal suspensions of cellulose nanocrystals (Moon et al. 2011). Recently, a new type of cellulose nanocrystals has been reported by Mathew et al. (2014) isolated through integrated process step of bioethanol plant. Cellulose nanocrystals (CNC) are rod-like nanoparticles which shows incredible properties for instance, aspect ratio (10–80), modulus (100–150 GPa), high crystallinity (up to 95%), low density (1.5–1.6 g cm^{-3}) and high surface area (up to few 100 m^2 g^{-1}) (Nagalakshmaiah et al. 2017; Mautner et al. 2016; Mathew et al. 2014; Habibi et al. 2010).

Reinforcing of CNC with polymer is the major investigated area in the composites field. CNC has great tendency to improve the mechanical properties of the end-user applications. Nevertheless, the processing methods of these composites are imperative. Various approaches were discussed in the literature regarding fabrication of CNC-reinforced composites. Melt processing technologies (batch processing, continuous melt processing), solvent casting, filtration. Spinning, impregnation, foaming, freeze-drying, sol-gel method, etc., are most reported technologies. Different processing routes give different properties of produced dimensional nanocomposites. Porous nanocomposites could be produced using freeze-drying,

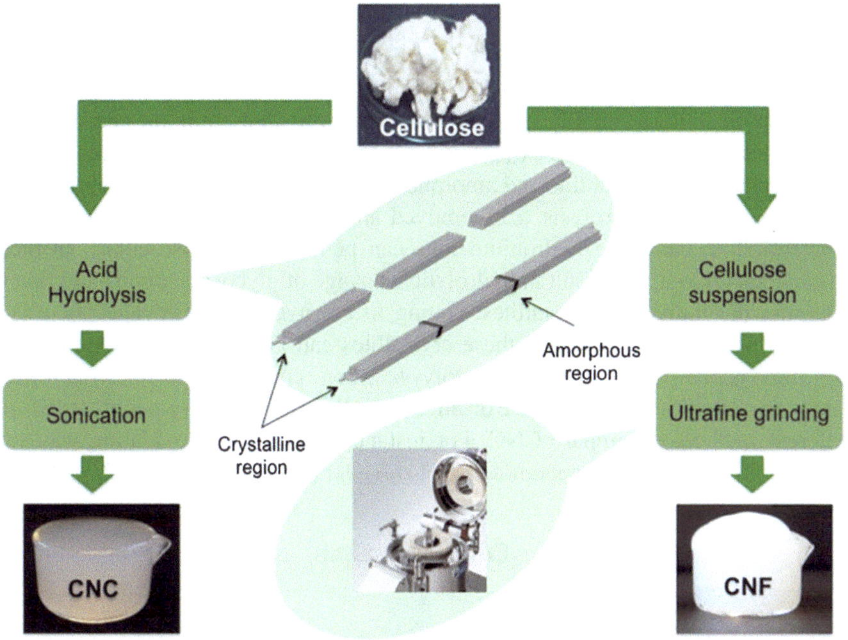

Fig. 1 Two reported routes for the isolation of cellulose crystals (CNC) and cellulose nanofibers (CNF) are acid hydrolysis and fine grinding. Copyright (2016), Lulea University of Technology, Sweden

foaming and sometime vacuum-filtration approach. Consequently, melt spinning, compounding, solvent casting leads to dense network formation (Karim et al. 2017b; Oksman et al. 2016).

Applications of CNC-based nanocomposites somehow depend on the networking properties of fabricated dimensional structures. This chapter describes the different preparation methods and properties of CNC precisely. Further, the different processing methods of nanocomposites and their performance in two specific emerging applications are deliberated in short.

2 Cellulose Nanocrystals: An Overview

Cellulose nanocrystals are generally referred as cellulose whiskers and were first produced by Rånby (1951). Sulfuric acid hydrolysis process is mostly used, which dissolves the amorphous parts of the cellulose by leaving the crystalline parts. Through this method, negative charged rod-like CNC with sulfate groups on the surface could be produced. The typical diameter of the CNC will vary between the 3 and 35 nm and length will be in 100 nm to few hundred nm. However, the size and morphology of the CNC completely depend on the source of the CNC produced.

2.1 Isolation Methods

The first step in the extraction of CNC is the purification of biomass to remove lignin, hemicellulose and other non-cellulosic materials present in the cell wall or to remove bacterial debris when extracting from bacterial cellulose. Obtained cellulose fibers contain both crystalline and amorphous regions in its structure (Ehmann et al. 2014). During acid hydrolysis, acid removed amorphous regions and leaves behind the crystalline part. The hydronium ions can penetrate cellulose chains in these amorphous regions and result in hydrolytic cleavage of glycosidic bonds and finally releasing the individual crystallites. Owing to the hydrolytic cleavage, there is a freedom of motion, and hence, these crystallites can grow. This can explain why the dimensions increase after hydrolysis when compared to the microfibril dimension (de Souza Lima and Borsali 2004). Researchers have been carried out different methods to prepare CNC. For instance, low toxicity oxidation was also reported with ammonium persulfate [$(NH4)_2S_2O_8$] (Scheme 2) in Fig. 2, yields the

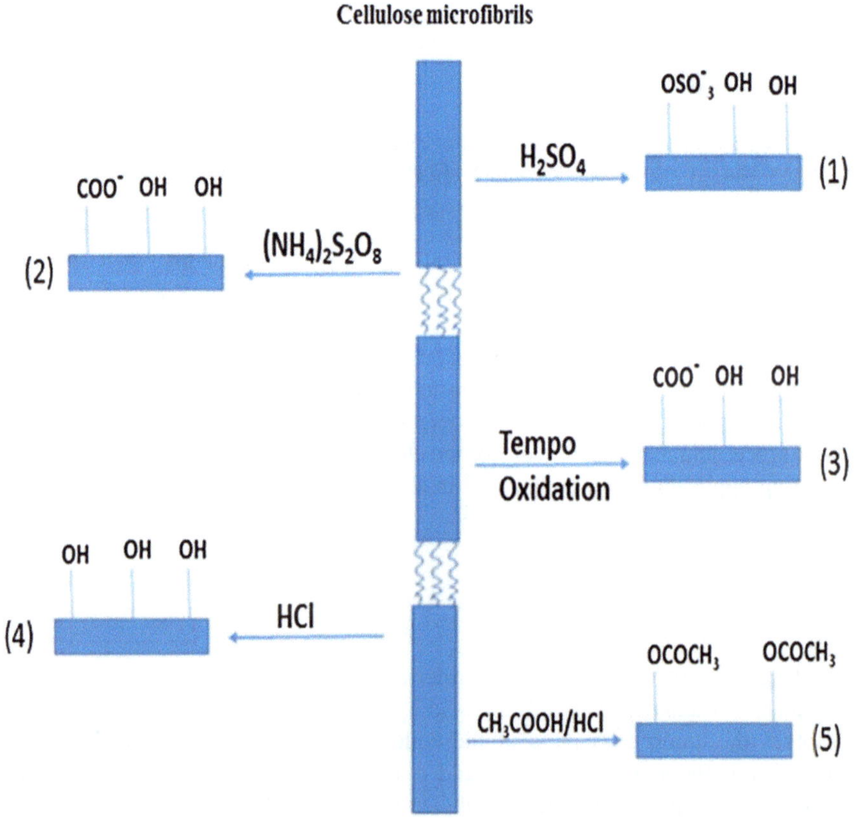

Fig. 2 Isolation of CNC from cellulose fibers by different methods. Adopted from International EPNOE Junior Scientists Meeting Future Perspectives in Polysaccharide Research et al. (2015)

carboxylate cellulose nanocrystals (Leung et al. 2011) and similar type of CNC was isolated by using the combination of (2, 2, 6, 6-Tetramethylpiperidin-1-yl) oxyl (TEMPO) and sodium hypochlorite. During this process, the surface hydroxyl groups of cellulose nanocrystals can be selectively transformed into carboxyl groups (Scheme 3) in Fig. 2, which may be useful for subsequent surface modification (Hirota et al. 2010). Hydrolysis with hydrochloric acid also used to preserve the hydroxyl groups (Scheme 4) of native cellulose but leads to less stable aqueous suspensions (Beck-Candanedo et al. 2005) compared to that of surface charged CNC. Furthermore, an acid mixture composed of hydrochloric and an organic acid (acetic acid) can be used to hydrolyze cellulose fibers inducing the hydrolysis and modification in one single-step process (Braun and Dorgan 2009), which offers the hydrophobic acetyl groups (Scheme 5) in Fig. 2 on the surface of cellulose nanocrystals. Recently, some novel systems have been developed for the isolation of cellulose nanocrystals, such as enzymatic hydrolysis, ionic liquid hydrolysis, gaseous acid hydrolysis, phosphoric acid and hydrobromic acid to yield CNC (Koshizawa 1960; Sadeghifar et al. 2011; Yu et al. 2013). However, most of the researchers extensively use sulfuric acid hydrolysis for the preparation of CNC.

CNC prepared from sulfuric acid hydrolysis is shown to be more stable than that are prepared using hydrochloric acid. This can be attributed to the acidic sulfate half-ester groups are introduced on the surface of the CNC that generates the electrical double-layer repulsion between the nanoparticles in the suspension. This phenomenon results in the less agglomeration and flocculation of the cellulose nanocrystals in the aqueous medium and also warrants their interaction with themselves and with the polymer matrix (Corrêa et al. 2010; Boluk et al. 2011).

Since 1950s, CNC was prepared from different sources like wood, cotton and tunicate. Off late, CNC is also extracted from the agricultural biomass, for example, chili leftover, coconut husk and fibers. The need of the materials that are compatible with environment is increasing, and hence, researchers are concentrating in extracting nanocellulose from the crop wastes like sugarcane bagasse, rice straw, oats straw and corn cob. This also allows the utilization of the crop wastes which otherwise are burnt in the fields. The CNC obtained from various sources are tabulated with their purification and the isolation process in Table 1.

3 Properties of CNC

The morphological, crystalline, thermal, rheological and mechanical properties of the isolated CNC are vital for the end-user applications. The morphology of the CNC will vary on the source of the isolated biomass. The crystallinity is one of the important factors for CNC, and it plays a vital role in the better mechanical performance of the materials. The size is also significant parameter as the aspect ratio (*L*/*D*) is depends on the length (*L*) and diameter (*D*) of the CNC. The dimensions and surface charge of the CNC are influenced by the hydrolysis time, concentration

Table 1 CNC extracted from different sources and their isolated process

Source	Purification technique	Preparation method	References
Ramie	Sodium hydroxide	H_2SO_4 hydrolysis	Peresin et al. (2010)
Cotton	Sodium hydroxide	H_2SO_4 hydrolysis	Lin and Dufresne (2014)
Lyocell fibers	Sodium hydroxide	Ammonium persulfate	Cheng et al. (2014)
Bacterial cellulose	Sodium hydroxide	H_2SO_4 hydrolysis	Hirai et al. (2009)
Wood	Hydrochloric acid and Sodium hydroxide	TEMPO/ Sonication	Hirota et al. (2010)
Wood	High-pressure homogenization (HPH)	Ball-milling	Nge et al. (2013)
Mulberry barks	Sodium hydroxide	H_2SO_4 hydrolysis	Li et al. (2009)
Coconut husk fibers	Sodium hydroxide Acid hydrolysis Bleaching with Sodium chlorite and glacial acetic acid	H_2SO_4 hydrolysis	Rosa et al. (2010)

of the acid, acid used for the preparation and acid-to-pulp ratio on the properties. These properties make them a highly compatible material for the nanocomposite as filler.

3.1 Morphological, Crystalline, Thermal and Mechanical Properties of CNC

The morphology of the obtained CNC can be studied by using various microscopic techniques such as transmission electron microscopy (TEM), scanning electron microscopy (SEM) and atomic force microscopy (AFM). The morphological inspection benefits to evaluate the surface morphology of the end CNC and with the same techniques the dimensions and subsequently aspect ratio. If the aspect ratio is high, the percolation network between the CNC will be high; therefore, it will lead to better mechanical performance. As explained above, the morphology of the CNC depends on the source. The TEM micrographs of the CNC isolated from different sources can be seen in Fig. 3. The dimensions and properties of the CNC depend on the source from it is extracted, the method of hydrolysis, hydrolysis conditions and ionic strength. The dimensions of CNC obtained from various sources listed in Table 2.

The crystallinity depends on the initial amorphous content present in the raw materials. By the removal of the amorphous materials (like lignin, hemicellulose etc.) using chemical treatments from the raw materials, remaining crystalline part

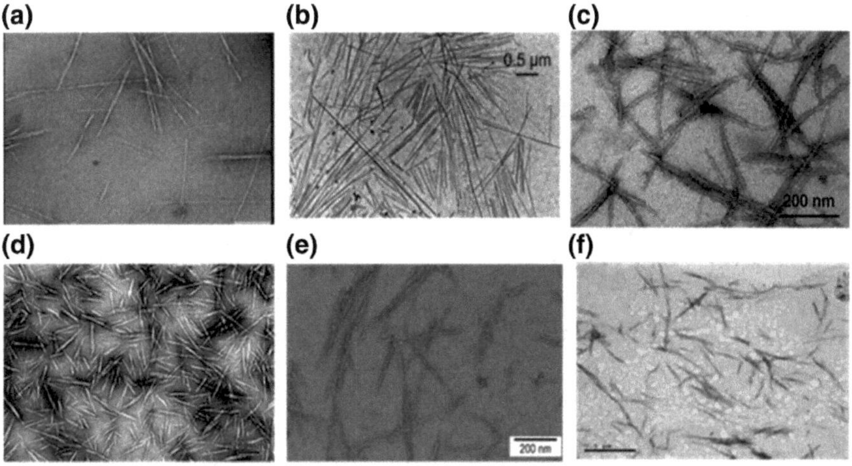

Fig. 3 Micrographs of CNC from **a** garlic straw, **b** tunicin, **c** Wood, **d** ramie, **e** tomato peel and **f** maize straw. Adopted from Park et al. (2010) with permission

Table 2 Dimensions and crystallinity of CNC extracted from different sources

Source	Length (nm)	Diameter (nm)	Crystallinity (%)	References
Wood	100–300	3–5	–	Beck-Candanedo et al. (2005)
Tunicin	100–1000	10–20	–	Favier et al. (1995)
Ramie	200–300	10–15	–	Habibi et al. (2008)
Garlic straw	480	6	68.8	Kallel et al. (2016)
Groundnut shells	111	5–18	74	Bano and Negi (2017)
Chili leftover	90–180	4–6	78.5	Nagalakshmaiah et al. (2016b)

could be responsible in the enhancement of crystallinity. Upon acid hydrolysis, amorphous regions will hydrolyze completely and keep the crystalline parts that will enhance the crystallinity for CNC. In general, the cellulose microfibers composed with alternative amorphous and crystalline regions. The crystallinity of the CNC is high compared to NFC since it not contains the amorphous regions. Basically, the crystallinity can be determined by using X-ray diffraction (X-RD). Typically, the crystallinity of the CNC will vary from 45 to 90% (depend so on the source) and the crystallinity of CNC extracted from different sources is cited in Table 2. However, the crystallinity will differ with measurement method as evaluated by Park et al. (2010).

Thermal property of CNC is key to process the polymer nanocomposites by melt processing. The thermal degradation of the CNC can be evaluated by the thermogravimetric analysis, which measures the loss of weight as a function of temperature. The isolated process is directly proportional to thermal stability. For instance, CNC extracted from hydrochloric acid hydrolysis shown the better thermal stability compared to the sulfuric acid hydrolyzed CNC, this is due to the sulfate groups present on the surface of CNC (Nagalakshmaiah et al. 2016b). The typical onset of thermal degradation arises between 200 and 300 °C. In order to improve the thermal stability, researchers often modify or graft the surface with different molecules. Different techniques have been used to measure the mechanical properties of CNC like AFM, Raman spectroscopy, inelastic X-ray scattering and X-ray diffraction analysis (Moon et al. 2011). However, evaluating the mechanical properties is very challenging due to different factors like crystalline structure (like cellulose-I and II), anisotropy, defects, size and percentage of crystallinity. Iwamoto et al. reported the elastic modulus of the CNC measured by using AFM tip-bending test and reported approximately 150 GPa (Iwamoto et al. 2009). In some other study, the Raman spectroscopy was used to calculate modulus of tunicate CNC and reported as 149 GPa (Šturcová et al. 2005). However, the fundamental understanding of the mechanical properties of CNC is very limited and challenging.

4 Processing of CNC-Based Nanocomposites

The reinforcement of nanomaterials into polymeric matrices has increased attention because of the high mechanical properties even at low filler content. Lately, using CNC as reinforcing filler with polymers is increasing due to their attractive mechanical and biodegradable properties. Favier et al. (1995) for the first time was reported CNC from tunicin reinforced with the styrene/butyl acrylate copolymer (poly[S-co-BuA]) matrix. The solvent casting composite film shows enhancing mechanical properties even with the low CNC content.

Different methods like electrospinning, thermo-pressing, melt extrusion and solvent casting have been reported for processing of the polymer nanocomposites based on CNC. However, the most extensively used methods are solvent casting, electro spinning and melt processing. The mechanical properties of the nanocomposites depend on three different factors: (i) The morphology and dimensions of CNC (ii) processing methods and (iii) the microstructure of matrix and matrix/filler interactions (Ramires and Dufresne 2011). The solvent casting process appears to be impressive mechanical performance compared to thermo-pressing and melt extrusion (Favier et al. 1995). This feature is mainly due to the low structural damage and CNC preserve their structural orientation in the matrix during casting evaporation. An illustrated image of the casting evaporation of CNC/ latex composites is shown in Fig. 4. It can be also attributed to their homogeneous dispersion in water due to their electrostatic repulsions between negatively charged CNC. However, this process is limiting to produce the composites at bulk scale. Moreover, this method can be applied to water-soluble polymers. The solvent casting evaporation is very tedious for water-insoluble polymers like polyethylene,

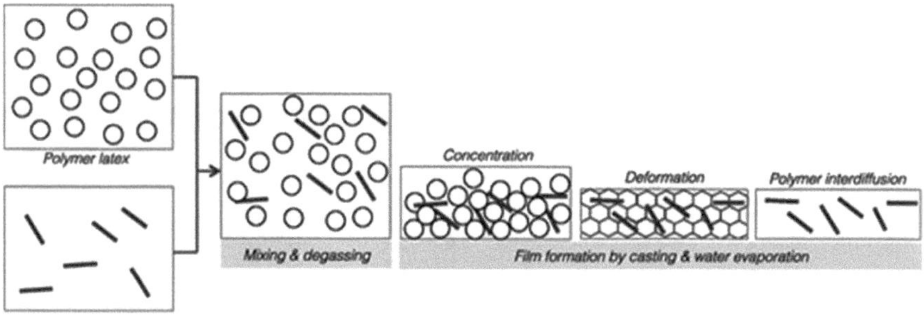

Fig. 4 Illustrated image of the solvent casting evaporation of CNC with latex particles, slightly changed. Adopted from Le Corre and Angellier-Coussy (2014)

polypropylene and polylactic acid. This is mainly due to the solvent exchange process of CNC to suitable with polymer solubility. The dispersion of CNC is very challenging in non-polar solvents due to the hydrophilic behavior. Moreover, this technique is non-industrial and non-economic and also limits the number of polymer matrices that can be used in association with cellulose nanocrystals as explained. Additionally, this process requires more energy and plenty of organic solvents. Hence, this method cannot be applied at industrial scale (Dufresne 2012).

Lately, the rapid developments in the commercial production of CNC require more industrial composite processing techniques that will enable industries to use their produced CNC. Such methods are melt-compounding techniques, such as melt extrusion. This process is well known as solvent-free process (green) and importantly this is industrially viable process and widely used in the polymer industry. However, few major issues are limiting to use this process. Mainly, due to the (i) irreversible agglomeration upon freeze-drying (ii) low thermal stability associated with sulfate groups (iii) inherent incompatibility between the cellulose and the polymer. Accordingly, this process is infrequently employed compared to the solvent casting and other methods for the preparation of CNC-based composites.

In order to avoid all three mentioned issues, researchers developed new methods like surface modification (Bondeson and Oksman 2007; Nagalakshmaiah et al. 2016a), surface adsorption (Pereda et al. 2014; Nagalakshmaiah et al. 2016c), premix blending (Mariano et al. 2015) and chemical grafting (Goffin et al. 2011a, b). The main objective of the above-mentioned methods is to promote the dispersion and compatibility of CNC with different polymer matrices. The appearance of CNC extruded nanocomposite films by different methods is shown in Fig 5. Goffin et al. reported the grafting of the lactide on the surface of CNC by ring-opening polymerization. The resulted composites can be seen in Fig. 4a. This process evident that the significant improvement in thermal stability and compatibility and enhance the mechanical properties could be obtained (Goffin et al. 2011b). Recently, premix encapsulation of CNC in polycarbonate (PC) was reported and the resulted PC+CNC capsules and ensued composites can be seen in Fig. 4b (Mariano et al. 2015). In some other study, ionic compatibilization of CNC with PP was reported, and this

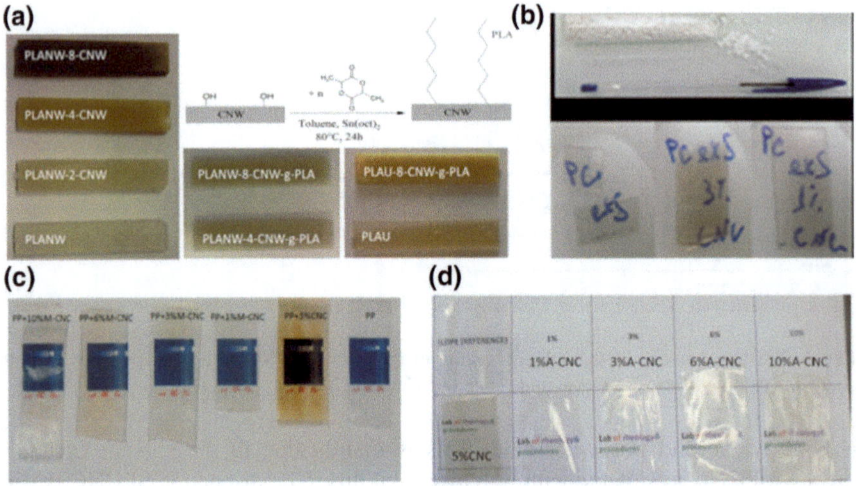

Fig. 5 Appearance of the CNC extruded composites with hydrophobic polymers. **a** PLA/g-CNW (Goffin et al. 2011b). **b** Encapsulated PC/CNC (Mariano et al. 2015). **c** PP with modified CNC (Nagalakshmaiah et al. 2016a). **d** Adsorbed CNC/LLDPE composites (Nagalakshmaiah et al. 2016c)

Table 3 Different methods reported on CNC-based polymer nanocomposites

Name of process	Processing method	Filler content (wt%)	Polymer matrices	References
Neat	Casting/melt extrusion	3–30	PEO	Alloin et al. (2011)
Neat	Melt extrusion	3–30	PVA	Mathew et al. (2011)
Physical adsorption	Melt extrusion	1–9	LDPE	Ben Azouz et al. (2012)
Neat	Melt extrusion	1–5	PLA	Fortunati et al. (2012)
Neat	Melt extrusion	1–15	PEVA	Sonia et al. (2013)
Adsorption/Grafting	Melt extrusion	1–20	PS	Lin and Dufresne (2013)
Premix or blends	Casting/melt extrusion	1–5	PLA+NR	Bitinis et al. (2013a, b)
Adsorption	Melt extrusion	1–9	LDPE	Pereda et al. (2014)
Modified	Melt extrusion	1–5	PLA	Arrieta et al. (2014)
Grafting	Casting/melt extrusion	1–5	LDPE/HDPE	Mokhena and Luyt (2014)
Grafting	Reactive extrusion	1	PLA	Yang et al. (2015)
Adsorption	Melt extrusion	6	PBS	Ludueña et al. (2016)
Grafting	Melt extrusion	5	PLLA	Spinella et al. (2016)

process enhances the thermal stability of the ensued composites can be seen in Fig. 4c (Nagalakshmaiah et al. 2016a). In another study, a simple physical adsorption of macromolecules on the surface of CNC and their reinforcement in linear low-density polyethylene (LLDPE) was reported (Nagalakshmaiah et al. 2016c). Despite having intensive studies the chemical grafting, premix processes are not viable to apply at industrial scale as mentioned above. Table 3 summarized the different methods reported the processing of the polymer nanocomposites-based on CNC.

5 Possible Emerging Applications

Nanocrystalline cellulose (NCC) reinforcement nanocomposites were prepared using different approaches as discussed in this chapter. Various operational dimensional products like biomedical scaffold for bone regeneration, lightweight packaging materials, building materials, insulating materials, electrical sensors, structural materials, energy storage materials and contaminants remediation filters/ membranes were reported and exploited in the literature (Afrin and Karim 2017; Karim et al. 2017a). There is an emerging demand (i) within the packaging industry, there is a trend toward sustainable lightweight packaging combined with a demand for biodegradable materials, particularly for food packaging applications. These demands are partly driven by legislation, e.g., to even ban non-biodegradable plastic materials. (ii) Within the information and communication industry, there is a constant need for biodegradable plastic materials. (iii) Within the automotive industry, there is a constant need to develop new materials that contribute to the reduction of greenhouse gas emissions as a result of the reduced weight of the vehicles, as well as new materials that can be recycled. (iv) Within the building industry, there is a need to reduce greenhouse gas emissions, either by an increased use of bio-based, recyclable and lightweight building materials or by decreasing the energy consumption during the lifetime of a building by the use of novel materials with improved insulation properties. (v) Within industries relying on formulations, such as the chemical, pharmaceutical, cosmetics, pulp and paper and food industries, there is a large incentive for bio-based, recyclable and safe products. Fortuitously, cellulose nanocrystal based nanocomposites fits very well for all these applications. Authors are not interested to discuss all mentioned applications of CNC-based nanocomposites but only two interesting and emerging applications are discussed below.

5.1 Water Remediation

The fabrication of nanoenabled composite membranes by incorporation of CNC is a new application area in the field of membrane technology. Dual approach might be used for the filtration of contaminants from water using CNC incorporated membranes: (a) physical nanofiltration/ultrafiltration process and (b) adsorption by functional entity (CNC). For the first time, a research group from Stony Brook University reported nanocellulose as functional entity for the capturing of pollutants present in water (Ma et al. 2011a, b, 2012). Therefore, affinity membranes were developed to permit the purification of contaminants based on physical/chemical properties or biological functions. Rather than operate purely operation on the sieving mechanism, affinity membrane based on the selectivity of the membrane to attach contaminants, by immobilizing specific ligands or produce specific functional groups onto the surface of membranes. Affinity membrane reflects technological advances in both fixed-bed liquid separation and membrane filtration and combines both the outstanding selectivity and the reduced pressure drops associated with filtration membranes.

Recently, Karim et al. (2014) used cellulose nanocrystal isolated form cellulose sludge as functional entity within the chitosan matrix for removal of model wastewater contaminated with dyes. Three targeted dyes listed very hazardous in EPA list were Victoria blue, Methyl Violet and Rhodamine 6G. Highest (98%) removal of Victoria blue followed by Methyl Violet and Rhodamine 6G were reported after 24 h incubation with freeze-dried nanocomposites affinity membranes.

In another study reported by the same group, cellulose nanocrystals isolated from an integrated bioethanol production process (CNC_{BE}) were used as functional material to increase the metal ions capturing efficiency of cellulose nanofiber (CNF) base material. Dip coating of base dimensional structure (CNF) in CNC_{BE} solution followed by drying was performed to capture the metal ions from mirror making industry. The produces affinity membranes showed remarkably very high removal efficiency (100%) of silver, copper and iron metal ions (Karim et al. 2016a). Furthermore, a comparative study was performed to see the efficiency of CNC isolated form cellulose sludge and through bioethanol route. Affinity membranes/filters fabricated using filtration have double-layered structure, top layer has functional CNC and bottom was made up of sludge fibers which have to make the membrane stable in wet environment and porous structure gives high permeability. Effluent of mirror making industry laden with heavy metal ions when treated with cellulose nanocomposites membranes showed high removal capacity (Karim et al. 2016b, 2017b).

5.2 *Scaffolds for Biomedical Applications*

In medical products and tissue engineering scaffolds, where cell growth and proliferation are the main requirements, nanostructured materials with controlled pore size and pore interconnectivity are considered advantageous. Porous nanocomposites based on nanocellulose for wound dressing materials, cartilage repair and bone regeneration were discussed in the literature, and some interesting examples are given here. Recently, electrospun mats based on chitosan/polyethylene oxide as matrix phase and cellulose nanocrystal (CNC) with different surface charges as a reinforcing phase were prepared. The crosslinked nanocomposite mats had a tensile strength of 58 MPa and a modulus of 3.1 GPa due to the good integration of CNC into the matrix. The mechanical stability, moisture stability and functional properties of the mats, such as the O_2/CO_2 permeability, water vapor transmission rate (WVTR) in the range of 1202–1879 g m^{-2} day^{-1} and compatibility toward adipose-derived stem cells, were found to be favorable for wound dressing (Naseri et al. 2015). Furthermore, this study was extended to understand the effect of charge on spinnability of solution for the production of nanocomposites (Naseri et al. 2016a). In another study, interpenetrating polymer network hydrogel based on sodium alginate and gelatin reinforced with cellulose nanocrystals form bioethanol process plant were prepared via a freeze-drying technique. The high porosity of the IPN hydrogels (>96%), high PBS uptake, and cytocompatibility toward mesenchymal stem cells are considered favorable for use as a substitute for cartilage (Naseri et al. 2016b).

6 Conclusion and Future Work

This chapter discussed the important aspects regarding the isolations of nanocrystals from various cellulosic biomass and how chemicals used for the isolation of these tiny entities affect the intrinsic properties of cellulose nanocrystals. Types and density of produced charge on the surface are two critical parameters responsible for the distribution and dispersion of these tiny particles within the polymer networks. Various techniques have been discussed in the literature for the characterization of these particles but still, researchers and scientists do need some more sophisticated techniques for better characterization and understanding at molecular level.

The application of these tiny particles at suspension and dimensional structures is reported in various reports, but the large-scale production is still limited. Recently, a sophisticated pilot scale plant at MoRe Research AB with collaboration with four other partners has been started. Up-scaling/processing of nanocomposites at industrial scale is still limited. Various understanding/efforts are still needed for commercialization of cellulose crystals-based nanocomposites, and it might be a future foal for researches.

References

Afrin S, Karim Z (2017) Green catalytic approach for isolation and surface modification of nanocellulose: necessity of enzymes over chemicals. Chem Bio Eng Rev 5:289–303

Alloin F, D'Aprea A, Dufresne A, Kissi NE, Bossard F (2011) Poly(oxyethylene) and ramie whiskers based nanocomposites: influence of processing: extrusion and casting/evaporation. Cellulose 18:957–973

Arrieta MP, Fortunati E, Dominici F, Rayón E, López J, Kenny JM (2014) PLA-PHB/cellulose based films: mechanical, barrier and disintegration properties. Polym Degrad Stab 107: 139–149

Bano S, Negi YS (2017) Studies on cellulose nanocrystals isolated from groundnut shells. Carbohydr Polym 157:1041–1049

Beck-Candanedo S, Roman M, Gray DG (2005) Effect of reaction conditions on the properties and behavior of wood cellulose nanocrystal suspensions. Biomacromolecules 6:1048–1054

Ben Azouz K, Ramires EC, Van den Fonteyne W, El Kissi N, Dufresne A (2012) Simple method for the melt extrusion of a cellulose nanocrystal reinforced hydrophobic polymer. ACS Macro Lett 1:236–240

Bitinis N, Fortunati E, Verdejo R, Bras J, Kenny JM, Torre L, López-Manchado MA (2013a) Poly (lactic acid)/natural rubber/cellulose nanocrystal bionanocomposites. Part II: properties evaluation. Carbohydr Polym 96:621–627

Bitinis N, Verdejo R, Bras J, Fortunati E, Kenny JM, Torre L, López-Manchado MA (2013b) Poly (lactic acid)/natural rubber/cellulose nanocrystal bionanocomposites Part I. Processing and morphology. Carbohydr Polym 96:611–620

Boluk Y, Lahiji R, Zhao L, McDermott MT (2011) Suspension viscosities and shape parameter of cellulose nanocrystals (CNC). Colloids Surf Physicochem Eng Asp 377:297–303

Bondeson D, Oksman K (2007) Dispersion and characteristics of surfactant modified cellulose whiskers nanocomposites. Compos Interfaces 14:617–630

Braun B, Dorgan JR (2009) Single-step method for the isolation and surface functionalization of cellulosic nanowhiskers. Biomacromolecules 10:334

Cheng M, Qin Z, Liu Y, Qin Y, Li T, Chen L, Zhu M (2014) Efficient extraction of carboxylated spherical cellulose nanocrystals with narrow distribution through hydrolysis of lyocell fibers by using ammonium persulfate as an oxidant. J Mater Chem A 2:251–258

Corrêa AC, de Morais TE, Pessan LA, Mattoso LHC (2010) Cellulose nanofibers from curaua fibers. Cellulose 17:1183–1192

de Souza Lima MM, Borsali R (2004) Rodlike cellulose microcrystals: structure, properties, and applications. Macromol Rapid Commun 25:771–787

Dufresne A (2012) Processing of polymer nanocomposites reinforced with cellulose nanocrystals: a challenge. Int Polym Process 27:557–564

Ehmann HMA, Mohan T, Koshanskaya M, Scheicher S, Breitwieser D, Ribitsch V, Stana-Kleinschek K, Spirk S (2014) Design of anticoagulant surfaces based on cellulose nanocrystals. Chem Commun 50:13070–13072

Favier V, Canova GR, Cavaillé JY, Chanzy H, Dufresne A, Gauthier C (1995) Nanocomposite materials from latex and cellulose whiskers. Polym Adv Technol 6:351–355

Fortunati E, Armentano I, Zhou Q, Iannoni A, Saino E, Visai L, Berglund LA, Kenny JM (2012) Multifunctional bionanocomposite films of poly(lactic acid), cellulose nanocrystals and silver nanoparticles. Carbohydr Polym 87:1596–1605

Goffin A-L, Raquez J-M, Duquesne E, Siqueira G, Habibi Y, Dufresne A, Dubois P (2011a) Poly (ε-caprolactone) based nanocomposites reinforced by surface-grafted cellulose nanowhiskers via extrusion processing: morphology, rheology, and thermo-mechanical properties. Polymer 52:1532–1538

Goffin A-L, Raquez J-M, Duquesne E, Siqueira G, Habibi Y, Dufresne A, Dubois P (2011b) From interfacial ring-opening polymerization to melt processing of cellulose nanowhisker-filled polylactide-based nanocomposites. Biomacromolecules 12:2456–2465

Habibi Y, Goffin A-L, Schiltz N, Duquesne E, Dubois P, Dufresne A (2008) Bionanocomposites based on poly(ε-caprolactone)-grafted cellulose nanocrystals by ring-opening polymerization. J Mater Chem 18:5002

Habibi Y, Lucia LA, Rojas OJ (2010) Cellulose nanocrystals: chemistry, self-assembly, and applications. Chem Rev 110:3479–3500

Hirai A, Inui O, Horii F, Tsuji M (2009) Phase separation behavior in aqueous suspensions of bacterial cellulose nanocrystals prepared by sulfuric acid treatment. Langmuir 25:497–502

Hirota M, Tamura N, Saito T, Isogai A (2010) Water dispersion of cellulose II nanocrystals prepared by TEMPO-mediated oxidation of mercerized cellulose at pH 4.8. Cellulose 17: 279–288

International EPNOE Junior Scientists Meeting Future Perspectives in Polysaccharide Research, Gericke M, Peršin Z, Kargl R, Cemef, European Polysaccharide Network of Excellence (2015) Book of abstracts. Faculty of Mechanical Engineering, Institute of Engineering Materials and Design, Laboratory for Characterisation and Processing of Polymers, Maribor

Iwamoto S, Kai W, Isogai A, Iwata T (2009) Elastic modulus of single cellulose microfibrils from tunicate measured by atomic force microscopy. Biomacromolecules 10:2571–2576

Kallel F, Bettaieb F, Khiari R, García A, Bras J, Chaabouni SE (2016) Isolation and structural characterization of cellulose nanocrystals extracted from garlic straw residues. Ind Crops Prod 87:287–296

Karim Z, Afrin S (2015) Nanocellulose as novel supportive functional material for growth and development of cells. Cell Devel Biol 4:2–7

Karim Z, Mathew AP, Mouzon J, Oksman K (2014) Nanoporous membranes with cellulose nanocrystals as functional entity in chitosan: removal of dyes from water. Carbohydr Poly 112:668–676

Karim Z, Claudpierre S, Grahn M, Oksman K, Mathew AP (2016a) Nanocellulose based functional membranes for water cleaning: tailoring of mechanical properties, porosity and metal ion capture. J Memb Sci 514:418–428

Karim Z, Grahn M, Oksman K, Mathew AP (2016b) High flux affinity membranes based on cellulose nanocomposite for removal of heavy metal ions from industrial effluent. RCS Adv 6:20644–20653

Karim Z, Afrin S, Husain Q, Danish R (2017a) Necessity of enzymatic hydrolysis for production and functionalization of nanocelluloses. Crit Rev Biotechnol 6:1–16

Karim Z, Hakalahti M, Tammelin T, Mathew A, Oksman K (2017b) Effect of in situ TEMPO surface functionalization of nanocellulose membranes on the adsorption of metal ions from aqueous solution. RSC Adv 7:5232–5241

Klemm D, Heublein B, Fink H-P, Bohn A (2005) Cellulose: fascinating biopolymer and sustainable raw material. Angew Chem Int Ed 44:3358–3393

Koshizawa T (1960) Degradation of wood and cotton linters in phosphoric acid. Jpn Tappi J 14(7): 455–458

Le Corre D, Angellier-Coussy H (2014) Preparation and application of starch nanoparticles for nanocomposites: a review. React Funct Polym 85:97–120

Leung ACW, Hrapovic S, Lam E, Liu Y, Male KB, Mahmoud KA, Luong JHT (2011) Characteristics and properties of carboxylated cellulose nanocrystals prepared from a novel one-step procedure. Small 7:302–305

Li R, Fei J, Cai Y, Li Y, Feng J, Yao J (2009) Cellulose whiskers extracted from mulberry: a novel biomass production. Carbohydr Polym 76:94–99

Lin N, Dufresne A (2013) Physical and/or chemical compatibilization of extruded cellulose nanocrystal reinforced polystyrene nanocomposites. Macromolecules 46:5570–5583

Lin N, Dufresne A (2014) Surface chemistry, morphological analysis and properties of cellulose nanocrystals with gradiented sulfation degrees. Nanoscale 6:5384–5393

Ludueña LN, Fortunati E, Morán JI, Alvarez VA, Cyras VP, Puglia D, Manfredi LB, Pracella M (2016) Preparation and characterization of polybutylene-succinate/poly(ethylene-glycol)/ cellulose nanocrystals ternary composites. J Appl Polym Sci 133:n/a-n/a

Ma H, Burger C, Hsiao BS, Chu B (2011a) Ultra-fine cellulose nanofibers: new nano-scale materials for water purification. J Mat Chem 21:7507–7510

Ma H, Burger C, Hsiao BS, Chu B (2011b) Ultrafine polysaccharide nanofibrous membrane for water purification. Biomacromolecules 12:970–976

Ma H, Burger C, Hsiao BS, Chu B (2012) Highly permeable polymer membranes containing channels for water purification. ACS Macro Let 1:723–726

Mariano M, El Kissi N, Dufresne A (2015) Melt processing of cellulose nanocrystal reinforced polycarbonate from a masterbatch process. Eur Polym J 69:208–223

Mathew AP, Gong G, Bjorngrim N, Wixe D, Oksman K (2011) Moisture absorption behavior and its impact on the mechanical properties of cellulose whiskers-based polyvinylacetate nanocomposites. Polym Eng Sci 51:2136–2142

Mathew AP, Oksman K, Karim Z, Liu P, Khan SA, Naseri N (2014) Process scale up and characterization of wood cellulose nanocrystals hydrolysed using bioethanol pilot plant. Indus Crop Produc 58:212–219

Mautner A, Maple HA, Kobkeatthawin T, Kokol V, Karim Z, Li K, Bismark A (2016) Phosphorylated nanocellulose papers for copper adsorption from aqueous solutions. Int J Environ Sci Technol 13:1861–1872

Mokhena TC, Luyt AS (2014) Investigation of polyethylene/sisal whiskers nanocomposites prepared under different conditions. Polym Compos 35:2221–2233

Moon RJ, Martini A, Nairn J, Simonsen J, Youngblood J (2011) Cellulose nanomaterials review: structure, properties and nanocomposites. Chem Soc Rev 40:3941

Nagalakshmaiah M, El Kissi N, Dufresne A (2016a) Ionic compatibilization of cellulose nanocrystals with quaternary ammonium salt and their melt extrusion with polypropylene. ACS Appl Mater Interfaces 8:8755–8764

Nagalakshmaiah M, kissi NE, Mortha G, Dufresne A (2016b) Structural investigation of cellulose nanocrystals extracted from chili leftover and their reinforcement in cariflex-IR rubber latex. Carbohydr Polym 136:945–954

Nagalakshmaiah M, Pignon F, El Kissi N, Dufresne A (2016c) Surface adsorption of triblock copolymer (PEO–PPO–PEO) on cellulose nanocrystals and their melt extrusion with polyethylene. RSC Adv 6:66224–66232

Nagalakshmaiah M, Nechyporchuk O, El Kissi N, Dufresne A (2017) Melt extrusion of polystyrene reinforced with cellulose nanocrystals modified using poly[(styrene)-co-(2-ethylhexyl acrylate)] latex particles. Eur Polym J 91:297–306

Naseri N, Mathew AP, Girandon L, Fröhlich M, Oksman K (2015) Porous electrospun nanocomposite mats based on chitosan-cellulose nanocrystals for wound dressing: effect of surface characteristics of nanocrystals. Cellulose 22:521–534

Naseri N, Mathew AP, Oksman K (2016a) Electrospinnability of bionanocomposites with high nanocrystal loadings: the effect of nanocrystal surface characteristics. Carbohydr Polym 147:464–472

Naseri N, Deepa B, Mathew AP, Girandon L, Oksman K (2016b) Nanocellulose based interpenetrating polymer network (IPN) hydrogels for cartilage applications. Biomacromolecules 14:3714–3723

Nge TT, Lee S-H, Endo T (2013) Preparation of nanoscale cellulose materials with different morphologies by mechanical treatments and their characterization. Cellulose 20:1841–1852

Oksman K, Aitomäki Y, Mathew AP, Siqueira G, Zhou Q, Butylina S, Tanpichai S, Zhou X, Hooshmand S (2016) Review of the recent developments in cellulose nanocomposite processing. Compos Part A 83:2–18

Park S, Baker JO, Himmel ME, Parilla PA, Johnson DK (2010) Cellulose crystallinity index: measurement techniques and their impact on interpreting cellulase performance. Biotechnol Biofuels 3:10

Pereda M, Kissi NE, Dufresne A (2014) Extrusion of polysaccharide nanocrystal reinforced polymer nanocomposites through compatibilization with poly(ethylene oxide). ACS Appl Mater Interfaces 6:9365–9375

Peresin MS, Habibi Y, Zoppe JO, Pawlak JJ, Rojas OJ (2010) Nanofiber composites of polyvinyl alcohol and cellulose nanocrystals: manufacture and characterization. Biomacromolecules 11:674–681

Ramires EC, Dufresne A (2011) A review of cellulose nanocrystals and nanocomposites. Tappi J 10:9–16

Rånby BG (1951) Fibrous macromolecular systems. Cellulose and muscle. The colloidal properties of cellulose micelles. Discuss Faraday Soc 11:158–164

Rosa MF, Medeiros ES, Malmonge JA, Gregorski KS, Wood DF, Mattoso LHC, Glenn G, Orts WJ, Imam SH (2010) Cellulose nanowhiskers from coconut husk fibers: effect of preparation conditions on their thermal and morphological behavior. Carbohydr Polym 81:83–92

Sadeghifar H, Filpponen I, Clarke SP, Brougham DF, Argyropoulos DS (2011) Production of cellulose nanocrystals using hydrobromic acid and click reactions on their surface. J Mater Sci 46:7344–7355

Sonia A, Priya Dasan K, Alex R (2013) Celluloses microfibres (CMF) reinforced poly (ethylene-co-vinyl acetate) (EVA) composites: dynamic mechanical, gamma and thermal ageing studies. Chem Eng J 228:1214–1222

Spinella S, Samuel C, Raquez J-M, McCallum SA, Gross R, Dubois P (2016) Green and efficient synthesis of dispersible cellulose nanocrystals in biobased polyesters for engineering applications. ACS Sustain Chem Eng 4:2517–2527

Šturcová A, Davies GR, Eichhorn SJ (2005) Elastic modulus and stress-transfer properties of tunicate cellulose whiskers. Biomacromolecules 6:1055–1061

Yang W, Dominici F, Fortunati E, Kenny JM, Puglia D (2015) Melt free radical grafting of glycidyl methacrylate (GMA) onto fully biodegradable poly(lactic) acid films: effect of cellulose nanocrystals and a masterbatch process. RSC Adv 5:32350–32357

Yu H, Qin Z, Liang B, Liu N, Zhou Z, Chen L (2013) Facile extraction of thermally stable cellulose nanocrystals with a high yield of 93% through hydrochloric acid hydrolysis under hydrothermal conditions. J Mater Chem A 1:3938

Nanofibrillated Cellulose-Based Nanocomposites

Hind Abdellaoui, Marya Raji, Hamid Essabir, Rachid Bouhfid and Abou el kacem Qaiss

Abstract Nanofibrillated cellulose (NFC), a form of nanocellulose, is currently recommended to be utilized in a wide of industrial applications like food packaging, printing, paper, biomedical, and nanocomposite materials. Their exploitation is not a coincidence, but a fruitful result of many studies showing that NFCs have exciting characteristics such as renewable, sustainable, recyclable, the high length-to-diameter ratio (aspect ratio), and high mechanical properties at the nanometric scale. This chapter is a boon to show the added value of NFCs and their applications in nanocomposites materials. To do this, this content deals with two parts: the first one focuses on the extraction of the NFCs from the cellulosic fiber, their structures, and the processes allowing to modify/treated nanocellulose surface to make it compatible with the polymer matrix. In the second part, focused on the manufacturing process of nanocomposites, their properties and the industrial applications are discussed in depth.

Keywords Nanofibrillated cellulose · Modify/treated nanocellulose Nanocomposites

1 Introduction

Nanocomposite materials based on petroleum sources and non-biodegradable polymers still have their ability to be the materials of this era (Essabir et al. 2017; Zari et al. 2018). However, the current trend is to develop a new generation of materials based on natural substances owing to many considerations where the environmental one is the most exciting and urgent (Raji et al. 2017a). This last one is realized by developing composites reinforced by natural elements such as lignocellulosic fibers which are widely spread in the earth, renewable, biologically

H. Abdellaoui · M. Raji · H. Essabir · R. Bouhfid · A. el kacem Qaiss (✉)
Moroccan Foundation for Advanced Science, Innovation and Research (MAScIR), Rabat, Morocco
e-mail: a.qaiss@mascir.com

M. L. Sanyang and M. Jawaid (eds.), *Bio-based Polymers and Nanocomposites*,
https://doi.org/10.1007/978-3-030-05825-8_4

recyclable, and endowed with interesting specific properties as indicated in the literature (Ku et al. 2011; Nunna et al. 2012; Singh et al. 2008).

Cellulose, contained in lignocellulosic fiber, is composed of nanosized microfibrils which are formed by crystalline and amorphous regions and can be extracted from several of sources like jute, kenaf, cotton, flax, and all natural fibers (Abdellaoui et al. 2015a; Le Duigou et al. 2010; Yahaya et al. 2015). Industrially, it has been used mainly for the implementation of cardboard and paper (Yahaya et al. 2015). The cellulose or nanocellulose, thanks to its nanometric size, is endowed with high mechanical strength, cheap, lightweight, and elevated aspect ratio. All those listed characteristics make the nanocellulose promising and highly recommend to use it as a reinforcement agent to develop a new generation of material that is nanocomposites.

The nanocellulose extraction or isolation (Essabir et al. 2016a), achieved either by mechanical or chemical treatment or combination of the two, makes it possible to distinguish between two categories: nanocrystalline cellulose (CNC) and nanofibrillated cellulose (NFC), in which the last one was introduced by Turbak et al. in the 1983s (Turbak et al. 1983) and will be detailed in this chapter.

Notwithstanding of the motivating characteristics of nanocellulose, some compatibility impediments prevent their perfect interfacial adhesion with the polymeric matrix (thermoset and thermoplastics) because of the hydrophilic nature of nanocellulose and hydrophobic one of the polymeric matrix. This incompatibility generates poor resistance to moisture absorption, reduced tensile strength and agglomerates formation during the manufacturing process. To remedy, modification or chemical treatment of the nanocellulose surface was performed by many researchers in which the characterization results confirmed the improvement of hydrophilicity of nanocellulose, ensuring good interfacial adhesion with polymeric matrix (Farias et al. 2017; Siró and Plackett 2010; Stocchi et al. 2007). This improvement influences positively the structural and mechanical properties of the final elaborated nanocomposites, allowing their use in a variety of industrial applications (Stocchi et al. 2007).

In this chapter, a crucial interest is focused on nanocomposites based on nanofibrillated cellulose and polymeric matrix. The first part consists of describing the structure of nanofibrillated cellulose, nanocellulose extraction processes, and the modification of its surface to make it compatible with the polymer matrix. The second part deals with the manufacturing process of nanocomposites based on nanofibrillated cellulose and their properties. In the end, some important industrial applications of nanofibrillated cellulose in nanocomposites materials are listed.

2 Opportunities and Challenge

The cellulose is estimated as the most abundant element in our earth (Yang et al. 2007), composed of semi-crystalline polysaccharide polymer and bulk material. Basically, nanometric celluloses can be extracted annually as mentioned in the literature (Raji et al. 2017a) from several natural resources like jute, sisal, flax, wood, cotton, eucalyptus, and coir fibers as well as non-plant sources as bacteria. In general, the cellulose fiber contains essentially a long chain of the ordered polymer of cellobiose unbranched chains of β-1, 4-linked glucose units, which act as a framework for the 3D polysaccharide structure. Each individual polysaccharide chain is bound with other in the microfibrils thanks to various strong intermolecular hydrogen bonds between hydroxyl groups of juxtaposed macromolecules, which in turn are grouped together for realizing macrofibrils.

For many industrial applications and taking account the environmental considerations, the cellulose materials have attracted more attention in the last decades, owing to their biodegradability, renewability, and high mechanical properties (specific properties) (Ben Azouz et al. 2012). Currently, the cellulose is exploited at too small dimensional scales, especially at a nanometric scale of 10^{-9} m, which is named nanocellulose (Vazquez et al. 2015). This latter is characterized by either rigid rod-shaped or flexible thin fibrils structure with 10–100 nm in length and 1–100 nm in diameter (Gacitua et al. 2005). As mentioned in literature, the researchers classify the nanocellulose by categories according to the shape of nanoparticles into three distinct categories: nanocrystalline cellulose (CNC) (Brinchi et al. 2013; Vazquez et al. 2015), bacterial cellulose (BC) (Albu et al. 2014; Eichhorn et al. 2010), and nanofibrillated cellulose (NFC) (Grumezescu 2017). Among them, the nanofibrillated cellulose (NFC) presents interesting advantages like greater aspect area (length-to-diameter: *L*/*d*), relatively high degree of crystallinity as well as important specific properties correlated to its weight, which make it perfect and ideal to exploit in different industrial fields like being nanofillers for nanocomposites and bio-nanocomposites production (Hietala et al. 2012; Song and Rojas 2013).

The primary objective for using nanofibrillated cellulose in nanocomposite materials based on the polymeric matrix (thermoplastic or thermoset) is to improve the mechanical properties of the final material (Brinchi et al. 2013). Actually, with the bio-tendency, the nanocomposites are estimated as a conditional replacement for the petroleum-based composites thanks to different and interesting properties granted by these discovered nanofillers (Alwani et al. 2014). The mechanical properties of nanocomposites will be quite significant if there is good interfacial adhesion between nanofillers and the polymeric matrix (Pracella et al. 2010; Sdrobi et al. 2012). In practical terms, the good interfacial adhesion is not assured because of the incompatibility between the hydrophobic matrix and the hydrophilic nanofibrillated cellulose (NFC), which causes a poor nanofiller–matrix interaction and low moisture resistance (Mir et al. 2013). Nonetheless, surface modification or treatment of nanocellulose may remedy this impediment by increasing the hydrophobicity of nanocellulose (Abdelmouleh et al. 2007; Singh and Samanta 2014). The chemical treatment is one of the most rewarding treatments used to improve the compatibility between nanocellulose and polymeric matrix allowing for the formation of the high dense hydrogen bonding network (Stocchi et al. 2007). Enhancing the mechanical properties of nanocomposites encourages the industrial and academical communities to develop more applications based on nanofibrillated cellulose and polymeric matrix (Erbas Kiziltas et al. 2015). Developing further these nanocomposites, their achievements and their applications is the purpose of this chapter.

3 Nanofibrillated Cellulose

Nanofibrillated cellulose, named also as cellulose nanofiber, microfibrillated cellulose (due to its length), or cellulose nanofibrils (CNFs) (Erbas Kiziltas et al. 2015; Niazi et al. 2017), is considered as sustainable nanocellulose with attractive advantages including biodegradability, good abundance, and renewability aspect as well as recyclability character (Eichhorn et al. 2010; Mohammed et al. 2015).

3.1 Structure of Nanofibrillated Cellulose (NFC)

The nanofibrillated cellulose (NFC) is micrometer long, flexible, and it has nanometer diameter of approximately 1–100 nm (Hietala et al. 2012; Song and Rojas 2013). They are highly entangled networks of nanofibers with both crystalline and amorphous cellulose domains (Shauddin et al. 2014). The nanofibrillated cellulose is obtained by high-pressure grinding of cellulose pulp suspension (Sdrobi et al. 2012). The NFCs are especially characterized by the degree of crystallinity parameter that is controlling their mechanical properties (Hakeem et al. 2011). Figure 1 illustrates the structure of nanofibrillated cellulose (NFC).

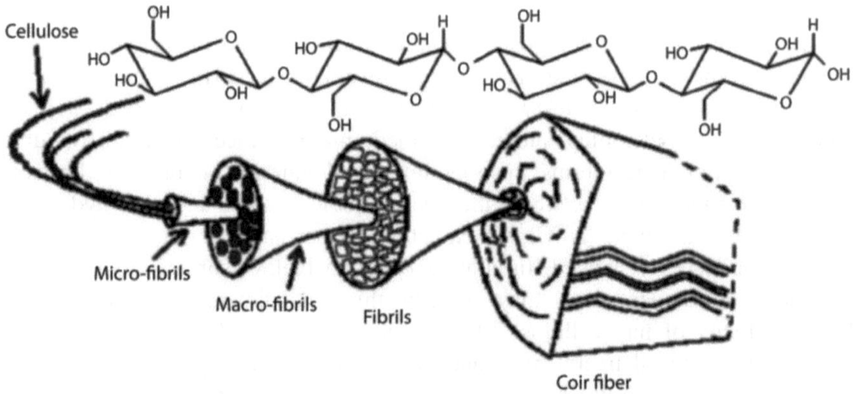

Fig. 1 Illustrative scheme of nanofibrillated cellulose (NFC) (Raji et al. 2017a)

3.2 *Preparation of Nanofibrillated Cellulose*

The nanofibrillated cellulose can be extracted or isolated from the cellulosic fiber by using a variety of methods including chemical pretreatment (alkaline-acid treatment), biological treatment (enzymatic treatment), mechanical treatment (crycrushing, grinding, high-intensity sonification, high-pressure homogenization, microfluidization), and a combination of both chemical and mechanical treatments (Thomas et al. 2015). The chemical pretreatment must be prior to the mechanical treatment as explained below.

3.2.1 Chemical Pretreatment Processes

Before the use of the mechanical treatment of cellulosic fiber, it is preferable to use the chemical pretreatment to facilitate the defibrillation of fiber and to reduce the energy of mechanical nanofibrillation process. Chemical pretreatment may be handled using alkaline-acid treatment and enzymatic hydrolysis for the purpose of increasing the surface area and enhancing the crystallinity degree (Raji et al. 2017a). In a study of Jyotishkumar et al (Thomas et al. 2015), they studied the effect of treatment or surface modification of jute fiber. Figure 2 shows the SEM results of raw jute fiber, alkali-treated fiber, bleached fiber, and acid hydrolyzed fiber. The results illustrate that the washing out and dissolving of hemicellulose and lignin start at the alkali treatment, and the entire removal was achievable at the bleaching step. Besides, the mechanical treatment (steam explosion) further banished the amorphous elements (lignin, hemicellulose, waxes, etc.) by defibrillation. Again, the SEM results indicate considerable changes in the morphology of jute fiber (size and smoothness) after acid hydrolysis (Thomas et al. 2015).

Alkaline Pretreatment Process

The alkaline pretreatment (pulping and bleaching), named also delignification (Azwa et al. 2013; Kumar et al. 2011), implies a degradation of lignin into solution of NaOH, which is removed by washing many times with water (Essabir et al. 2013, 2016a). This process leads to degradation of the linkage between lignin and carbohydrates where the entire elimination of the lignin could not be done without removing the carbohydrates.

Bleaching Treatment

Bleaching process is elected to remedy this matter, which consists to banish entirely the rest of lignin and impurities (Ait Laaziz et al. 2017). This process would be realized with hydrogen peroxide, ozone, chlorine dioxide (Raji et al. 2017a).

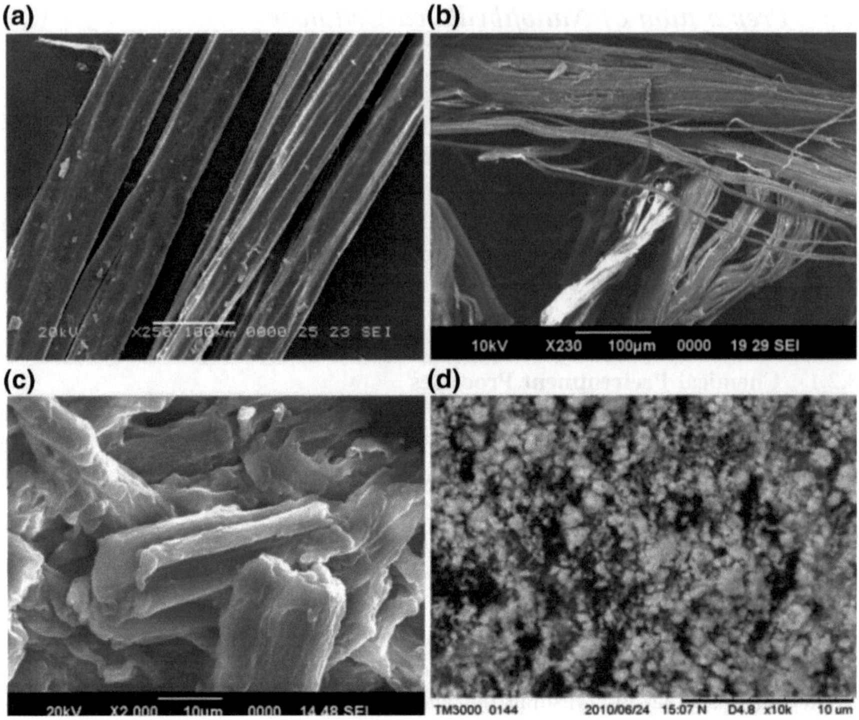

Fig. 2 Jute fiber at different processing steps: **a** raw jute fiber, **b** alkali-treated jute fiber, **c** bleached jute fiber, **d** acid hydrolyzed jute fiber (Thomas et al. 2015)

Acid Pretreatment Process

The acid pretreatment process is used to banish the amorphous regions of cellulosic fibers and enhances the crystallinity of fibers (Jönsson and Martín 2016). It can be handled using sulfuric acid (H_2SO_4) and hydrochloric acid (HCl). The properties of obtained nanofibrillated cellulose (NFC) may be influenced by acid concentration, temperature, and time of process.

Enzyme Pretreatment Process

It is a biological pretreatment characterized by a low energy requirement, it is also a safe and cost-effective process (Hall et al. 2011). Microorganisms/enzymes such as laccace are used to degrade or modify the lignin and hemicellulose contents from cellulosic fiber without disturbing cellulose content (Abdellaoui and Echaabi 2014).

3.2.2 Mechanical Treatment Process

The mechanical process, made after the chemical pretreatment, permits the defibrillation of cellulosic fiber (Hall et al. 2011). There are many mechanical processes to convert cellulosic fiber to nanofibrillated cellulose (NFC) like grinding, crycrushing, microfluidization, high-pressure homogenization, high-intensity sonification as detailed below.

High-Pressure Homogenization

The high-pressure homogenization is considered as an efficient mechanical method to refine cellulosic fiber; especially, it does not involve addition of any organic solvents (Tian et al. 2016). This process consists of passing a cellulose suspension at high pressure. The parameters, pressure, impact, high velocity and shear forces on fluid, imply a high degree of microfibrillation of cellulosic fibers. However, this process is still onerous because of the high energy consumption (Thomas et al. 2015).

Microfluidization

Similarly to the high-pressure homogenization process, microfluidizers imply the use of an intensifier pump to improve pressure to defibrillate cellulosic fiber (Wang et al. 2015). The size of nanofibrillated cellulose (NFC) is defined depending on the number of passes via the homogenizer (Thomas et al. 2015).

High-Intensity Sonication

During this process, the nanofibrillated cellulose is isolated by oscillating power with hydrodynamic forces of ultrasound, where the molecules absorb the ultrasonic energy of high-intensity waves and then involve the formation, expansion, and implosion of microscopic gas bubbles (Menon et al. 2017).

Grinding

This process permits also to break up cellulose into nanofibrillated cellulose. The mechanism is based on the shear forces to decompose the hydrogen bonds and cell wall structures of cellulosic fibers. Generally, ten cycles are recommended to create uniform size of nanofibrillated cellulose (Qu et al. 2017).

Crycrushing

In this process, the cellulosic fibers are kept in water in which the cellulose absorbs water in its cavity. The resulted cellulose is immersed in liquid nitrogen and crushed by mortar and pestle. A high impact force applied on frozen cellulosic fiber promotes the rupture, leading to the conversion of nanofibrillated cellulose (NFC) (Thomas et al. 2015).

3.3 Modification of Nanofibrillated Cellulose

Incorporating the nanofibrillated cellulose in polymeric matrix cannot be always rewarding because of the hydrophilic nature of nanocellulose (Ahmad et al. 2015). For that, obtaining nanocomposites with high interfacial adhesion and better specific properties may be achieved by enhancing the hydrophobicity of nanocellulose (Mohkami and Talaeipour 2011). Different treatment processes have been investigated in the purpose of reducing the hydrophilic character of nanocellulose and improving the interfacial adhesion with polymeric matrix. Among the most used treatments, some are silylation, acetylation, mercerization, coupling agents (3-amino-propyltriethoxysilane, titanate agent coupling, etc.) are used to chemically modifyc the functional hydroxyl groups on nanocellulose via covalent bonding (Thomas et al. 2015).

3.3.1 Grafting

The grafting is one of the most used techniques to increase the hydrophobicity of nanocellulose by two principal ways: "grafting onto" and "grafting from."

- Grafting onto: during this process, the pre-synthesized polymer chains are attached onto cellulose surface hydroxyl groups by coupling agents (Araújo et al. 2008).
- Grafting from: from immobilized initiator, the in situ surface-initiated polymerization happens to make polymer chains on the substrate (Raji et al. 2017a).

3.3.2 Acetylation

Acid anhydrides or acylchlorides are among the agents used in the acetylation process (Taj et al. 2007). The acetylation may be made in homogenous and heterogeneous ways by the presence of acetic anhydride, acetic acid, and a nip of catalyst (sulfuric acid or perchloric acid) (Hamour et al. 2015; Saheb and Jog 1999). In homogenous way, the moderately acetylated molecules start to spread in acetylating medium after being completely soluble (Abdellaoui et al. 2015b).

3.3.3 Silylation

The silylation technique permits the stabilization of the distribution of nanocellulose in a suspension (Xie et al. 2010). Many coupling agents are used for silylation like alkyl molecules (*N*-butyl, *N*-octyl, *N*-dodecyl, and isopropyl), where they are randomly dispersed at the nanofibrillated cellulose surface. The degree of silylation can affect the morphology of obtained nanocellulose, where dispersion in a high silylation solvent leads to a disintegration of the crystals and loss of original morphology (Abdellaoui and Echaabi 2014).

3.3.4 Surface-Active Agent/Surfactant

Also, the surfactant is used to minimize the hydrophilicity of nanocellulose. In the literature, many surfactants are exploited like isocyanates, maleate polypropylene, alkenyl succinic anhydride, where the modified cellulose changes the surface tension dynamics of nanocomposites (Abdellaoui and Echaabi 2014).

4 Nanofibrillated Cellulose-Based Nanocomposites

Nanofibrillated cellulose (NFC), owing to its high specific properties (mechanical, thermal and physical), elevated surface area, and low environment impact, has aroused great interest as filler or reinforcement for the polymeric matrix (Grumezescu 2017; Erbas Kiziltas et al. 2015). However, the NFC has a strong hydrophilic surface that inhibits the good interfacial adhesion between this reinforcement and synthetic matrix (Hietala et al. 2012). Therefore, it is highly recommended to use chemical-treated/modified nanocellulose in nanocomposites in order to enhance the specific characteristics of resulting nanocomposites (Abdelmouleh et al. 2007). Figure 3 explains the dispersion and distribution of nanoparticles into the matrix.

Thermoset and thermoplastic are the most known and exploited matrix in the composite materials thanks to their interesting characteristics and ease of manufacturing (Khanam et al. 2015). The thermoplastics like low-density polyethylene (LDPE) (Hoidy and Al-mulla 2013; Sdrobi et al. 2012), high-density polyethylene (HDPE) (Essabir et al. 2016b; Raji et al. 2016), polypropylene (PP), and others (Raji et al. 2018) are used especially for mass-produced composites. Concerning the thermoset as like epoxy (Abdellaoui et al. 2017), polyester (Benyahia et al. 2013; Gupta et al. 2016), phenol (Solyman et al. 2017), are essentially performed to produce high performance composites. The literature confirms that those synthetic matrices have been greatly modified by adding nanofibrillated cellulose (NFC) as reinforcement.

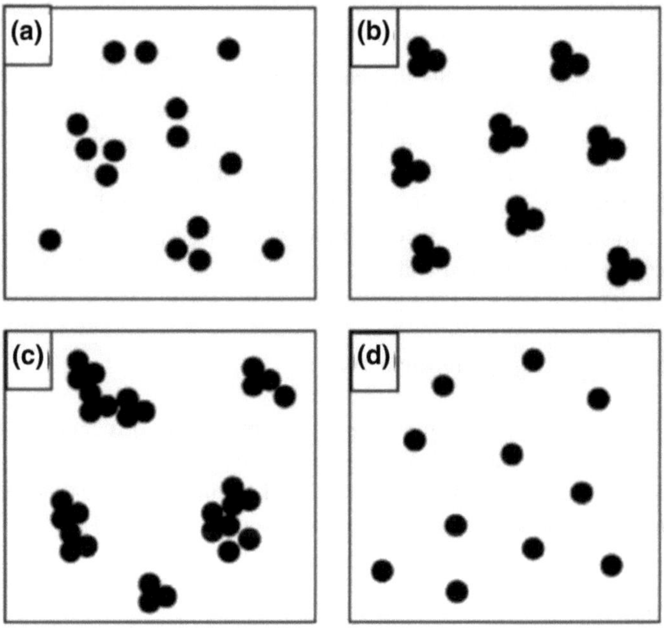

Fig. 3 Dispersion/distribution of nanoparticles into the polymeric matrix. **a** Good dispersion, poor distribution, **b** poor dispersion, good distribution, **c** pour dispersion, poor distribution, and **d** good dispersion, good distribution (Hedayati et al. 2011)

4.1 Preparation of Nanofibrillated Cellulose Base Nanocomposites

Practically, solution casting, melt compounding, electrospinning are the most used processes to elaborate nanofibrillated cellulose-reinforced nanocomposites. The properties of elaborated nanocomposites depend precisely on the nanocellulose morphology and size, good dispersion/distribution of nanocellulose/NFC into the matrix, nature of the matrix, and the used manufacturing techniques. Several studies were carried out on the elaboration of NFC-based nanocomposite. As an example, the study of Barari et al, where they are broached the fabrication of NFC bio-based epoxy by a liquid composite molding process (LCM), it's noted an improvement of mechanical and tribological properties compared to virgin composite without NFC (El Makssoudi et al. 2014).

Generally, to prepare NFC-based thermoset nanocomposite, the dried NFC must be mixed with thermoset polymer matrix. The obtained mixture is poured into a mold with desired dimensions. Then, the silicon oil may be used to facilitate the removal of the composite after manufacturing; the resulted product was cured in cold or hot press machine. Some important manufacturing processes are briefly enumerated in this chapter.

4.1.1 Electrospinning

In this process, an electrically charged jet of polymer melt is produced out of an orifice, and an elevated voltage of supply is applied between the orifice and collecting target. The jet of polymer melt solidifies, and correlated fibers are collected on the collecting target. Figure 4 illustrates this process.

4.1.2 Casting Evaporation

The most motivating in this process is that easy to manufacture without using sophisticated equipment. During this process, the solvent is evaporated from nanocomposites film after the dispersion casting. Nanofibrillated cellulose (NFC) must be homogeneously distributed and dispersed in a polymeric matrix to obtain high-strength nanocomposites. We may discern three systems based on polymer used like the hydrosoluble system, emulsion system, and non-hydrosoluble system (Abdellaoui and Echaabi 2014).

4.1.3 Melt Compounding

Practically, this is an extraction and impregnation method, implying the addition of the matrix using the thermomechanical mixing (compounding). It highly recommended that the polymer (thermoplastic) be in melting state or with a low viscosity to guaranty a complete impregnation. This method is composed of two steps: solutions mixing and melt compounding. The first step, solution mixing, permits to fabricate masterbatches of polymer/nanocellulose, where the combination was

Fig. 4 Electrospinning process

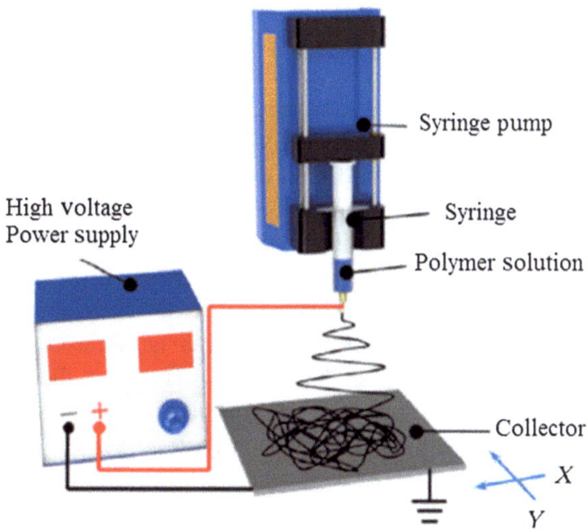

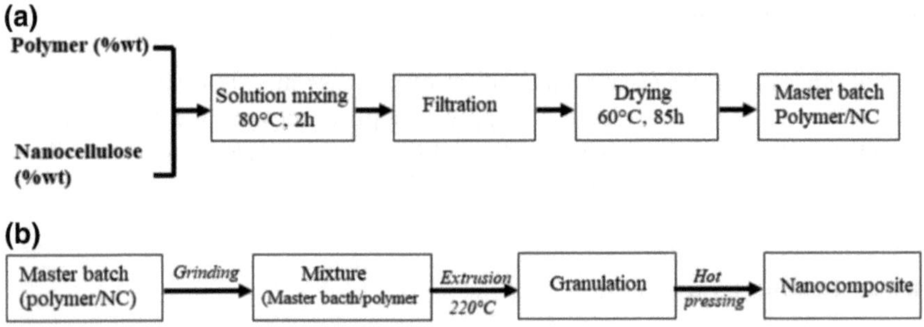

Fig. 5 Melt compounding process, **a** masterbatch preparation, **b** nanocomposite preparation

mixed into solution with stirring at 80 °C, filtered, and dried for obtaining the masterbatch based on polymer/NC as explained in Fig. 5a. In the second step, the obtained masterbatch was diluted with neat polymer by melt processing using a twin-screw mixer. Then, the material was pelletized and compression molded to elaborate samples for characterization (Raji et al. 2017b). Figure 5b indicates the steps of nanocomposite preparation.

4.2 Nanocomposite Properties

The properties of nanofibrillated cellulose-based nanocomposites were determined through nanocomposite characterization whether mechanical, rheological or thermal. The nanofibrillated cellulose material is most important and most investigated, owing to their mechanical properties, aspect ratio, and nanoscale dimension. Excellent specific properties can be obtained by mixing NFC with polymeric matrix even at small amount (low content of filler). In this respect, interesting results are obtained through several research works which we will be careful to briefly mention some of them.

Gantayat et al. (2017) studied the mechanical properties of carbon nanofiber CNF-based epoxy nanocomposites. The nanofibrils were functionalized by a mixed acid chemical method, and the nanocomposites were realized with different filler concentration as shown in Fig. 6. The results illustrate that the tensile stress increases monotonically with increase in the carbon nanofibrils content until it reaches 20% wt. with respect to neat epoxy resin. This improvement is probably due to strong interfacial adhesion between CNF and epoxy. However, the nanocomposite starts to degrade with further increasing of carbon nanofiber content (3%wt).

In another study, Saba et al. (2017) studied the effect of nanofibrillated cellulose on the thermal and dynamic mechanical analysis (DMA) of nanocomposites as a function of temperature. The nanocomposites based on nanofibrillated cellulose and epoxy were prepared by hand layup method at different filler contents (0.5, 0.75, and 1 wt%). Thermogra- vimetric methods (DTG) presented in Fig. 7, the results indicate

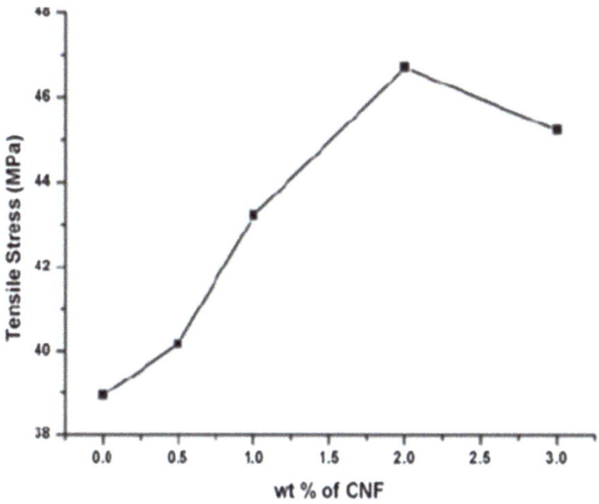

Fig. 6 Tensile stress of CNF-based epoxy nanocomposite (Raji et al. 2017b)

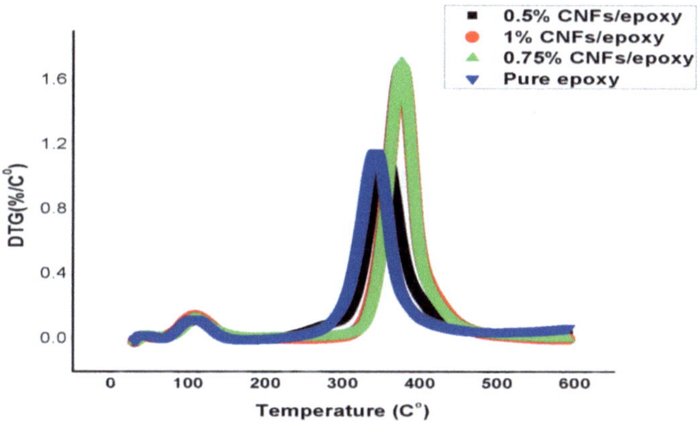

Fig. 7 DTG curves for NCF-based epoxy nanocomposites (Saba et al. 2017)

an increase in the thermal stability of the elaborated nanocomposites with low content of filler and, nanocomposites at 0.75wt% content confer high resistance stability compared to others. Concerning the DMA results, it is noted that the nanocomposites at 0.75wt% content confer highest value of storage modulus with respect to nanocomposites with 0.5 and 1wt% content.

Again, some researchers have used nanofibrillated cellulose for reinforcing thermoplastic matrix. Niazi et al. (2017) have worked on the effect of the addition of phosphorylated nanocellulose (PCNF) on the mechanical and thermal properties of poly(vinyl alcohol) (PVA) nanocomposites. The tensile properties were studied for three different relative humidities (0, 53, and 93% RH) as explicit in Fig. 8. It is noted that at 93% RH, the elastic modulus increases from 0.12 to 0.82 MPa with

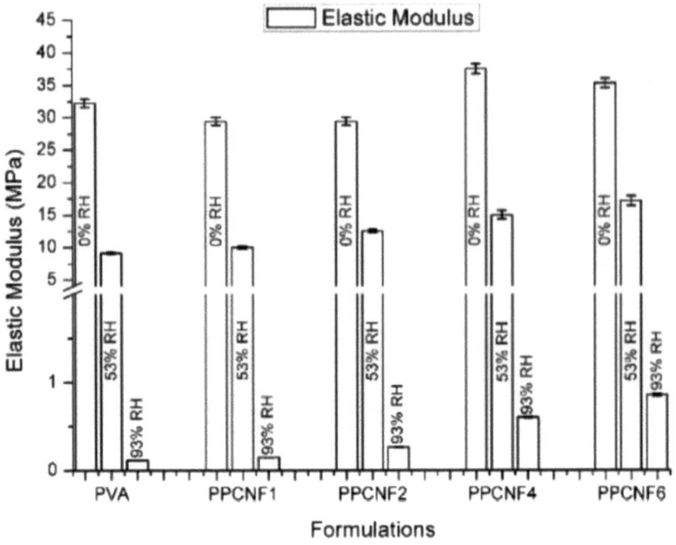

Fig. 8 Tensile modulus of PVA/nanocellulose membrane at 0% RH, 53% RH, and 93% RH (Saba et al. 2017)

the increase of filler concentration (6% wt.). This result is due to higher water uptake after four days at 53 and 93% RH and is an indication to apply nanocomposites with phosphorylated nanocellulose and PVC matrix as membrane for CO_2 separation.

Other researchers used nanocomposites based on natural matrix like the work of Jyotishkumar et al (Thomas et al. 2015), in which they studied nanocomposites based on jute nanocellulose and natural rubber (NR) latex. The nanocellulose fibers were extracted from raw jute fiber by steam explosion process. The mechanical results present that the elastic modulus of elaborated nanocomposites increased from 1.3 to 3.8 MPa with the addition of jute nanocellulose fiber (3 wt%). This enhancement is mainly attributed to the possible restriction of polymer chain mobility in the surrounding of cross-linked cellulose–cellulose tridimensional network. Figure 9 presents the stress measurement for three different nanocomposites and raw natural rubber.

5 Nanocomposite Applications

Recently, the nanocellulose gained tremendous interest for the purpose of replacing the petroleum-based reinforcements, despite their commercialization is still in the embryonic state, owing to their high rigidity, nanometric dimension as well as the high uniformity and fewer defects. The nanocellulose is mainly responsible for

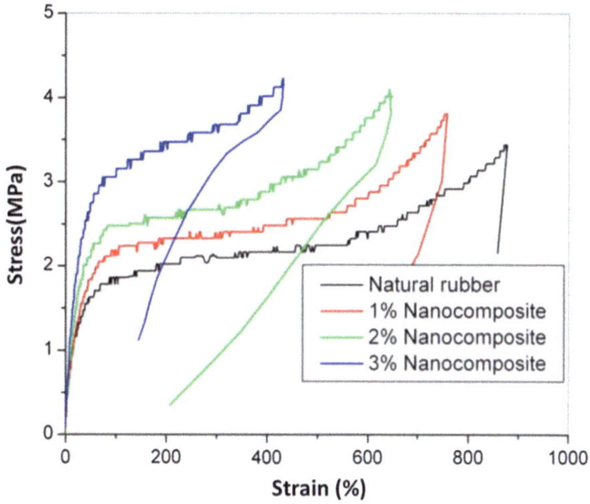

Fig. 9 Stress versus strain for jute nanocellulose fiber-based natural rubber nanocomposites (Thomas et al. 2015)

developing interesting properties for nanocomposites, where they are being used in medical, electronics, construction, automotive, packaging, and wastewater. We can classify the nanocellulose application into novel application, high volume, and low value as detailed below.

- High volume: automotive interiors, automotive body components, building and construction, etc.
- Low value: aerogels, oil and gas industry, construction, air and water filtration, and industrial viscosity modifiers.
- Novel application: exceptionally is about some sophisticated technologies like photonic structures, light-emitting diodes, industrial and medical additive manufacturing. Likewise, this category includes nanomaterials and nanocomposites thanks to their weight, strength, and rheological/optical properties. Figure 10 illustrates several fields where the nanocellulose can be greatly applied (Abdellaoui and Echaabi 2014).

Industrially, some companies of electroacoustic devices have used nanocellulose as a membrane for high-quality sound (Kim et al. 2018). Besides, they are used for high-quality electronic paper (e-paper). Nowadays, the nanofibrillated cellulose has emerged in the industrial community for a huge variety of applications like as printing (Li and Panigrahi 2004), medical applications (Doktor et al. 2011), and packaging in food applications (Saba et al. 2014) due to the high demands on product safety, low cost, and environmental considerations (Pracella et al. 2010). In food industry, nanofibrillated cellulose is requested for their high mechanical properties and sufficient barrier to oxygen, microorganisms, water vapor light, and other contaminants to avoid food deterioration (Vatai 2010). In biomedical industry, they are solicited for healthcare applications (cosmetics, biomedicine, etc.), owing to their cyto/hemo-compatibility, biodegradability as well as non-toxic aspect for people (Khalil et al. 2015).

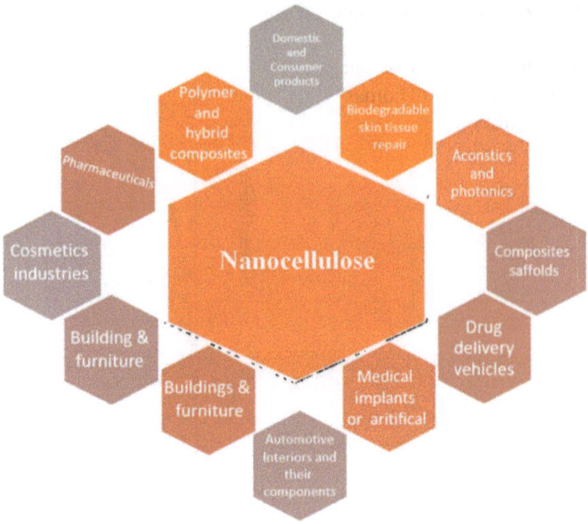

Fig. 10 Several nanocellulose applications

6 Conclusion

In this chapter, we reviewed the nanofibrillated cellulose (NFC) structure, its preparation, and its surface modification to be able to reinforce the polymeric matrix. Besides, we reviewed the nanocomposites nanofibrillated cellulose-based nanocomposite processing, the resulted properties, and their applications. The NFC-based nanocomposites have the potential to be a green materials in vast amount of industrial applications, owing to their exciting mechanical properties, low weight, cheapness, biodegradability, abundance, and high crystallinity. These promising characteristics will only have effect if the surface of the nanocellulose is modified to be compatible with the hydrophobic nature of the polymeric matrices, ensuring good interfacial adhesion and increased mechanical properties of the elaborated nanocomposites.

References

Abdellaoui H, Echaabi J (2014) Rheological models for modeling the viscoelastic behavior in liquid composite molding processes (LCM) review. J Reinf Plast Compos 33(8):1–19

Abdellaoui H, Bensalah H, Echaabi J, Bouhfid R, el kacem Qaiss A (2015a) Fabrication, characterization and modelling of laminated composites based on woven jute fibres reinforced epoxy resin. Mater Des 68:104–113

Abdellaoui H, Bouhfid R, Echaabi J, el kacem Qaiss A (2015b) Experimental and modeling study of viscoelastic behaviour of woven dried jute under compressive stress. J Reinf Plast Compos 34(5):405–420

Abdellaoui H, Bensalah H, Raji M, Rodrigue D, Bouhfid R, el kacem Qaiss A (2017) Laminated epoxy biocomposites based on clay and jute fibers. J Bionic Eng 14(2):379–389

Abdelmouleh M, Boufi S, Belgacem MN, Dufresne A (2007) Short natural-fibre reinforced polyethylene and natural rubber composites: effect of silane coupling agents and fibres loading. Compos Sci Technol 67(7–8):1627–1639

Ahmad F, Choi HS, Park MK (2015) A review: natural fiber composites selection in view of mechanical, light weight, and economic properties. Macromol Mater Eng 300(1):10–24

Ait Laaziz S, Raji M, Hilali E, Essabir H, Rodrigue D, Bouhfid R (2017) Bio-composites based on polylactic acid and Argan nut shell: production and properties. Int J Biol Macromol 104:30–42

Albu MG, Vuluga Z, Panaitescu DM, Vuluga DM, Căşărică A, Ghiurea M (2014) Morphology and thermal stability of bacterial cellulose/collagen composites. Cent Eur J Chem 12(9):968–975

Alwani MS, Abdul Khalil HPS, Sulaiman O, Islam MN, Dungani R (2014) An approach to using agricultural waste fibres in biocomposites application: thermogravimetric analysis and activation energy study. BioResources 9(1):218–230

Araújo JR, Waldman WR, De Paoli MA (2008) Thermal properties of high density polyethylene composites with natural fibres: coupling agent effect. Polym Degrad Stab 93(10):1770–1775

Azwa ZN, Yousif BF, Manalo AC, Karunasena W (2013) A review on the degradability of polymeric composites based on natural fibres. Mater Des 47:424–442

Ben Azouz K, Ramires EC, Van den Fonteyne W, El Kissi N, Dufresne A (2012) Simple method for the melt extrusion of a cellulose nanocrystal reinforced hydrophobic polymer. ACS Macro Lett 1(1):236–240

Benyahia A, Merrouche A, Rokbi M, Kouadri Z (2013) Study the effect of alkali treatment of natural fibers on the mechanical behavior of the composite unsaturated polyester-fiber Alfa abstract. 21ème Congrès Français de Mécanique, 1–6

Brinchi L, Cotana F, Fortunati E, Kenny JM (2013) Production of nanocrystalline cellulose from lignocellulosic biomass: technology and applications. Carbohyd Polym 94(1):154–169

Doktor G, Fakult N, Ashraf H, Asran S, Sayed A, Gutachter K, Michler GH (2011) Electrospinning of polymeric nanofibers and nanocomposite materials: fabrication, physico-chemical characterization and medical applications

Eichhorn SJ, Dufresne A, Aranguren M, Marcovich NE, Capadona JR, Rowan SJ et al (2010) Review: current international research into cellulose nanofibres and nanocomposites. J Mater Sci 45(1):1–33

El Makssoudi A, Abdellaoui H, El Ouatib R, Tahiri M (2014) Development of composite materials based on expanded perlite and plastic wastes. Mechanic-chemical properties. In: 2nd annual international conference on chemistry, chemical engineering and chemical process (CCECP 2014), pp 38–46

Erbas Kiziltas E, Kiziltas A, Bollin SC, Gardner DJ (2015) Preparation and characterization of transparent PMMA-cellulose-based nanocomposites. Carbohyd Polym 127:381–389

Essabir H, Hilali E, Elgharad A, El Minor H, Imad A, Elamraoui A, Al Gaoudi O (2013) Mechanical and thermal properties of bio-composites based on polypropylene reinforced with Nut-shells of Argan particles. Mater Des 49:442–448

Essabir H, Bensalah MO, Rodrigue D, Bouhfid R, el kacem Qaiss A (2016a) Biocomposites based on Argan nut shell and a polymer matrix: effect of filler content and coupling agent. Carbohyd Polym 143:70–83

Essabir H, Boujmal R, Bensalah MO, Rodrigue D, Bouhfid R, el kacem Qaiss A (2016b) Mechanical and thermal properties of hybrid composites: oil-palm fiber/clay reinforced high density polyethylene. Mech Mater 98:36–43

Essabir H, Raji M, Essassi EM, Rodrigue D, Bouhfid R, el kacem Qaiss A (2017) Morphological, thermal, mechanical, electrical and magnetic properties of ABS/PA6/SBR blends with Fe_3O_4 nano-particles. J Mater Sci Mater Electron 28(22):17120–17130

Farias D, Cordeiro R, Canabarro BR, Scholz S, Sim RA (2017) Surface lignin removal on coir fibers by plasma treatment for improved adhesion in thermoplastic starch composites [João Gabriel Guimarães de Farias a, Rafael Cordeiro Cavalcante a]. Carbohydr Polym 165:429–436

Gacitua EW, Ballerini AA, Jinwen Z (2005) Polymer nanocomposites: synthetic and natural fillers. Maderas Ciencia Y Tecnología 7(3):159–178

Gantayat S, Rout D, Swain SK (2017) Structural and mechanical properties of functionalized carbon nanofiber/epoxy nanocomposites. Mater Today Proc 4(8):9060–9064

Grumezescu AM (2017). Food packaging nanotechnology in the agri-food industry, vol 7. Elsevier Inc, Netherlands, p 805

Gupta G, Gupta A, Dhanola A, Raturi A (2016) Mechanical behavior of glass fiber polyester hybrid composite filled with natural fillers. In: IOP conference series: materials science and engineering, vol 149. pp 12091

Hakeem KR, Mohammad J, Alothman Othman Y (2011) Agricultural biomass based potential materials. Springer International Publishing, Switzerland

Hall M, Bansal P, Lee JH, Realff MJ, Bommarius AS (2011) Biological pretreatment of cellulose: enhancing enzymatic hydrolysis rate using cellulose-binding domains from cellulases. Biores Technol 102:2910–2915

Hamour N, Boukerrou A, Djidjelli H, Maigret JE, Beaugrand J (2015) Effects of MAPP compatibilization and acetylation treatment followed by hydrothermal aging on polypropylene alfa fiber composites. Int J Polym Sci

Hedayati M, Salehi M, Bagheri R, Panjepour M, Maghzian A (2011) Ball milling preparation and characterization of poly (ether ether ketone)/surface modi fi. Powder Technol 207(1–3):296–303

Hietala M, Mathew AP, Oksman K (2012) Bionanocomposites of thermoplastic starch and cellulose nanofibers manufactured using twin-screw extrusion. Eur Polym J 1–7

Hoidy WH, Al-mulla EAJ (2013) Study of preparation for co-polymer nanocomposites using PLA/LDPE/CTAB modified clay. Iraqi Nat J Chem 49:61–72

Jönsson LJ, Martín C (2016) Pretreatment of lignocellulose: formation of inhibitory by-products and strategies for minimizing their effects: review. Biores Technol 199:103–112

Khalil HPSA, Bhat AH, Bakar AA, Tahir PM, Zaidul ISM, Jawaid M (2015) Cellulosic nanocomposites from natural fibers for medical applications: a review. In: Handbook of polymer nanocomposites. Processing, performance and application: volume C: Polymer nanocomposites of cellulose nanoparticles. Springer, Berlin, pp 475–511

Khanam PN, Ponnamma D, AL-Madeed MA (2015) Electrical properties of graphene polymer nanocomposites. In: Graphene-based polymer nanocomposites in electronics, Springer series on polymer and composite materials, pp 25–47

Kim H, Hong J, Pyo S (2018) Acoustic characteristics of sound absorbable high performance concrete. Appl Acoust 138(April):171–178

Ku H, Wang H, Pattarachaiyakoop N, Trada M (2011) A review on the tensile properties of natural fiber reinforced polymer composites. Compos B Eng 42(4):856–873

Kumar R, Obrai S, Sharma A (2011) Chemical modifications of natural fiber for composite material. Pelagia Res Libr 2(4):219–228

Le Duigou A, Davies P, Baley C (2010) Interfacial bonding of flax fibre/poly(l-lactide) bio-composites. Compos Sci Technol 70(2):231–239

Li X, Panigrahi S (2004) Flax fiber-reinforced composites and the effect of chemical treatments on their properties. Appl Eng Agric 25(3):1–11

Menon MP, Selvakumar R, Kumar PS, Ramakrishna S (2017) Extraction and modification of cellulose nanofibers derived from biomass for environmental application. Roy Soc Chem 7:42750–42773

Mir SS, Nafsin N, Hasan M, Hasan N, Hassan A (2013) Improvement of physico-mechanical properties of coir-polypropylene biocomposites by fiber chemical treatment. Mater Des 52:251–257

Mohammed L, Ansari MNM, Pua G, Jawaid M, Islam MS (2015) A review on natural fiber reinforced polymer composite and its applications. Int J Polym Sci 2015:1–15

Mohkami M, Talaeipour M (2011) Investigation of the chemical structure of carboxylated and carboxymethylated fibers from waste paper via XRD and FTIR analysis. BioResources 6 (2):1988–2003

Niazi MBK, Jahan Z, Berg SS, Gregersen ØW (2017) Mechanical, thermal and swelling properties of phosphorylated nanocellulose fibrils/PVA nanocomposite membranes. Carbohyd Polym 177:258–268

Nunna S, Chandra PR, Shrivastava S, Jalan AK (2012) A review on mechanical behavior of natural fiber based hybrid composites. J Reinf Plast Compos 31(11):759–769
Pracella M, Haque MMU, Alvarez V (2010) Functionalization, compatibilization and properties of polyolefin composites with natural fibers. Polymers 2(4):554–574
Qu T, Zhang X, Gu X, Han L, Ji G, Chen X, Xiao W (2017) Ball milling for biomass fractionation and pretreatment with aqueous hydroxide solutions. Am Chem Soc 5:7733–7742
Raji M, Essabir H, Essassi EM, Rodrigue D, Bouhfid R, el kacem Qaiss A (2016) Morphological, thermal, mechanical, and rheological properties of high density polyethylene reinforced with illite clay. Polym Polym Compos 16(2):101–113
Raji M, Essabir H, Bouhfid R, el kacem Qaiss A (2017a) Impact of chemical treatment and the manufacturing process on mechanical, thermal, and rheological properties of natural fibers-based composites. In: Handbook of composites from renewable materials. Wiley, Hoboken, pp 225–252
Raji M, Essabir H, Rodrigue D, Bouhfid R, el kacem Qaiss A (2017b) Influence of graphene oxide and graphene nanosheet on the properties of polyvinylidene fluoride nanocomposites. Polym Polym Compos. https://doi.org/10.1002/pc.24292
Raji M, Mekhzoum MEM, Rodrigue D, el kacem Qaiss A, Bouhfid R (2018) Effect of silane functionalization on properties of polypropylene/clay nanocomposites. Compos Part B. https://doi.org/10.1016/j.compositesb.2018.04.013
Saba N, Tahir P, Jawaid M (2014) A review on potentiality of nano filler/natural fiber filled polymer hybrid composites. Polymers 6:2247–2273
Saba N, Safwan A, Sanyang ML, Mohammad F, Pervaiz M, Jawaid M, Sain M (2017) Thermal and dynamic mechanical properties of cellulose nanofibers reinforced epoxy composites. Int J Biol Macromol 102:822–828
Saheb DN, Jog JP (1999) Natural fiber polymer composites: a review. Adv Polym Technol 18 (4):351–363
Sdrobi A, Darie RN, Totolin M, Cazacu G, Vasile C (2012) Low density polyethylene composites containing cellulose pulp fibers. Compos B Eng 43(4):1873–1880
Shauddin SM, Shaha CK, Khan MA (2014) Effects of fiber inclusion and γ radiation on physico-mechanical properties of jute caddies reinforced waste polyethylene composite. J Polym Biopolym Phys Chem 2(4):91–97
Singh TJ, Samanta S (2014) Characterization of natural fiber reinforced composites-bamboo and sisal: a review. IJRET: Int J Res Eng Technol 3(7):187–195
Singh S, Mohanty AK, Sugie T, Takai Y, Hamada H (2008) Renewable resource based biocomposites from natural fiber and polyhydroxybutyrate-co-valerate (PHBV) bioplastic. Compos A Appl Sci Manuf 39(5):875–886
Siró I, Plackett D (2010) Microfibrillated cellulose and new nanocomposite materials: a review. Cellulose 17(3):459–494
Solyman WSE, Nagiub HM, Alian NA, Shaker NO, Kandil UF (2017) Synthesis and characterization of phenol/formaldehyde nanocomposites: studying the effect of incorporating reactive rubber nanoparticles or Cloisite-30B nanoclay on the mechanical properties, morphology. J Radiat Res Appl Sci 10(1):72–79
Song J, Rojas OJ (2013) Approaching super-hydrophobicity from cellulosic materials: a review. Paper Chem 28(2):216–238
Stocchi A, Lauke B, Vázquez A, Bernal C (2007) A novel fiber treatment applied to woven jute fabric/vinylester laminates. Compos A Appl Sci Manuf 38(5):1337–1343
Taj S, Munawar MA, Khan S (2007) Natural fiber-reinforced polymer composites: review. Proc Pakistan Acad Sci 44:129–144
Thomas MG, Abraham E, Jyotishkumar P, Maria HJ, Pothen LA, Thomas S (2015) Nanocelluloses from jute fibers and their nanocomposites with natural rubber: preparation and characterization. Int J Biol Macromol 81:768–777
Tian C, Yi J, Wu Y, Wu Q, Qing Y, Wang L (2016) Preparation of highly charged cellulose nanofibrils using high-pressure homogenization coupled with strong acid hydrolysis pretreatments. Carbohyd Polym 136:485–492

Turbak A, Snyder F, Sandberg K (1983) Microfibrillated cellulose, a new cellulose product: properties, uses, and commercial potential. In: Sarko A (ed) Proceedings of the ninth cellulose conference, Applied polymer symposium, vol 37. Wiley, New York, pp 815–827

Vatai G (2010) Separation technologies in the processing of fruit juices. In: Separation, extraction and concentration processes in the food, beverage and nutraceutical industries. Woodhead Publishing series in food science, technology and nutrition, pp 381–395. Woodhead Publishing Limited, UK

Vazquez A, Foresti M, Moran J, Cyras V (2015) Extraction and production of cellulose nanofibers. In: Handbook of polymer nanocomposites. Processing, performance and application, pp 81–118

Wang W, Sabo RC, Mozuch MD, Kersten P, Jin JYZY (2015) Physical and mechanical properties of cellulose nanofibril films from bleached eucalyptus pulp by endoglucanase treatment and microfluidization. J Polym Environ 23:551–558

Xie Y, Hill CAS, Xiao Z, Militz H, Mai C (2010) Silane coupling agents used for natural fiber/polymer composites: a review. Compos Part A Appl Sci Manuf 41(7):806–819

Yahaya R, Sapuan SM, Jawaid M, Leman Z, Zainudin ES (2015) Effect of fibre orientations on the mechanical properties of kenaf–aramid hybrid composites for spall-liner application. Defence Technol 12(1):52–58

Yang H, Yan R, Chen H, Lee DH, Zheng C (2007) Characteristics of hemicellulose, cellulose and lignin pyrolysis. Fuel 86(12–13):1781–1788

Zari N, Raji M, El Mghari H, Bouhfid R, el kacem Qaiss A (2018) Nanoclay and polymer-based nanocomposites: materials for energy efficiency. In: Polymer-based nanocomposites for energy and environmental applications. Woodhead, UK, pp 75–103

Bacterial Cellulose Nanocomposites

N. Pa'e, I. I. Muhamad, Z. Hashim and A. H. M. Yusof

Abstract Bacterial cellulose (BC) is a biopolymer with high purity of cellulose and excellent mechanical properties. Increased interest in the use of natural polymer makes BC as an excellent alternative for plant cellulose. Although both celluloses consist of unbranched pellicle with chemically equivalent structure, bacterial cellulose exhibits greater properties and potential in wider applications. The structure of bacterial cellulose that consists only glucose monomer and nanosized cellulose fibres secreted by the bacteria induces it to have high water-holding capacity, high crystallinity, high degree of polymerization and high mechanical strength. Furthermore, the characterization of BC can be certainly altered by incorporation with materials that are not essential for the bacterial growth into the fermentation medium. This unique property of BC opens a new gate for the development of new cellulose nanocomposites with desired properties by the incorporation of selective suitable materials. The BC nanocomposites produced opens new opportunity for various usages of BC in different fields of application in the pharmaceutical, chemical, medical and wastewater treatment plants.

Keywords Bacterial cellulose · Bacterial cellulose composites
Nanocomposites · Biopolymer composites

N. Pa'e · I. I. Muhamad · Z. Hashim · A. H. M. Yusof
Department of Bioprocess and Polymer Engineering, School of Chemical and Energy Engineering, Faculty of Engineering, Universiti Teknologi Malaysia, 81310 Johor Bahru, Johor, Malaysia

I. I. Muhamad (✉)
Cardiac Biomaterials Cluster, IJN-UTM Cardiovascular Engineering Center, FBME, UTM, Level 2, Block B, Building V01, 81310 Johor Bahru, Johor, Malaysia
e-mail: idaidayu@utm.my

M. L. Sanyang and M. Jawaid (eds.), *Bio-based Polymers and Nanocomposites*,
https://doi.org/10.1007/978-3-030-05825-8_5

1 Introduction

Cellulose from bacteria has more advantages than the cellulose found in plants. These advantages provide plenty room for the use of the bacterial cellulose (BC) in various fields. The most significant advantage of the BC compared to plant cellulose is its high purity. The BC from fermentation is produced in a pure form without lignin and hemicelluloses (Tyagi and Suresh 2015). It is highly hydrophilic with high mechanical strength. As BC is 100% pure and produced in hydrophilic matrix forms, its extensive fibrils and high mechanical strength are maintained throughout the formation. The BC can be produced using different methods and substrates whereby the properties can be modified based on the application. Soluble or insoluble materials added in the fermentation medium, such as enzymes and metallic compound, can directly be incorporated in the cellulose network during the synthesis (Serafica et al. 2002). These properties open the opportunities in many fields mainly in medical field, wastewater treatment, paper and also audio industries.

Addition of other materials that are not needed in the growth of bacteria in the fermentation medium is proven to affect the yield and properties of BC (Ruka et al. 2013). This modification was done either by using in situ or ex situ techniques (Shah et al. 2013). Several researchers have shown the ability to improve the production of BC, while others have successfully altered the properties of the cellulose by adding certain substrates or by manipulating the operating conditions. For example, Park et al. (2013) reported that the addition of magnetite nanoparticles and polyaniline enhanced the thermal stability of BC while da Silva et al. (2016) reported that the addition of polyethylene glycol improved hydrophilicity properties of BC.

This unique ability of BC allowing a series of potential applications as a new novel BC composite can be produced with modified properties based on its function. These applications range from high mechanical strength of hydrogel with addition of genipin (Dayal and Catchmark 2016; Nakayama et al. 2004), incorporation of aloe vera as wound dressing (Saibuatong and Phisalaphong 2010), antimicrobial film with addition of silver nanoparticles (Yang et al. 2012; Wu et al. 2014) to the use of BC as membranes such as cellulose acetate membranes reinforced with BC sheet (Gindl and Keckes 2004). Development of electromagnetic nanocomposite with the incorporation of magnetite nanoparticles and polyaniline was reported by Park et al. (2013). Juncu et al. (2015) suggested the novel uses of BC for drug delivery with the addition of carboxymethyl cellulose. Furthermore, Shanshan et al. (2012) demonstrated that the production of BC membrane in *N*-methylmorpholine *N*-oxide had better mechanical and barrier properties.

Undeniably, conservation of the forest, particularly trees, is essential in managing the global warming. However, excessive use of trees for cellulose-based products such as paper, biofuels and construction materials has been continuously depleting the forest resources (deforestation) which lead to global warming problem. At the same time, natural polymers such as cellulose have attracted many attentions due to the effect of environmental pollution of the synthetic polymer (Pei et al. 2013). Previously, plant-derived celluloses were widely used. For example, cellulose fibres have been used as eco-composite plastics in agricultural fields (González-Sánchez et al. 2014), biosorbent for heavy metal removal (O'Connell et al. 2008), nanofiller for biofilm (Salehudin et al. 2014; Slavutsky and Bertuzzi 2014) and other newly developed degradable composites (Piccinno et al. 2015). However, in recent time, BC is started to be used because of its excellent and promising properties (Ashori et al. 2012). Table 1 shows the development of BC and its application starting from dessert or *nata* in the 1990s followed by various other applications throughout the year.

2 Bacterial Cellulose (BC)

BC is a source of cellulose other than plant cellulose. It is produced by bacteria from many genera such as *Acetobacter*, *Achromobacter*, *Agrobacterium and Sarcina*. However, from that many lists of cellulose producers, the only species that is well known with its capability to produce cellulose in large quantity is *Acetobacter* genus. While within that species, *A. xylinus* or *Acetobacter xylinum* (*A. xylinum*) is the one that is being used extensively in research and studies (Jonas and Farah 1997).

A. xylinum produces cellulose from glucan chains. The chains extrude into the fermentation medium from *A. xylinum* pores. These processes are repeated until a bundle of microfibrils is gathered and forms the BC. Medium for BC fermentation can be from any kind of carbon sources or sugars. *Acetobacter* needs oxygen to produce the BC. Therefore, the production will occur mostly at the surface of the liquid (Schramm and Hestrin 1954, Zahan et al. 2014; Hsieh et al. 2016).

The BC pellicle is pure and extremely hydrophilic. Therefore, it needs no treatment which makes its original high mechanical strength retained. These unique characteristics of BC open many rooms for new applications as the properties of BC can be changed by manipulating the fermentation process. The BC has more advantages compared to plant cellulose. The most significant advantage is its purity. The BC has no hemicelluloses or lignin as in plant cellulose; thus, less processing steps are required. Besides, its structure is stronger and finer compared to plant cellulose which is due to longer and finer fibre length of BC. Figure 1 shows a microstructure of BC and plant cellulose at 5000 times magnification.

2.1 Synthesis of Bacterial Cellulose

There are several bacteria which are able to produce cellulose such as the strain from genera *Acetobacter*, *Agrobacterium*, *Pseudomonas*, *Rhizobium* and *Sarcina*. It has superior properties such as ultrafine network structure, high biodegradability

Table 1 Development in bacterial cellulose research

Development	Details	References
1950s Research on bacterial strain, fermentation medium and fermentation condition	Synthesis of cellulose by *Acetobacter xylinum*	Hestrin and Schramm (1954)
	Improved cellulose production using *Acetobacter xylinum* mutant	De Wulf et al. (1996)
	Improved production of cellulose with Acetobacter SP.LMG 1518 in submerged culture	Vandamme et al. (1998)
	Production of BC from fructose in continuous culture	Naritomi et al. (1998)
	Optimization of fermentation condition of *Acetobacter xylinum* in shaking culture	Son et al. (2001)
	Increased production of BC in synthetic media under shaking condition	Son et al. (2003)
	Production of BC from persimmon vinegar	Kim et al. (2006)
1990s Research on characterization of BC	Mechanical properties of bacterial sheets	Yamanaka et al. (1989)
	Characterization of BC produced by *Acetobacter pasteurianus* strain	Bertocchi et al. (1997)
	Characterization on mechanical properties of BC and chitosan blends	Wu et al. (2004)
	Microbial cellulose structure in stationary and agitated culture	Czaja et al. (2004)
	Characterization of water in BC	Gelin et al. (2007)
2000 Research on new application of BC	*Nata de coco* as dessert	Budhiono et al. (1999)
	Electronic paper displays made from microbial cellulose	Shah and Brown (2005)
	BC as wound healer	Czaja et al. (2006)
	Antimicrobial films from BC	Gao et al. (2014)
	BC for skin repair materials	Fu et al. (2011)
	BC as carrier for drug delivery system	Amin et al. (2012)
2009 Research on modification of BC and its application	Modification of BC using nano aloe vera for wound dressing	Saibuatong and Phisalaphong (2010)
	Hybrid BC nanocrystals and silver nanoparticles	George et al. (2014)
	Modification of BC using magnetic field	Fijalkowski et al. (2015)

(continued)

Table 1 (continued)

Development	Details	References
	Modification of BC–alginate composites for scaffold in tissue engineering	Kirdponpattara et al. (2015)
	Surface modification of BC using trimethylsilylation for oil–water separation	Sai et al. (2015)
	Modification of BC structure under ultrasonic irradiation	Paximada et al. (2016)

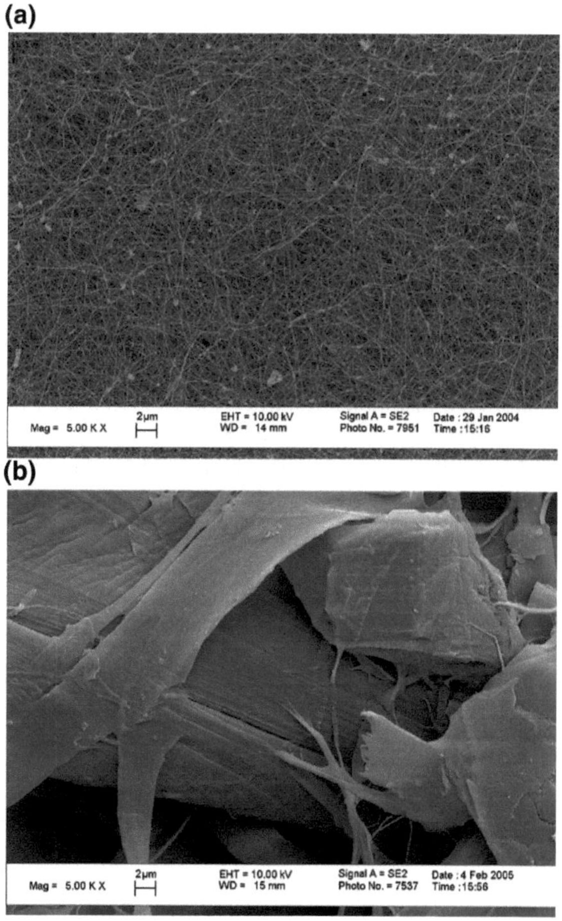

Fig. 1 SEM images of **a** bacterial cellulose and **b** plant cellulose at 5000 times magnification

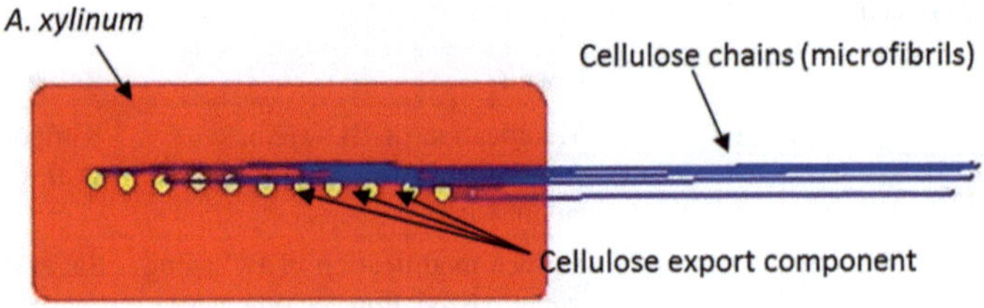

Fig. 2 Secretion of microfibrils by *Acetobacter* cells

and high mechanical strength as compared to plant cellulose (Tsuchida and Yoshinaga 1997). It is expected to be an alternative for biodegradable biopolymer. According to Lee et al. (2002), cellulose synthesized by the bacterium *A. xylinum* and plant cellulose is highly crystalline. However, the arrangements of glucosyl units of the crystallites make them differ to each other.

A. xylinum is acetic acid-producing bacterium which is widely used as the model system to study the enzymes and genes involved in cellulose biosynthesis. This species of gram-negative bacteria has high capability to produce cellulose by converting carbon source such as glucose or sucrose into cellulose. It is well known that *A. xylinum* is an obligate aerobe and forms cellulose at the air/liquid interface in undisturbed cultures. Figure 2 demonstrates the formation of microfibrils by *A. xylinum.*

Multiple cellulose chain or microfibrils are synthesized in the interior of bacteria cell and spun out of BC export component or nozzles (Iguchi et al. 2011). The cellulose synthesis will continue until a limited condition such as when carbon sources are insufficient or when the BC filled the discs in fermentation using rotary discs reactor (Pa'e 2009; Khairul et al. 2016).

Most cellulose-producing acetic acid bacteria can convert glucose into gluconic acid and ketogluconic acid. The enzyme responsible for the conversion of glucose to gluconic acid is membrane-bound glucose dehydrogenase (GDH) (Hwang et al. 1999). This will simply remove glucose from the fermentation medium and avoid the formation of cellulose (Toru et al. 2005).

2.2 *Application of Bacterial Cellulose*

BC has similar chemical structure as plant cellulose, but it has higher purity without lignin or hemicelluloses. Therefore, using BC is more economic compared to plant cellulose because it can skip many stages in the production of pure plant cellulose (Jonas and Farah 1997), thus lowering the production cost. Moreover, its distinguishing properties such as ultrafine nanofibre, biodegradability and high mechanical test (Amin et al. 2012) make it as promising materials for many applications.

The interest on applications of BC grows rapidly since the 1990s. Those applications started with the BC usage in its ordinary form and expand to BC composites *via* the modification of BC using other materials.

3 Modification of Bacterial Cellulose for the Production of Bacterial Cellulose Nanocomposites

The most attractive feature of BC production is the ability to control and modify the properties of the cellulose product while it is being synthesized. Additions of other materials that are not required for bacterial growth to the medium can alter the yield and properties of the cellulose produced (Ruka et al. 2013). With this ability, the chemical composition (Shirai et al. 1994) and properties of BC such as hydrophilicity can be adjusted (Martínez-Sanz et al. 2013). Those improvements are required to enhance the capabilities of BC in order to fulfil the demand in different fields (Shah et al. 2013). The capacity to control and adjust the cellulose characteristic during the synthesis allows the development of new BC composites with better properties than would be conceivable by post-synthetic processing of other sources of cellulose.

Modification of BC to produce BC composites can be done using numerous methods depending on the additives or materials used. The method likewise changes with the required applications. In general, there are two fundamental approaches in modifying BC: ex situ and in situ modification methods.

3.1 Ex Situ Modification of Bacterial Cellulose

Ex situ modification method can be explained as the modification of pure BC after the cellulose is harvested (Shah et al. 2013). This involves the impregnation of BC with other materials to produce composites. The interaction might be physical or occur through definite hydrogen bonding between the BC and reinforcement materials. Liquid substances and tiny solid particles can easily penetrate and become engrossed inside the porous BC matrix. A schematic diagram showing the synthesis of BC composites using this method is provided in Fig. 3.

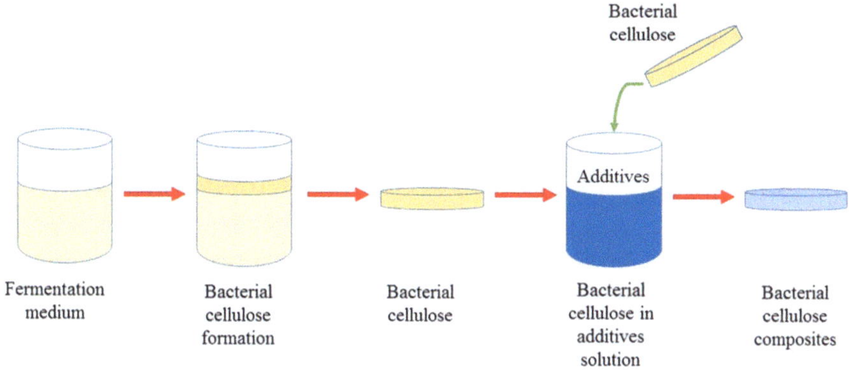

Fig. 3 Modification of bacterial cellulose using ex situ method

The properties of BC are differently affected by the additives used. The major hurdle associated with the ex situ modification method is the size and nature of the additives. Specifically, only submicron to nanosized materials can be impregnated into BC matrix. This is because larger particles cannot enter the BC pores, and hydrophobic materials are not able to combine with BC. Moreover, the structural arrangement of the BC fibril is not always uniform. Therefore, penetrating materials might not be homogeneously distributed inside the BC matrix. Accordingly, there is a need to identify new BC composite synthesis routes to resolve this problem. Table 2 lists previous works on producing BC composites using ex situ modification methods.

3.2 In Situ Modification of Bacterial Cellulose

The in situ method utilizes the addition of other materials as additives to the BC during its synthesis, which then becomes part of the cellulose structure. Using this method, additives were added to BC fermentation medium at the beginning of the process. A schematic diagram showing the synthesis of BC composites using this method is provided in Fig. 4.

Microfibrils of BC become denser with time and produce a web-shaped structure (Horii et al. 1997; Tang et al. 2010) that can trap various materials added to the medium (Ul-Islam et al. 2012). The encaged materials become part of the BC fibril network, resulting in BC composites. Bacterial composites can be synthesized via static fermentation (Wu et al. 2014; Saibuatong and Phisalaphong 2010), agitated culture (Cheng et al. 2009; Yan et al. 2008) and in vessels with rotating discs called rotary disc reactor (Pa'e et al. 2013; Serafica et al. 2002).

In situ modification of BC is a widely used approach that employs a wide range of modifications in agitation equipment and operating methods. However, this technique has certain limitations that prevent the synthesis of many BC composites. For example, several important antimicrobial agents such as titanium oxide cannot be added directly to the media because of their toxic effects on microorganisms. Moreover, BC composite synthesized through agitation culture cannot be applied as a gel or sheet in biomedical applications. Table 3 lists previous works on producing BC composites using in situ modification methods.

3.3 Characterization of Bacterial Cellulose Nanocomposites

BC composites can be characterized in many different ways. These important properties will vary from application to application. Microstructure study of the BC composites will provide information about the appearance of the cellulose produced under microscopic scale. Field emission scanning electron microscope is usually used to study the microstructure of the BC. The FESEM is a versatile analytical

Table 2 Previous works on ex situ modification of bacterial cellulose

Materials added	Impact	References
Polyethylene glycol	• Produced composite with high hydrophilicity which is suitable for biomedical devices • No significant effect to crystalline structure	da Silva et al. (2016)
Conductive polymer complex: poly (3,4-ethylenedioxythiophene)-poly (4-styrenesulphonate) (PEDOT/PSS)	• Development of composites that are suitable for electrodes and temperature sensors • Strong absorption in the red and near-infrared spectral region	Aleshin et al. (2015)
Carboxylmethyl cellulose (CMC)	• Produced CMC-BC composites that are suitable for drug release • Higher swelling rate up to 20% compared to CMC alone without BC	Juncu et al. (2015)
Ionic conducting polymer (ICP): *N*-hydroxyethyl acrylamide and 3-sulphopropyl methacrylate potassium salt	• Produced optically transparent BC/ICP composite with electro-conductive capability • Improved the electrical conductivity of the composite • No further significant increase on optical transparency for impregnated of 2 ICP salt	Jeon et al. (2014)
Polylactic acid (PLA), polycaprolactone (PCL), cellulose acetate (CA) and poly(methyl methacrylate) (PMMA)	• Produced composites with enhanced mechanical properties • Increased Young's modulus for all polymers added with increase of BC in the composites • High surface-area-to-volume ratio	Pircher et al. (2014)
Poly (vinyl alcohol) and chitosan	• Produced composite for release of ibuprofen sodium salt • Decrease of swelling ability with the increase of BC in the composites • Low cumulative release of ibuprofen sodium salt	Pavaloiu et al. (2014)
Polylactic acid (PLA)	• Produced composites films with high water barrier • Decrease of water permeability by 48%	Martínez-Sanz et al. (2013)
Polyaniline (PANI)	• Produced BC composite films with better thermal stability and optimal weight loss (20%) due to degradation • Polymerization of PANI on BC produced dark green film indicating colour of the aniline monomers	Park et al. (2013)

(continued)

Table 2 (continued)

Materials added	Impact	References
Polyurethane resin	• Produced composites that are able to emit light • Suitable as a substrate for organic light-emitting diode (OLED) • Higher degradation temperature up to 350 °C and 8% increase of reflective index compared to polyurethane resin	Ummartyotin et al. (2012)

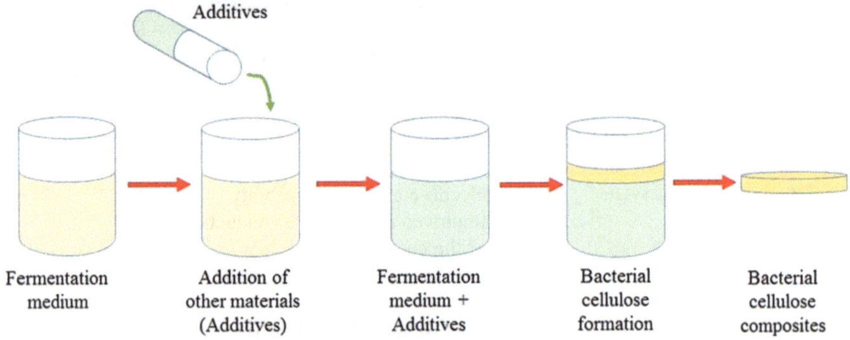

Fig. 4 Modification of bacterial cellulose using in situ method

ultrahigh resolution that extends imaging and analytical resolution beyond previously achievable limits. The equipment is well suited for characterization of BC since it provides high magnification up to 30,000 magnification. Since BC consists of multiple nanocellulose networks, equipment with high magnification like FESEM will provide better overview of the samples. It also has a unique in-lens detector for clear topographic imaging in high vacuum mode.

The possible interactions between the components in the BC composites were evaluated by X-ray diffractometer (XRD) and Fourier transform infrared spectroscopy (FTIR). XRD is routinely employed to characterize the phase composition and per cent crystallinity for organic or inorganic materials. It is important to study the crystallinity patterns of BC and BC composites since it influence many of its characteristics including tensile strength, opacity and its swelling ability.

FTIR identifies chemical bonds in a molecule by producing an infrared absorption spectrum. The spectra produce a profile of the sample, a distinctive molecular fingerprint that can be used to screen and scan samples for many different components. Therefore, FTIR is an effective analytical instrument for detecting functional groups in the BC composites and characterizing the difference between native BC.

Table 3 Previous works on in situ modification of bacterial cellulose

Materials added	Impact	References
Carboxylmethyl cellulose, pectin, gelatine, cornstarch and corn steep liquor	• Produced BC composites with maintained basic network structure with addition of all additives • Increased mechanical properties when pectin, gelatine and carboxylmethyl cellulose were present • Decreased crystallinity with carboxylmethyl cellulose and gelatine as additives	Dayal and Catchmark (2016)
Silver nanoparticle	• Produced antimicrobial wound dressing similar to a commercial silver-containing dressing • No obvious difference in crystallinity • Good antimicrobial effect against *E. coli* and *S. aureus*	Wu et al. (2014)
Poly(3-hydroxybutyrate) (PHB)	• Addition of PHB altered properties of BC film produced • Resulted in decrease of crystallinity by 34% and tensile strength (36%) but increased the cellulose yield	Ruka et al. (2013)
Polyaniline (PANI)	• Produced conductive films from BC-PANI composites • Increased 42% of BC conductivity and electrical sensitivity with addition of PANI	Pa'e et al. (2013)
Magnetite nanoparticles (MNPs)	• Produced BC composite films that have electromagnetic properties • MNP was successfully embedded in the fibre structure • Introduction of MNP enhanced thermal stability of the composite films with degradation temperature of 273 °C	Park et al. (2013)

Tensile properties specify in what way that the materials respond to forces applied in tension. A tensile test is an important test performed to define the modulus of elasticity, elongation and Young's modulus of native BC and BC composites. In this test, the specimen is carefully prepared and loaded in a very controlled manner while measuring the applied load and the elongation of the specimen over some distance. The results from the testing are used for the selection of materials for various purposes and application considering how it will react under different types of forces.

4 Application of Bacterial Cellulose Nanocomposites in Different Fields

Modification of BC had been done to enhance the properties of native BC and impart some additional properties for certain specific application. A variety of additives materials that had been used lead to the development of many new composite materials design for application on different fields.

4.1 Application in Biomedical Field

The modification of the bacterial cellulose can occur during or after fermentation by introducing selected bioactive material as additive in order to produce bacterial cellulose nanocomposites. This unique nanostructured matrix of composite materials are widely used in biomedical applications (Liyaskina et al. 2017). Modifying bacterial cellulose results in a composite material with better properties such as good mechanical properties and high moisture-keeping properties. These features make modified bacterial cellulose an excellent dressing material for treating different kinds of wounds, burns and ulcers. BC and BC nanocomposites are mostly used in the medical field, including wound healing materials (Legeza et al. 2004; Ul-Islam et al. 2012; Kim et al. 2011), artificial skin, blood vessels (Klemm et al. 2001; Charpentier et al. 2006; Arias et al. 2016), scaffolds for tissue engineering (Watanabe et al. 1998) and drug delivery (Amin et al. 2012; Müller et al. 2013). Table 4 lists different uses of BC nanocomposites in the medical field.

Table 4 Application of BC nanocomposites in biomedical field

Field	Application	Example of additives	References
Biomedical	• Antimicrobial wound dressing	• Chitosan	Lin et al. (2013)
		• Silver nanoparticles	Maneerung et al. (2008)
		• Montmorillonite	Ul-Islam et al. (2012)
	• Scaffold for tissue engineering	• Alginate	Kirdponpattara et al. (2015)
		• Calcium phosphates	Busuioc et al. (2016)
		• Silk fibroin	Barud et al. (2016)
	Hydrogel for controlled drug release	• Carboxymethyl cellulose	Juncu et al. (2015)
		• Graphene oxide	Luo et al. (2017)

4.2 Application in Electronic Device

The uses of BC nanocomposites have been explored as conducting materials for application in electric and electronic devices. These include sensors, electronic, papers and display devices (Muller et al. 2012; Lee et al. 2012). Previous researchers have revealed the production of BC nanocomposites with conducting properties with addition of CNTs (Yoon et al. 2006), graphene (Feng et al. 2012) and graphite nanoplatelets (Zhou et al. 2013) which are well known as conducting nanomaterial. One of the examples of BC conducting polymer composite was prepared by Muller et al. (2011). In their research, pyrrole was added into BC matrix through in situ oxidative polymerization resulted to 1 S/cm conductance for the composites. Another BC conducting composite was also successfully produced from polyaniline (Pa'e et al. 2013; Muller et al. 2011). The BC–polyaniline nanocomposites prepared were reported to have good conducting properties, thus suitable for the uses as electronic device such as sensor, signal receiver and fuel cell and to guide enviable cell function for tissue engineering applications. Table 5 shows some application of BC nanocomposites in electric and electronic fields.

4.3 Application in Wastewater Treatment

In wastewater treatment, BC with its unique properties is suitable to be used as adsorbent for heavy metal removal (Wang et al. 2015; Lu et al. 2010). This includes

Table 5 Application of BC nanocomposites in biomedical field

Field	Application	Example of additives	References
Conducting materials and electronic devices	• Electronic paper and E-book	• Refined cellulose	Mormino and Bungay (2003)
		• Dichromate dyes	Shah and Brown (2005)
	• Batteries	• Palladium	Evans et al. 2003
		• Polypyrrole	Zhang et al. (2015)
		• Molybdenum disulphide	Zhang et al. (2016)
	• Electric and electronic instruments	• Titanium dioxide	Gutierrez et al. (2012)
		• Graphite	Kiziltas et al. (2016)
		• Polyaniline	Hu et al. (2011)
	• Biosensor	• Gold	Li et al. (2011)
		• Platinum	Foresti et al. (2017)

Table 6 Bacterial cellulose as biosorbent for heavy metal removal in wastewater treatment

Biosorbent	Preparation method	Adsorbate	References
BC coated with polyethylenimine	Ex situ	Copper Lead	Wang et al. (2015)
Amino-BC	Ex situ	Lead Cadmium Copper	Lu et al. (2014)
Ammonium sulphamate-BC	Ex situ	Chromium	Lu et al. (2013)
Spherical iron oxide-BC composite	Ex situ	Lead Manganese Chromium	Zhu et al. (2011)
Carboxymethylated-BC	In situ	Copper Lead	Chen et al. (2009)

high tensile strength, porous structure and high water-holding capacity. Besides that, BC has large surface area with many hydroxyl groups in the chain that make it effective for the separation of heavy metals ions (Lu et al. 2013). Most importantly, BC is simple to produce and the modification can be done using various materials and methods. Furthermore, the physiochemical properties of this BC can be controlled by changes in growth conditions or by chemical modification to obtain desired functionality (Pa'e et al. 2011; Sokolnicki et al. 2006; Serafica et al. 2002). These features along with its biocompatibility and low production make BC ideal to be used as eco-friendly biosorbent for heavy metal removal.

However, the BC itself as biosorbent has several disadvantages such as low adsorption capacity, poor selectivity and high hydrophilicity which makes it swell easily in water. Therefore, new functional group is added to BC to modify its properties and improve the activity of BC on adsorption of heavy metal ions in wastewater treatment. Table 6 lists the previous works on the use of BC as biosorbent for heavy metal removal.

5 Conclusion

BC is a unique material; hence, research and development in BC has expanded fast. This expansion of the BC exploration is due to several advantages such as BC's properties which are suitable for numerous application, easy production process and economical viability. From the foregoing review of the literature, the progress in BC studies continues with its application as nanocomposites. Various synthetic and organic materials were added in situ or ex situ to produce nanocomposites materials with certain properties. At present, the focus is essentially moving towards the reasonable uses of BC and BC nanocomposites as a based material for various medical devices, electronic devices and biosorbent for heavy metal removal.

Various uses in many fields are expected to increase the demand on BC and BC nanocomposites production. Thus, efforts that focus on large-scale production need to be highlighted next.

References

Aleshin AN, Berestennikov AS, Krylov PS, Shcherbakov IP, Petrov VN, Trapeznikova IN, Mamalimov RI, Khripunov AK, Tkachenkob AA (2015) Electrical and optical properties of bacterial cellulose films modified with conductive polymer PEDOT/PSS. Synth Met 199:147–151

Amin MCM, Abadi AG, Ahmad N, Katas H, Jamal JA (2012) Bacterial cellulose film coating as drug delivery system: physicochemical, thermal and drug release properties. Sains Malaysiana 41(5):561–568

Arias SL, Shetty AR, Senpan A, Echeverry-Rendón M, Reece LM, JAllain JP (2016) Fabrication of a functionalized magnetic bacterial nanocellulose with iron oxide nanoparticles. J Vis Exp 26:111

Ashori A, Sheykhnazari S, Tabarsa T, Shakeri A, Golalipour M (2012) Bacterial cellulose/silica nanocomposites: preparation and characterization. Carbohydr Polym 90:413–418

Barud HGO, Barud HS, Cavicchioli M, do Amaral TS, de Oliveira Junior OB, Santos DM, de Oliveira Almeida Petersen AL, Celes F, Borges VM, de Oliveira CI, de Oliveira PF, Furtado RA, Tavares DC, Ribeiro SJL (2016) Preparation and characterization of a bacterial cellulose/silk fibroin sponge scaffold for tissue regeneration. Carbohydr Polym 128:41–51

Bertocchi C, Delneri D, Signore S, Weng Z, Bruschi CV (1997) Characterization of microbial cellulose from a high-producing mutagenized Acetobacter pasteurianus strain. Biochem Biophys Acta 1336:211–217

Budhiono A, Rosidi B, Taher H, Iguchi M (1999) Kinetics aspects of bacterial cellulose formation in nata de coco culture system. Carbohydr Polym 40:137–143

Busuioc C, Stroescu M, Stoica-Guzun A, Voicu G, Jinga SI (2016) Fabrication of 3D calcium phosphates based scaffolds using bacterial cellulose as template. Ceram Int 42(14):15449–15458

Charpentier PA, Maguire A, Wan WK (2006) Surface modification of polyester to produce a bacterial cellulose-based vascular prosthetic device. Appl Surf Sci 252:6360–6367

Chen S, Zou Y, Yan Z, Shen W, Shi S, Zhang X, Wang H (2009) Carboxymethylated-bacterial cellulose for copper and lead ion removal. J Hazard Mater 161:1355–1359

Cheng KC, Catchmark JM, Demirci A (2009) Enhanced production of bacterial cellulose by using a biofilm reactor and its material property analysis. J Biol Eng 3(12)

Czaja W, Romanovicz D, Brown RM Jr (2004) Structural investigations of microbial cellulose produced in stationary and agitated culture. Cellulose 11(3):401–411

Czaja W, Krystynowicz A, Bielecki S, Brown RM Jr (2006) Microbial cellulose—the natural power to heal wound. Biomaterials 27:145–151

da Silva R, Sierakowski MR, Bassani HP, Zawadzki SF, Pirich CL, Ono L, de Freitas RA (2016) Hydrophilicity improvement of mercerized bacterial cellulose films by polyethylene glycol. Int J Biol Macromol 86:599–605

Dayal MS, Catchmark JM (2016) Mechanical and structural property analysis of bacterial cellulose composites. Carbohydr Polym 144:447–453

De Wulf P, Joris K, Vandamme EJ (1996) Improved cellulose formation by an *Acetobacter xylinum* mutant limited in keto gluconate synthesis. J Chem Technol Biotechnol 67:665–672

Evans BR, O'Neill HM, Malyvanh VP, Lee I, Woodward J (2003) Palladium-bacterial cellulose membranes for fuel cells. Biosens Bioelectron 18:917–923

Feng Y, Zhanga X, Shena Y, Yoshino K, Feng W (2012) A mechanically strong, flexible and conductive film based on bacterial cellulose/graphene nanocomposite. Carbohydr Polym 87:644–649

Fijałkowski K, Żywicka A, Drozd R, Niemczyk A, Junka AF, Peitler D, Kordas M, Konopacki M, Szymczyk P, Fray ME, Rakoczy R (2015) A modification of bacterial cellulose through exposure to the rotating magnetic field. Carbohydr Polym 133:52–60

Foresti ML, Vázquez A, Boury B (2017) Applications of bacterial cellulose as precursor of carbon and composites with metal oxide, metal sulfide and metal nanoparticles: a review of recent advances. Carbohydr Polym 157:447–467

Fu L, Zhang Y, Zhang J, Yang G (2011) Bacterial cellulose for skin repair materials. In: Fazel-Rezai R (ed) Biomedical engineering—frontiers and challenges. In-Tech.Rijeka, Croatia

Gao C, Yan T, Du J, He F, Luo H, Wan Y (2014) Introduction of broad spectrum antibacterial properties to bacterial cellulose nanofibers via immobilising ε-polylysine nanocoatings. Food Hydrocolloids 36:204–211

Gelin K, Bodin A, Gatenholm P, Mihranyan A, Edwards K, Strømme M (2007) Characterization of water in bacterial cellulose using dielectric spectroscopy and electron microscopy. Polymer 48(26):7623–7631

George J, Kumar R, Sajeevkumar VA, Ramana KV, Rajamanickam R, Abhishek V, Nadanasabapathy SS (2014) Hybrid HPMC nanocomposites containing bacterial cellulose nanocrystals and silver nanoparticles. Carbohydr Polym 105:285–292

Gindl W, Keckes J (2004) Tensile properties of cellulose acetate butyrate composites reinforced with bacterial cellulose. Compos Sci Technol 64(15):2407–2413

González-Sánchez C, Martínez-Aguirre A, Pérez-García B, Martínez-Urreaga J, de la Orden MU, Fonseca-Valero C (2014) Use of residual agricultural plastics and cellulose fibers for obtaining sustainable eco-composites prevents waste generation. J Clean Prod 83:228–237

Gutierrez J, Tercjak A, Algar I, Retegi A, Mondragon I (2012) Conductive properties of TiO_2/bacterial cellulose hybrid fibres. J Colloid Interface Sci 377(1):88–93

Hestrin S, Schramm M (1954) Synthesis of cellulose by *Acetobacter xylinum.* 2. Preparation of freeze-dried cells capable of polymerizing glucose to cellulose. Biochem J 58(2):345–352

Horii F, Yamamoto H, Hirai A (1997) Microstructural analysis of microfibrils of bacterial cellulose. Macromol Symp 120:197–205

Hsieh JT, Wang MJ, Lai JT, Liu HS (2016) A novel static cultivation of bacterial cellulose production by intermittent feeding strategy. J Taiwan Inst Chem Eng 63:46–51

Hu W, Chen S, Yang Z, Liu L, Wang H (2011) Flexible electrically conductive nanocomposite membrane based on bacterial cellulose and polyaniline. J Phys Chem B 115:8453–8845

Hwang JW, Yang YK, Hwang JK, Pyun YR, Kim YS (1999) Effects of pH and dissolved oxygen on cellulose production by *Acetobacter xylinum* BRCS in agitated culture. J Biosci Bioeng 88(2):183–188

Iguchi M, Huang HC, Chen LC, Lina SB, Chen HH (2011) Nano-biomaterials application: In situ modification of bacterial cellulose structure by adding HPMC during fermentation. Carbohydr Polym 83:979–987

Jeon S, Yoo YM, Park JW, Kim HJ, Hyun J (2014) Electrical conductivity and optical transparency of bacterial cellulose based composite by static and agitated methods. Curr Appl Phys 14(12):1621–1624

Jonas R, Farah LF (1997) Production and application of microbial cellulose. J Polym Degrad Stab 59:101–106

Juncu G, Stoica-Guzun A, Stroescu M, Isopencu G, Jinga SI (2015) Drug release kinetics from carboxymethyl cellulose-bacterial cellulose composite films. Int J Pharm 510(2):485–492

Khairul AZ, Norhayati P, Ida IM (2016) An evaluation of fermentation period and discs rotation speed of rotary discs reactor for bacterial cellulose production. Sains Malaysiana 45(3): 393–400

Kim SY, Kim JN, Wee YJ, Park DH, Ryu HW (2006) Production of bacterial cellulose by *Gluconacetobacter* sp. RKY5 isolated from persimmon vinegar. Appl Biochem Biotechnol 129–132:705–715

Kim J, Cai Z, Lee HS, Choi GS (2011) Preparation and characterization of a bacterial cellulose/chitosan composite for potential biomedical application. J Polym Res 18:739–744

Kirdponpattara S, Khamkeaw A, Sanchavanakit N, Pavasant P, Phisalaphong M (2015) Structural modification and characterization of bacterial cellulose–alginate composite scaffolds for tissue engineering. Carbohydr Polym 132:146–155
Kiziltas EE, Kiziltas A, Rhodes K, Emanetoglu NW, Blumentritt M, Gardner DJ (2016) Electrically conductive nano graphite-filled bacterial cellulose composites. Carbohydr Polym 136:1144–1151
Klemm D, Schumann D, Udhardt U, Marsch S (2001) Bacterial synthesized cellulose—artificial blood vessels for microsurgery. Prog Polym Sci 26:1561–1603
Lee RL, Paul JW, Willem HZ, Isak SP (2002) Microbial cellulose utilization: fundamentals and biotechnology. J Microbiol Mol Biol Rev 66(3):506–577
Lee BH, Kim HJ, Yang HS (2012) Polymerization of aniline on bacterial cellulose and characterization of bacterial cellulose/polyaniline nanocomposite films. Curr Appl Phys 12:75–80
Legeza VI, Galenko-Yaroshevskii VP, Zinov'ev EV, Paramonov BA (2004) Effects of new wound dressings on healing of thermal burns of the skin in acute radiation disease. Bull Exp Biol Med 138:311–315
Li Z, Zhu BJ, Yang JX, Peng K, Zhou BH, Xu RQ, Hu WL, Chen SY, Wang HP (2011) Method for manufacture of bacterial cellulose hydrogel cold pack. CN Patent No 201020239963.4
Lin WC, Lien CC, Yeh HJ, Yu CM, Hsu SH (2013) Bacterial cellulose and bacterial cellulose–chitosan membranes for wound dressing applications. Carbohydr Polym 94(1):603–611
Liyaskina E, Revin V, Paramonova E, Nazarkina M, Pestov N, Revina N, Kolesnikova S (2017) Nanomaterials from bacterial cellulose for antimicrobial wound dressing. J Phys Conf Ser 784:(1)
Lu M, Li YY, Guan XH, Wei DZ (2010) Preparation of bacterial cellulose and its adsorption of Cd^{2+}. J Northeast Univ 31(8):1196–1199
Lu M, Guan XH, Xu X, Wei D (2013) Characteristic and mechanism of Cr(VI) adsorption by ammonium sulfamate-bacterial cellulose in aqueous solutions. Chin Chem Lett 24:253–256
Lu M, Zhang YM, Guan XH, Xu X, Gao T (2014) Thermodynamics and kinetics of adsorption for heavy metal ions from aqueous solutions onto surface amino-bacterial cellulose. Trans Nonferrous Metals Soc China 24:1912–1917
Luo H, Ao H, Li G, Li W, Xiong G, Zhu Y, Wan Y (2017) Advanced nano- and bio-materials: a pharmaceutical approach bacterial cellulose/graphene oxide nanocomposite as a novel drug delivery system. Curr Appl Phys 17(2):249–254
Maneerung T, Tokura S, Rujiravanit R (2008) Impregnation of silver nanoparticles into bacterial cellulose for antimicrobial wound dressing. Carbohydr Polym 72:43–51
Martínez-Sanz M, Lopez-Rubio A, Lagaron JM (2013) High-barrier coated bacterial cellulose nanowhiskers films with reduced moisture sensitivity. Carbohydr Polym 98:1072–1082
Mormino R, Bungay H (2003) Composites of bacterial cellulose and papermade with a rotating disk bioreactor. Appl Microbiol Biotechnol 62:503–506
Muller D, Rambo CR, Recouvreux DOS, Porto LM (2011) Chemical in situ polymerization of polypyrrole on bacterial cellulose nanofibers. Synth Met 161:106–111
Muller D, Mandelli JS, Marins JA, Soares BG (2012) Electrically conducting nanocomposites: preparation and properties of polyaniline (PAni)-coated bacterial cellulose nanofibers (BC). Cellulose 19:1645–1654
Müller A, Ni Z, Hessler N, Wesarg F, Müller FA, Kralisch D, Fischer D (2013) The biopolymer bacterial nanocellulose as drug delivery system: investigation of drug loading and release using the model protein albumin. J Pharm Sci 102:579–592
Nakayama A, Kakugo A, Gong JP, Osada Y, Takai M, Erata T, Kawano S (2004) High mechanical strength double-network hydrogel with bacterial cellulose. Adv Funct Mater 14:1124–1128
Naritomi T, Kouda T, Yano H, Yoshinaga F (1998) Effect of ethanol on bacterial cellulose production from fructose in continuous culture. J Ferment Bioeng 85(6):598–603
O'Connell DW, Birkinshaw C, O'Dwyer TF (2008) Heavy metal adsorbents prepared from the modification of cellulose: a review. Biores Technol 99:6709–6724
Pa'e N (2009) Rotary discs reactor for enhanced production microbial cellulose. Master thesis, Universiti Teknologi Malaysia, Skudai

Pa'e N, Zahan KA, Muhamad II (2011) Production of biopolymer from *acetobacter xylinum* using different fermentation methods. Int J Eng Technol (IJET-IJEN) 11(5):90–98

Pa'e N, Zahan KA, Muhamada II, Kok FS (2013) Modified fermentation for production of bacterial cellulose/polyaniline as conductive biopolymer material. Jurnal Teknologi 62(2):21–23

Park M, Cheng J, Choi J, Kim J, Hyun J (2013) Electromagnetic nanocomposite of bacterial cellulose using magnetite nanoclusters and polyaniline. Colloids Surf B 102:238–242

Pavaloiu RD, Stoica-Guzun A, Stroescu M, Jinga SI, Dobre T (2014) Composite films of poly (vinyl alcohol)–chitosan–bacterial cellulose for drug controlled release. Int J Biol Macromol 68:117–124

Paximada P, Dimitrakopoulou EA, Tsouko E, Koutinas AA, Fasseas C, Mandala IG (2016) Structural modification of bacterial cellulose fibrils under ultrasonic irradiation. Carbohydr Polym 150:5–12

Pei Y, Yang J, Liu P, Xu M, Zhang X, Zhang L (2013) Fabrication, properties and bioapplications of cellulose/collagen hydrolysate composite films. Carbohydr Polym 92:1752–1760

Piccinno F, Hischier R, Saba A, Mitrano D, Seeger S, Som C (2015) Multi-perspective application selection: a method to identify sustainable applications for new materials using the example of cellulose nanofiber reinforced composites. J Clean Prod 112:1199–1210

Pircher N, Veigel S, Aignea N, Nedelecc JM, Rosenau T, Liebner F (2014) Reinforcement of bacterial cellulose aerogels with biocompatible polymers. Carbohydr Polym 111:505–513

Ruka DR, Simon GP, Deana KM (2013) In situ modifications to bacterial cellulose with the water insoluble polymer poly-3-hydroxybutyrate. Carbohydr Polym 92:1717–1723

Sai H, Fu R, Xing L, Xiang J, Li Z, Li F, Zhang T (2015) Surface modification of bacterial cellulose aerogels web-like skeleton for oil/water separation. ACS Appl Mater Interfaces 7(13): 7373–7381

Saibuatong O, Phisalaphong M (2010) Novo aloe vera—bacterial cellulose composite film from biosynthesis. Carbohydr Polym 79:455–460

Salehudin MH, Salleh E, Muhamad II, Mamat SNH (2014) Starch-based biofilm reinforced with empty fruit bunch cellulose nanofibre. Mater Res Innovations 18:322–325

Schramm M, Hestrin S (1954) Factors affecting production of cellulose at the air/liquid interface of a culture of *Acetobacter xylinum*. Microbiology 11:123–129

Serafica G, Mormino R, Bungay H (2002) Inclusion of solid particles in bacterial cellulose. Appl Microbiol Biotechnol 58:756–760

Shah J, Brown RM Jr (2005) Towards electronic paper displays made from microbial cellulose. Appl Microbiol Biotechnol 66(4):352–355

Shah N, Ul-Islam M, Khattak WA, Park JK (2013) Overview of bacterial cellulose composites: a multipurpose advanced material. Carbohydr Polym 98:1585–1598

Shanshan G, Jianqing W, Zhengwei J (2012) Preparation of cellulose films from solution of bacterial cellulose in NMMO. Carbohydr Polym 87(2):1020–1025

Shirai A, Takahashi M, Kaneko H, Nishimura S, Ogawa M, Nishi N, Tokura S (1994) Biosynthesis of a novel polysaccharide by *Acetobacter xylinum*. Int J Biol Macromol 16 (6):297–300

Slavutsky MA, Bertuzzi MA (2014) Water barrier properties of starch films reinforced with cellulosenanocrystals obtained from sugarcane bagasse. Carbohydr Polym 110:53–61

Sokolnicki AM, Fisher RJ, Harrah TP, Kaplan DL (2006) Permeability of bacterial cellulose membranes. J Membr Sci 272(1–2):15–27

Son HJ, Heo MS, Kim YG, Lee SJ (2001) Optimization of fermentation conditions for the production of bacterial cellulose by a newly isolated *Acetobacter* sp. A9 in shaking cultures. Biotechnol Appl Biochem 33(1):1–3

Son HJ, Kim HG, Kim KK, Kim HS, Kim YG, Lee SJ (2003) Increased production of bacterial cellulose by *Acetobacter* sp. V6 in synthetic media under shaking culture conditions. Biores Technol 86(3):215–219

Tang W, Jia S, Jia Y, Yang H (2010) The influence of fermentation conditions and post-treatment methods on porosity of bacterial cellulose membrane. World J Microbiol Biotechnol 26:125–131

Toru S, Kazunori T, Masaya K, Tetsuya M, Takaaki N, Shingeru M, Kenji K (2005) Cellulose production from glucose using a glucose dehydrogenase gene (gdh)-deficient mutant of *Gluconacetobacter xylinus* and its use for bioconversion of sweet potato pulp. J Biosci Bioeng 99(4):415–422

Tsuchida T, Yoshinaga F (1997) Production of bacterial cellulose by agitation culture system. J Pure Appl Chem 69(11):2453–2458

Tyagi N, Suresh S (2015) Production of cellulose from sugarcane molasses using *Gluconacetobacter* intermedius SNT-1: optimization & characterization. J Clean Prod 112:71–80

Ul-Islam M, Khan T, Park JK (2012) Nanoreinforced bacterial cellulose–montmorillonite composites for biomedical applications. Carbohydr Polym 89(4):1189–1197

Ummartyotin S, Juntaro J, Sain M, Manuspiya H (2012) Development of transparent bacterial cellulose nanocomposite film as substrate for flexible organic light emitting diode (OLED) display. Ind Crops Prod 35:92–97

Vandamme EJ, De Baets S, Vanbaelen A, Joris K, De Wulf P (1998) Improved production of bacterial cellulose and its application potential. Polym Degrad Stab 59:93–99

Wang J, Lu X, Ng PF, Lee K, Fei B, Xin JH, Wu J (2015) Polyethylenimine coated bacterial cellulose nanofiber membrane and application as adsorbent and catalyst. J Colloid Interface Sci 440:32–38

Watanabe K, Tabuchi M, Morinaga Y, Yoshinaga F (1998) Structural features and properties of bacterial cellulose produced in agitated culture. Cellulose 5:187–200

Wu YB, Yu SH, Mi FL, Wu CW, Shyu SS, Peng CK, Chao AC (2004) Preparation and characterization on mechanical and antibacterial properties of chitosan/cellulose blends. Carbohydr Polym 57(4):435–440

Wu J, Zheng Y, Song W, Luan J, Wen X, Wu Z, Chen X, Wang Q, Guo S (2014) In situ synthesis of silver-nanoparticles/bacterial cellulose composites for slow-released antimicrobial wound dressing. Carbohydr Polym 102:762–771

Yamanaka S, Watanabe K, Kitamura N, Iguchi M, Mitsuhashi S, Nishi Y, Uryu M (1989) The structure and mechanical properties of sheets prepared from bacterial cellulose. J Mater Sci 24:3141–3145

Yan Z, Chen S, Wang H, Wang B, Jiang J (2008) Biosynthesis of bacterial cellulose/multi-walled carbon nanotubes in agitated culture. Carbohydr Polym 74:659–665

Yang G, Xie J, Hong F, Cao Z, Yang X (2012) Antimicrobial activity of silver nanoparticle impregnated bacterial cellulose membrane: effect of fermentation carbon sources of bacterial cellulose. Carbohydr Polym 87:839–845

Yoon SH, Jin HJ, Kook MC, Pyun YR (2006) Electrically conductive bacterial cellulose by incorporation of carbon nanotubes. Biomacromolecules 7:1280–1284

Zahan KA, Pa'e N, Muhamad II (2014) Process parameter for fermentation in rotary discs reactor for optimum microbial cellulose production using response surface methodology. Bioresources 9(2):1858–1872

Zhang Z, Zhang J, Zhao X, Yang F (2015) Core-sheath structured porous carbon nanofiber composite anode material derived from bacterial cellulose/polypyrrole as an anode for sodium-ion batteries. Carbon 95:552–559

Zhang F, Tang Y, Yang Y, Zhang X, Lee CS (2016) In-situ assembly of three-dimensional MoS_2 nanoleaves/carbon nanofiber composites derived from bacterial cellulose as flexible and binder-free anodes for enhanced lithium-ion batteries. Electrochim Acta 211:404–410

Zhou T, Chen D, Jiu J, Nge TT (2013) Electrically conductive bacterial cellulose composite membranes produced by the incorporation of graphite nanoplatelets in pristine bacterial cellulose membranes. Express Polym Lett 7:756–766

Zhu H, Jia S, Wan T, Jia Y, Yang H, Li J, Yan L, Zhong C (2011) Biosynthesis of spherical Fe_3O_4/Bacterial cellulose nanocomposites as adsorbents for heavy metal ions. Carbohydr Polym 86:1558–1564

Properties of Nano-composites Based on Different Clays and Polyamide 6/Acrylonitrile Butadiene Styrene Blends

Marya Raji, Elmokhtar Essassi, Hamid Essabir, Denis Rodrigue, Abou el kacem Qaiss and Rachid Bouhfid

Abstract In the last years, serval researches have been focused on organophilic clay as reinforcements for polymer matrices. In this respect, the aim of this chapter is to valorize mineral resources; montmorillonite clay was modified using hexadecyltrimethylammonium bromide (CTAB) and then used as reinforcement in a thermoplastic copolymer matrix to compare with pristine montmorillonite and commercially organo-modified montmorillonite (Cloisite 20A). The nano-composites were prepared by melt compounding using a blend of polyamide 6 (PA6) with acrylonitrile butadiene styrene (ABS) as the matrix. Scanning electron microscopy (SEM), Fourier transform infrared spectroscopy (FTIR), thermogravimetric analysis (TGA), X-ray diffraction (XRD) as well as mechanical and rheological tests were carried out to understand the properties of these nano-composites at different particle contents. The results obtained clearly showed that the Moroccan montmorillonite was successfully modified and its addition in the selected matrix substantially improved the properties of the resulting nano-composites.

Keywords Montmorillonite · Polyamide 6 · Acrylonitrile butadiene styrene
Surface modification · Nano-composites

M. Raji · H. Essabir · A. Qaiss · R. Bouhfid (✉)
Moroccan Foundation for Advanced Science, Innovation and Research (MAScIR), Laboratory of Polymer Processing, Institute of Nanomaterials and Nanotechnology (NANOTECH), Rabat, Morocco
e-mail: r.bouhfid@mascir.com

M. Raji · E. Essassi
Laboratory of Organic Chemistry and Heterocyclic, Faculty of Science, Mohammed V University, Rabat, Morocco

D. Rodrigue
Department of Chemical Engineering and CERMA, Université Laval, Quebec City G1V0A6, Canada

M. L. Sanyang and M. Jawaid (eds.), *Bio-based Polymers and Nanocomposites*,
https://doi.org/10.1007/978-3-030-05825-8_6

1 Introduction

In recent years, polymer nano-composite has been a topic of high scientific research and industrial development (Zeng et al. 2005), especially for several applications like automotive (Raji et al. 2016b), aerospace (Raji et al. 2017b), construction (Essabir et al. 2016b), and packing (Zari et al. 2018). This interest is related to their excellent specific properties such as mechanical properties (Raji et al. 2016a) and thermal stability (Raji et al. 2017a). This kind of materials can be produced using either by thermoplastics (Laaziz et al. 2017), thermosets, or elastomers (Abdellaoui et al. 2017) filled by low content of nano-sized particles (less than 100 nm) (Gacitua et al. 2005). The incorporation of these small particles into a polymer matrix can also decrease the cost of materials (Wang et al. 2013) and its impact on the environment (Essabir et al. 2016a). The nanoparticles can be synthetic or natural (Bensalah et al. 2017; Boujmal et al. 2017). Typical examples of natural nanoparticles are layered materials such as graphene (Pretschuh et al. 2014), silicate (Vuluga et al. 2014), and clays (Bhattacharya and Aadhar 2014)), or fiber-like materials such as carbon nano-tubes and nano-fibers (Nuzzo et al. 2014), sepiolite (Mejía et al. 2014; Zotti et al. 2014), or cellulose nano-fibers (Erden and Ho 2017).

Inorganic resources as nanoparticle fillers are being exploited substantially as an alternative to organic and synthetic materials because of their high in-plane strength and stiffness, rich intercalation chemistry as well as their abundance in nature and availability. Among them, the use of clay nanoparticles is attracting ever-growing interest due to its large aspect ratio and their unique layered structure (Leszczy et al. 2007) and high surface area (Amendola et al. 2012). The excellent properties of polymer nano-composite reinforced with clays (Singla et al. 2012) led to substantial mechanical (Usuki 2002) and thermal stability (Gao 2004) improvement. They also decrease gas and liquid permeability (Bergaya and Lagaly 2013). Moreover, they were shown to improve flame retardancy (Laoutid et al. 2013), without affecting the optical properties of the base polymer (Navrátilová et al. 2007). But their main advantage is to improve the rigidity of a neat polymer (Wang et al. 2013).

2 Clay Minerals: Structure, Modifications, and Properties

2.1 Clay Structure

The word "clay" has ambivalent definitions. On the first hand, in terms of simple definitions relating to geological knowledge, it is used to define any fine-grained fraction of rocks, sediments, and soil particles less than 2 μm in equivalent spherical diameter. On the other hand and from the chemical knowledge, this word includes a large group of hydrated phyllosilicates. Roughly speaking, clay minerals are essentially microcrystalline secondary minerals based on hydrous magnesium or aluminum silicates and carry negative charges that have sheet-like structures with

Table 1 Classification of planar 2:1 phyllosilicates (Guggenheim et al. 2006)

Layer charge[a]	Interlayer material	Group	Octahedral character	Example of species
0	None	Talc-Pyrophyllite	Trioctahedral Dioctahedral	Talc-pyrophyllite
0.4–1.2	Hydrated exchangeable cations	Smectite	Trioctahedral Dioctahedral	Saponite Montmorillonite
1.2–1.8	Hydrated exchangeable cations	Vermiculite	Trioctahedral Dioctahedral	Vermiculite Vermiculite
1.7–2.0	Non-hydrated monovalent (>50%) cations	True(Flexible) Mica	Trioctahedral Dioctahedral	Phlogopite Muscovite
1.2–1.7	Non-hydrated mono- or divalent cations	Interlayer-deficient Mica	Trioctahedral Dioctahedral	Illite Wonesite
2.6–4.0	Non-hydrated or divalent (>50%) cations	Brittle Mica	Trioctahedral Dioctahedral	Clintonite Margarite
variable	Hydroxide sheet	Chlorite	Trioctahedral Dioctahedral Di, Trioctahedral	Clinochlore Donbassite Cookeite

[a]Approximate negative net charge per anionic $O_{22}(OH)_4$ formula unit

very fine particle size and a general chemical formula (Ca, Na, H)(Al, Mg, Fe, $Zn)_2(Si, Al)_4O_{10}(OH)_2–xH_2O$, where x represents the variable amount of water.

Generally, the different structures of clay minerals are basically composed of two basic units: the silica tetrahedron “SiO_2” and alumina octahedron “AlO_6” at different ratios; (*i*) In 1:1 ratio, this structure is based on one octahedral sheet linked to one tetrahedral sheet as Kaolinite (González and Del 2006), Halloysite (Zhao et al. 2013), etc. (*ii*) In 2:1 ratio, this structure is created from two tetrahedral sheets sandwiching an octahedral sheet that is classified into separate groups on the basis of the layer charge and interlayer material (Tian and Tagaya 2007); see Table 1.

Montmorillonite (MMT) as one of the most popular inorganic fillers, a member of the smectite family, is classified into the general family of 2:1-layered silicates (Carrado 2000), identified with the chemical formula $M_x(Al_2{-}xMg_x)Si_4O_{10}(OH)_2 \cdot nH_2O$ (Laoutid et al. 2013). Its crystal structure is composed of nanometer-thick layers based on alumina octahedral sheets sandwiched between two silica tetrahedral sheets (Majeed et al. 2013). They are also characterized by the presence of cations in between layers (Na^+, K^+, Ca^+, etc.) (Tamura et al. 2008). This type of clay is characterized by its easy availability, low density, and its low cost-effectiveness (Jeong et al. 2011).

2.2 Problematic

Montmorillonite like as all types of clay is naturally hydrophilic (Oliveira and Machado 2013). This character makes them difficult to mix and interact with most polymer matrices which are hydrophobic (Nekhlaoui et al. 2014). For these reasons, montmorillonite must be surface-treated before their introduction into a polymer matrix to make high-quality nano-composites by ensuring good dispersion and interfacial adhesion (de Paiva et al. 2008).

2.3 MMT Modifications

Generally, clay surface can be modified by organic compounds based on the exchange of sodium cations located within the interlayer's or in the galleries with organic cations (Donescu et al. 2008) (Fig. 1). The most commonly used are alkyl ammonium (Tamura et al. 2008), sulfonium (Majeed et al. 2013), and phosphonium (Nazir et al. 2016) ions. This substitution changes the hydrophilic silicate surface to an organophilic one (Raji et al. 2016b). It also increases the gallery spacing (*d*-spacing) between the layers (Kumar et al. 2010; Mansoori et al. 2012). For example, Cloisite 20A is a commercially organo-modified montmorillonite intercalated with dimethyl, dehydrogenated tallow quaternary ammonium (2M2HT), where HT is the hydrogenated tallow ($\sim$65% C18, $\sim$30% C16, and $\sim$5% C14) (Bhat et al. 2008; Diagne et al. 2006; Tjong 2006). But the modification depends on the cation-exchange capacity of the layers (de Paiva et al. 2008), as well as the content and structure of the surfactant and the way the surfactant molecules organize themselves in the clay (Fujimori et al. 2010).

The synthesis of organo-modified montmorillonite was carried out by intercalating the clay layers with a long-chain organic surfactant. The preparation was done by solubilization of hexadecyltrimethylammonium bromide ($[H_3C\ (H_2C)_{15}–N^+–(CH_3)_3]\ Br^-$) in ethanol. This surfactant solution was then slowly added to the clay suspension obtained by the dispersion of (MMT-Na) in water:ethanol (1:1) at 80 °C under continued high-intensity stirring. Finally, centrifugation was used to separate the organo-modified clay. The particles were washed five times with a water:ethanol (1:1) solution to remove any residues. The general reaction for MMT modification is shown in Eq. 1:

$$Na^+{-}Clay + HOOC{-}R{-}NH_3 + Cl^- \rightarrow HOOC{-}R{-}NH_3^+{-}Clay + NaCl \quad (1)$$

2.4 *MMT Properties*

2.4.1 Structural Properties

- ***X-ray diffraction (DRX)***

Wide-angle X-ray diffraction (WAXD) was used to identify the chemical composition and crystallographic structure of the clays. The measurements were carried out on a Bruker D8 Discover using the Cu Kα radiation (λ = 0.154184 nm) and a GADDS detector. The samples were consolidated in an aluminum holder and scanned at 45 kW from 2° to 10° of 2θ. The diffraction patterns were analyzed using the X'Pert High Score software (Fig. 2).

Figure 2 shows the XRD pattern of MMT-Na and organo-modified clay MMT-CTAB. Both clays display a broadband in the region between 2θ = 6.16° and 2θ = 4°, respectively. The interlayer spacing corresponding of the peak of MMT-Na around 1.43 nm which increased to 2.2 nm for the organo-modified clay is getting close to that of Cloisite 20A (*d* = 2.84 nm calculated using the large band at 3.1°), according to Bragg's equation (Kumar et al. 2010). It can be concluded that, after clay modification by the organic surfactant, intercalation between the silicates layers occurred. This intercalation constitutes an efficient way to increase surface hydrophobicity which is a fundamental prerequisite for good compatibility between the polymer matrix and the montmorillonite surface (Agag et al. 2001). It also increases the distance between each layer (*d*-spacing).

- ***Fourier transforms infrared spectra (FTIR)***

Fourier transform infrared spectra (FTIR) of the different clays and nano-composites were recorded using an ABB Bomem FTLA 2000-102 spectrometer (ATR: SPECAC GOLDEN GATE). The spectra were obtained by an accumulation of 16 scans and a resolution of 4 cm^{-1}.

The FTIR spectra of MMT-Na, Cloisite 20A, and MMT-CTAB are presented in Fig. 3. The bands around 3600 and 3400 cm^{-1} are indicative of montmorillonite in all spectrums, which can be attributed to O–H stretching for the silicate and water

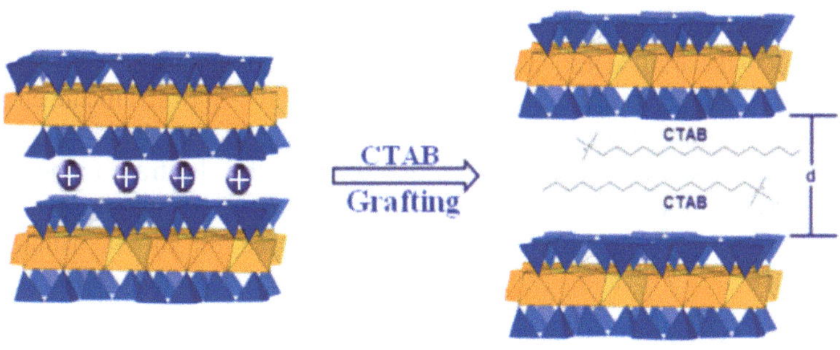

Fig. 1 Schematic of synthetic procedure for MMT-CTAB

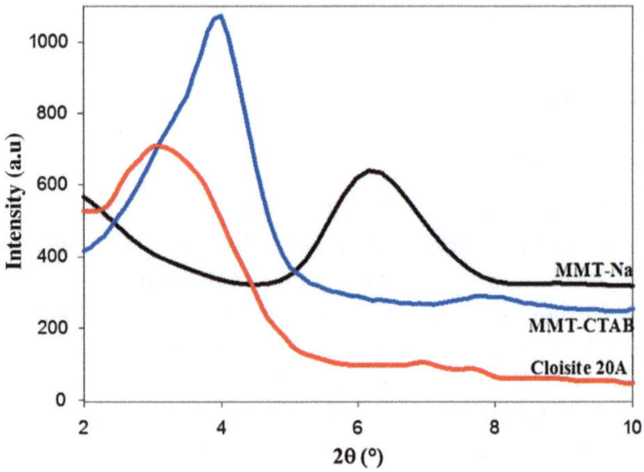

Fig. 2 XRD patterns of MMT-Na and modified MMT

(Kumar et al. 2010). In the FTIR spectrum of MMT-CTAB appears a band in the region of 1645 cm^{-1} is attributed to the –OH bending mode of the adsorbed water (Kumar et al. 2010). The characteristic band at 1134 cm^{-1} is assigned to Si–O stretching and out-of-plane Si–O stretching mode of montmorillonite (Yürüdü et al. 2005). The band at 1004 cm^{-1} is ascribed to Si–O stretching (in-plane) vibration for layered silicates (Xie et al. 2011). The IR bands at 945, 814, and 719 cm^{-1} are attributed to Al–Al–OH, Al–Fe–OH, and Al–Mg–OH bending vibrations, respectively (Navrátilová et al. 2007; Xie et al. 2011; Xue et al. 2007). Moreover, the –OH bending mode band of adsorbed water at 1645 cm^{-1} in the MMT-Na spectrum has a lower intensity because the absorbed water in MMT-CTAB is more limited due to the presence of the organic cations (Kumar et al. 2010). In addition, the bands at 1470 and at 1416 cm^{-1} are assigned to H–C–H and C–N stretching vibrations, respectively (Kumar et al. 2010). Figure 3 also presents two weaker bands as a doublet at 3054 and at 3028 cm^{-1} which can be associated to the C–H bonds of aromatic stretching vibrations normally present in CTAB (Gavrilko et al. 2013; Aowda et al. 2011). Finally, the presence of CTAB in the MMT-CTAB spectrum is confirmed, which implies that CTAB has been intercalated into the MMT-Na structure (nano-space inside montmorillonite).

In the FTIR spectrum of Cloisite 20A (Fig. 3), the transmittance bands at 1040 and 916 cm^{-1} are due to the stretching vibration of Si–O–Si from silicate and to the Al–OH–Al deformation of aluminates, respectively. Another band at 726 cm^{-1} belongs to Si–O stretching of quartz and silica. Finally, the transmittance bands between 600 and 400 cm^{-1} can be associated with the Al–O stretching and Si–O bending vibrations. In addition to the bands observed for montmorillonite, the bands located at 2930 and 2856 cm^{-1} can be assigned to the asymmetric and

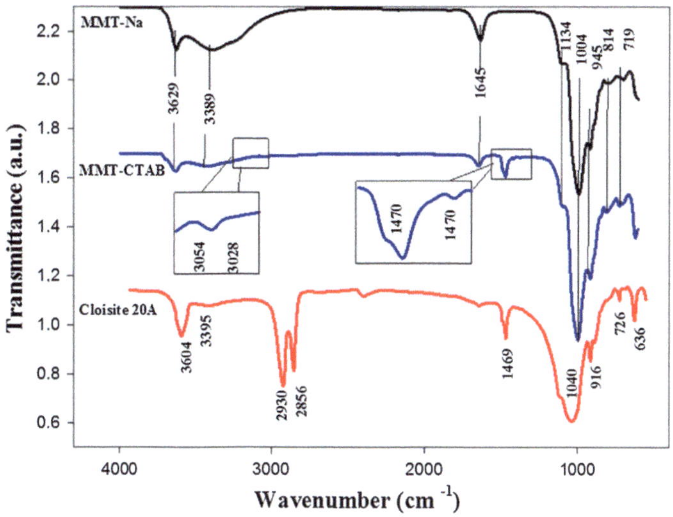

Fig. 3 FTIR spectra of MMT-Na, organo-modified clay (MMT-CTAB), and Cloisite 20A particles

symmetric vibrations of the methylene groups (CH_2) from surfactant molecules (long aliphatic chains), and the band at 1469 cm^{-1} may be ascribed to the asymmetric deformation vibrations of $-CH_3$ of the aliphatic chain of the surfactant (Navrátilová et al. 2007).

2.4.2 Morphological Properties

- ***Scanning electron microscopy (SEM)***

Montmorillonite powder was subjected to scanning electron microscopy (SEM) to investigate the morphology of their granules. Small amount of montmorillonite powder was poured on the carbon tape, and then, the excess was blown with air gun to ensure that small pieces of the powder remain on the tape.

The shapes of the Montmorillonite were studied by scanning electron microscopy (SEM). A general view using the micrographs of the clay powder can be observed in Fig. 4. Nevertheless, the stronger nanoparticles intermolecular interactions, that is, a cohesive tension make an attempt to the clay particles to aggregate and to form the agglomerates (Akin and Tihminlioglu 2018). Figure 4 shows that the montmorillonite possesses a plate-like morphology with nanometric size.

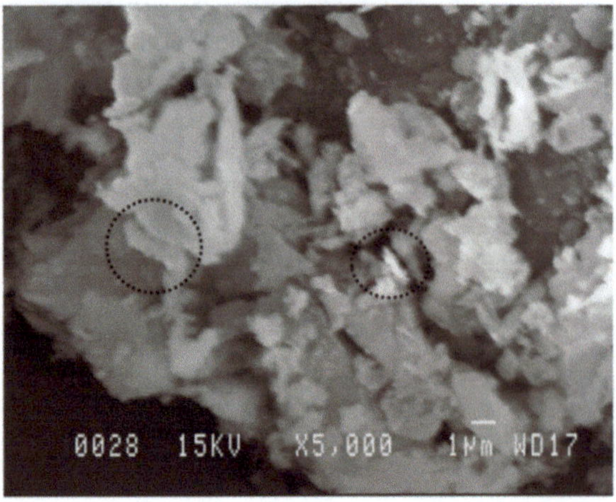

Fig. 4 SEM images of montmorillonite clay

2.4.3 Thermal Properties

- ***Thermogravimetric analysis (TGA)***

Thermal stability plays an essential role in determining the technical applications of composite materials. For this reason, TGA was used to compare the degradation profiles of the different types of clay used (MMT-Na, MMT-CTAB, and Cloisite 20A) and the results are presented in Fig. 5.

Figure 5 shows that the thermal decomposition of MMT-Na occurs in two steps: The first one around 82 °C (14% weight loss) is due to free water and interlayer water residing between the aluminosilicate layers and related to the hydration spheres of the cations (Carastan et al. 2013), while the second weight loss between 500 and 700 °C (2% weight loss) is due to dehydroxylation of the aluminosilicate lattice (Xie et al. 2011). On the other hand, the decomposition curves of the organoclay can be divided into four parts: (1) The free water region below 200 °C

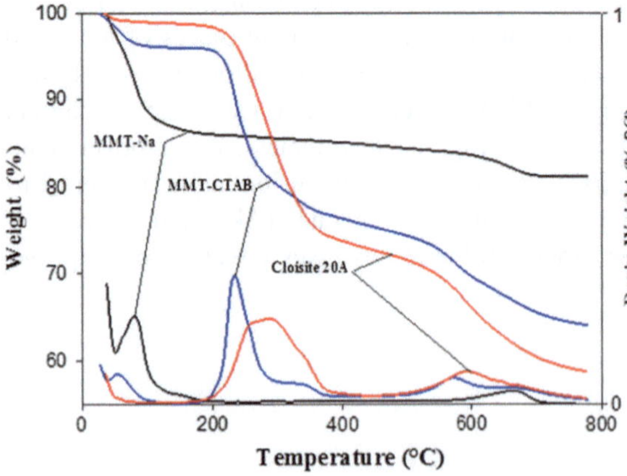

Fig. 5 TGA and DTG curves for the different clays used (Cloisite 20A, MMT-CTAB, and MMT-Na)

is characterized by a weight loss of 5% in MMT-CTAB which is lower than of MMT-Na (Pavlidou and Papaspyrides 2008). (2) The region where the CTAB and all organic substances thermally decompose is around of 231 °C with a total weight loss of 23% (Wu and Chen 2004) such region is also present for Cloisite 20A at 288 °C (26% loss) (Hoidy and Al-mulla 2013). This again confirms the intercalation of MMT-Na by CTAB as determined by XRD. (3) A weight loss of 3% due to water in the interlayer space is usually removed around 336 °C for MMT-CTAB (Wu and Chen 2004). Finally, (4) both of evaporation and complete decomposition of organic and inorganic moieties in the surfactant (Zhang et al. 2013), as well as dehydroxylation of the aluminosilicate lattice, was observed around 593 °C for Cloisite 20A (weight loss 3%) and around 600 °C for MMT-CTAB (weight loss 2%), respectively, which are lower than for MMT-Na. This behavior is associated to the presence of organic substances having a direct effect on decomposition (Zhang et al. 2013), depending on the number of the carbon chains acting as energy sources leading to further decomposition (Kokal et al. 2011). FTIR results also confirmed the presence of organic molecules in MMT-CTAB and Cloisite 20A. One can, therefore, conclude that MMT-CTAB exhibits excellent thermal stability compared to MMT-Na and close to that of Cloisite 20A, confirming again that intercalation of MMT-Na by the surfactant was successful (Pavlidou and Papaspyrides 2008).

The grafted yield (%), which corresponds to the number of intercalated molecules which effectively participated in the coupling reaction, can be determined using the weight loss between 200 and 800 °C corresponding to surfactant degradation (MMT-CTAB and Cloisite 20A). The results show that the amount is around 21 and 30%, respectively.

3 Organo-Modified Clay Nano-composite Preparation

3.1 Problematic

As for as any composite materials, the properties of nano-composites mostly depend on the dispersion distribution of the particles in the polymer matrix. For this reason, various methods were developed to compound montmorillonite such as: in situ polymerization (Hakeem et al. 2014), solution blending (He et al. 2010), and melt blending (Raji et al. 2016a). But most processes use the latter because it is cost-effective and can produce homogeneous compounds with good dispersion/distribution at large scale (Šupová et al. 2011).

3.2 Nano-composites Preparation

The objective of this chapter is to compare the morphological, thermal, rheological, and mechanical properties of polymers nano-composite based on different nanoparticles: pristine and modified montmorillonite, as well as commercially available Cloisite 20A. The nano-composites are based on unmodified montmorillonite (MMT-Na), organo-modified montmorillonite (MMT-CTAB) by adding hexadecyltrimethylammonium bromide (CTAB) as a surfactant, and three different montmorillonite clays: unmodified clay (PA6/MMT-Na), organoclay (PA6/MMT-CTAB) composite and commercially clay (Cloisite 20A) nano-composite. In all cases, the nanoparticles were first dispersed into polyamide 6 (masterbatches) and then compounded with acrylonitrile butadiene styrene (ABS) to get the final nano-composites by means of a simple melt compounding via a Leistritz ZSE-18 twin-screw extruder (Leistritz Extrusiontechnik GmbH, Germany), under the following conditions: a constant barrel temperature of 240 °C with 220 °C for the die. In general, different amounts (1, 2, 3, 4, and 5% wt.) of each nano-clay were incorporate into ABS using three different masterbatches (10% wt. nanoparticles) obtained by the combination of PA6 with either MMT-Na, MMT-CTAB, and Cloisite 20A. After extrusion and pelletizing in a precision grinder (FRITSCH Pulverisette 19) into pieces of 2–3 mm, the compounds were injection molded using an Engel e-Victory injection molding machine with a 40 tons platen capacity. The process temperature was fixed at 240 °C in the barrel and at 220 °C at the nozzle, while the mold was maintained at 45 °C (Bensalah et al. 2017). Then, the (ABS/PA6/MMT-Na), (ABS/PA6/MMT-CTAB) and (ABS/PA6/Cloisite 20A) nano-composites were tested using different characterization methods as described next. A schematic representation of the preparation methods is given in Fig. 6.

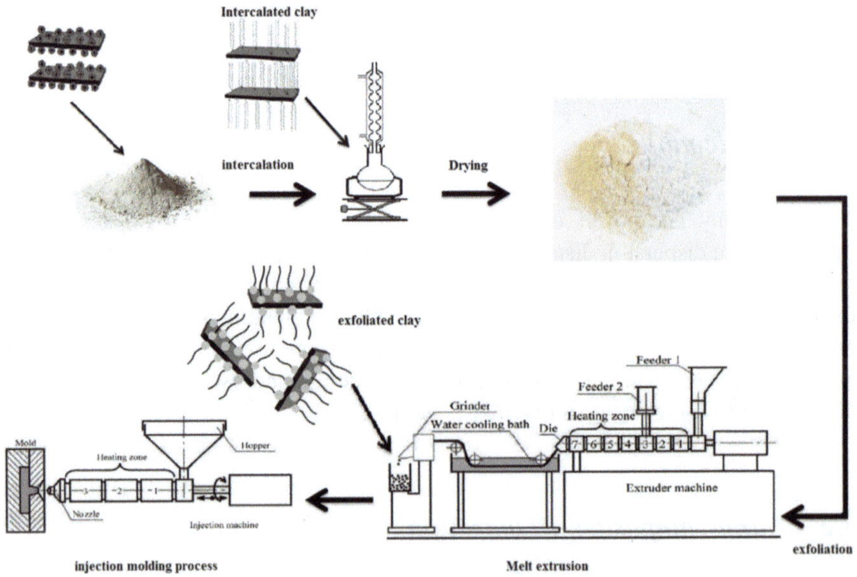

Fig. 6 Schematic representation of the preparation method for the polymer clay nano-composites produced

4 Nano-composites Properties

4.1 Morphological Properties

- ***Scanning electron microscopy (SEM)***

Scanning electron microscopy (SEM) constitutes an efficient tool to investigate the morphology of the nano-sized clays dispersion/distribution in the copolymer matrix. To obtain clean and accurate fractures, all the composites were cry fractured in liquid nitrogen.

Figure 7 shows the SEM images of the cryo-fracture surface of the nano-composites of ABS-PA6/Cloisite 20A, ABS-PA6/MMT-CTAB, and of ABS-PA6/MMT-Na containing 3 wt% clay content. The micrographs of clay nano-composites presented in Fig. 7a, b, c show typical co-continuous morphologies, an ideal model for high performance multi-component polymeric materials, in which the ABS (50%) domains presented by the white regions dispersed in the PA6 (47%) matrix, illustrated by the black part, these two immiscible polymers are combined together to form one system (Essabir et al. 2017; Li and Shimizu 2005). It's clearly observed that the gray parts correspond to the organoclay particles are located in PA6 phase, this phenomenon is due to the significant effect of exfoliated clay platelets on the rheological behavior of polymers, more precisely, the preferential location of the organoclay in the PA6 phase can significantly increase its viscosity and prevent the coalescence of the dispersed phase during extrusion (Yan et al. 2012). The selective localization of the clay existing in the PA6 phase is due to the different affinities between the organoclay and the two polymers (Yan et al. 2012). The clay layers are easily exfoliated by PA6 molecular chains compared to ABS molecular chains because of the higher polarity of the PA6 chains (Li and Shimizu 2005).

It's seen also from the Fig. 4 that the clay particles in the PA6 polymer formed a small aggregates, with the average of nanometer-size particles was around of 260, 195 and 459 nm correspond to Cloisite 20A, MMT-CTAB, and MMT-Na, respectively, this variation of the aggregates size, clearly indicates that the particles in MMT-Na clay before the treatment was in closely attached or in aggregates form due to the intermolecular forces and then in the presence of cationic surfactant, the clay particles are highly exfoliated into small aggregates, this may be due to the sodium ions in the interlayer spaces of clay can be exchanged for an amino acid such as CTAB (Zýková et al. 2009). In addition, it's clear that the organoclay composites surface appear to be covered by PA6 polymer. So, the interfacial adhesion in the presence of cationic surfactant into the interlayer spaces has been improved.

It's clear on Fig. 4 that the clays particles are uniformly dispersed in the matrix with small amount of agglomerates in the case of the MMT-Na composite, attesting that the use of melt compounding process to manufacture different types of composites is evident and enabled better particles distribution.

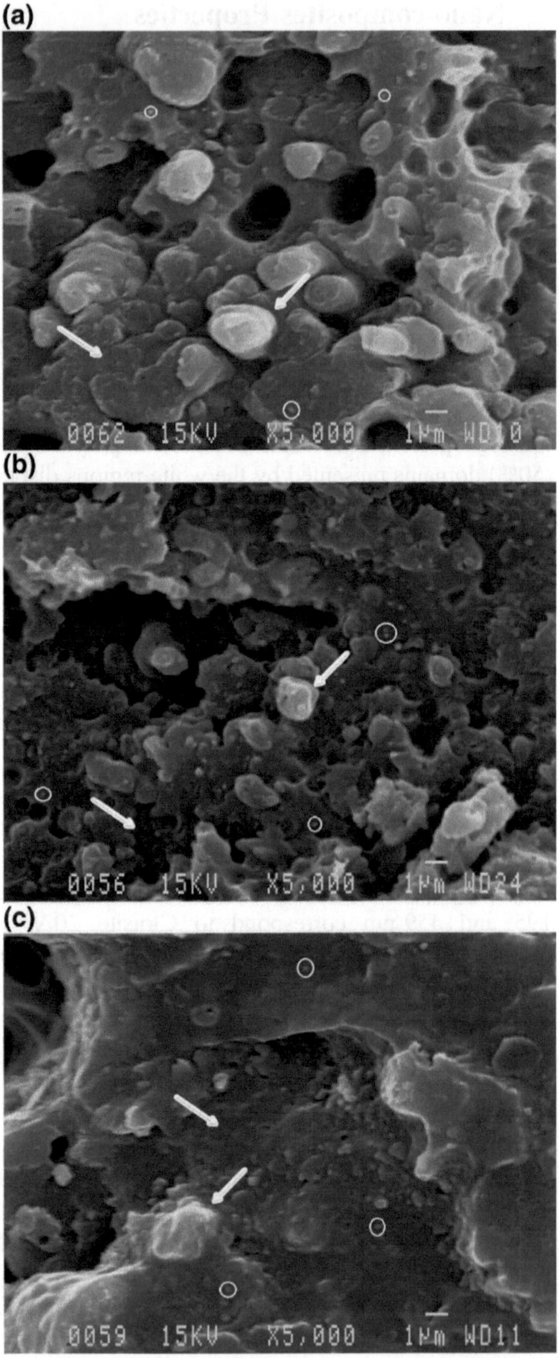

Fig. 7 SEM images of composites based on ABS-PA6 and of **a** 3 wt% of Cloisite 20A, **b** 3 wt% of MMT-CTAB, and **c** 3 wt% of MMT-Na

4.2 Thermal Properties

- ***Thermogravimetric analysis*** **(*TGA*)**

The thermal degradation of the different clays and nano-composites was evaluated by thermogravimetric analysis (TGA) using a model Q5000 from TA Instruments. Roughly 20 mg of each sample was placed in a platinum pan and heated under air from room temperature to 800 °C at a heating rate of 10 °C/min to yield the onset decomposition temperature.

Typical thermogravimetric curves of the nano-composite filled with different amounts of clay nanoparticles are presented in Fig. 8. All the curves exhibited a one-step degradation which is attributed to the random radical scission mechanism of polyolefin's degradation (Albano et al. 1999). The addition of clay increased the degradation temperature of ABS/PA6/Clay nano-composite because clay has a much higher decomposition temperature than the polymer matrix (Santos et al. 2013). Furthermore, the thermal stability of ABS/PA6/Clay nano-composite increased with clay loading.

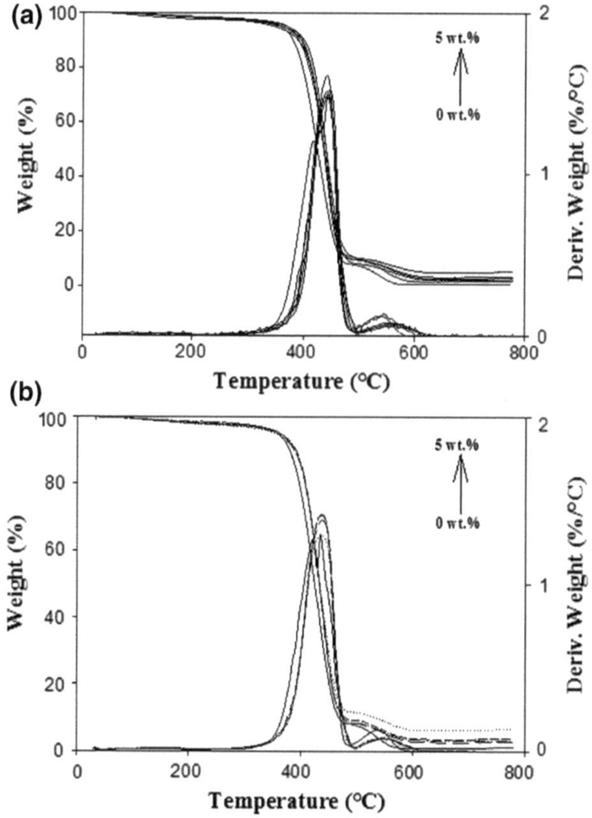

Fig. 8 TGA and DTG curves for: **a** the different nano-composites based on MMT-Na, and **b** the different nano-composite based on Cloisite 20A

Table 2 Temperature corresponding to a weight loss of 5% (T5%), 10% (T10%), 15% (T15%), and 20% (T20%) from TGA measurements of ABS/PA6/MMT nano-composite with various MMT-Na and Cloisite 20A contents

	Decomposition temperature for MMT-Na (°C)				Decomposition temperature for Cloisite 20A (°C)			
Clay content (wt%)	T5%	T10%	T15%	T20%	T5%	T10%	T15%	T20%
0	356	380	390	397	356	380	390	397
1	356	390	401	408	356	388	400	407
2	357	394	405	412	357	388	400	407
3	369	397	406	412	364	389	401	408
4	367	394	406	413	360	389	401	408
5	368	399	410	417	364	391	401	408

The temperatures corresponding to different weight loss are listed in Table 2 and it is clear that a direct relation between these temperatures and nano-clay loading exists. Furthermore, the highest thermal stability was observed for the MMT-Na and Cloisite 20A at 5% wt. and their T5, T10, T15, and T20% are about 12, 19, 20, 20 °C and 8, 11, 11, 11 °C higher than for the neat matrix, respectively. This can be explained by the effect of clays acting as thermal insulators and mass transport barriers to improving thermal stability (Bidsorkhi et al. 2014; Tartaglione et al. 2008). Thermal stability improvement is a direct effect of the excellent thermal barrier effect provided by the layered clays within the matrix. So, the better thermal stability of the nano-composites can be attributed to the clay layered structure limiting the mobility of small molecules produced during thermal decomposition. Furthermore, the thermal stability of MMT-Na is higher than Cloisite 20A due to a large number of grafted carbon chains present in the surfactant. These carbon chains acted as energy sources helping the samples to further decompose (Kokal et al. 2011).

4.3 *Mechanical Properties*

Dynamic mechanic analysis (DMA) and dynamic mechanical thermal analysis (DMTA) were performed using a Rheometer Solid Analyzer (RSA) according to ASTM D4092-01 (ASTM D 4092 – 01 2013). Due to the stiffness of the studied materials, a dual cantilever configuration was used. In DMA, a sinusoidal mechanical excitation is applied from which the force, deformation, and phase angle shift between the force and deformation were obtained as a function of frequency. The frequency sweeps ranged from 0.015 to 15 Hz with a strain of 0.002 (linear region of the materials). Similar conditions were used for dynamic mechanical thermal tests (DMTA) by heating the samples from 20 to 200 °C at a heating rate of 10 °C/min.

- ***Dynamic mechanical analysis (DMA)***

The complex modulus of the MMT-Na, MMT-CTAB, and Cloisite 20A nano-composites has been plotted as a function of nanoparticle content in Fig. 9. It can be seen that the complex modulus of MMT-Na, MMT-CTAB, and Cloisite 20A increased significantly with clay addition. This behavior can be associated to reinforcing effects through the formation of efficient and strong interactions between the polymer chains and the clay galleries at the interface which enhanced the compatibility between the phases (modified filler) (Hakeem et al. 2014). This can also be attributed to the incorporation of rigid particles changing the local molecular dynamics (lower polymer chain mobility) (Laaziz et al. 2017). Moreover, the frequency has also an effect on the composites response since the materials are viscoelastic. It is clear from Fig. 9 that E^* increases with increasing frequencies for the range of conditions tested (0.015–15 Hz).

- ***Dynamic mechanical thermal analysis (DMTA)***

The glass transition temperature of the composite, as obtained from the maximum of the damping factor ($\tan\delta = G''/G'$) curves using dynamic mechanical thermal analysis, is presented in Fig. 10. A clear decrease from 0 up to 4% wt. is observed: about 1.6 °C for Cloisite 20A and MMT-CTAB. Then, a slight increase is observed at 5% wt. suggesting that the mobility of the bulk polymer chains is restricted by the presence of the nanoparticles and related to their content (Santos et al. 2013). For MMT-Na, the glass transition temperature Tg decreases continuously by about 1.0 °C from 0 to 5% wt. The addition of low nano-clay content slightly decreased the glass transition temperature of the composite because the silicate layers can act as nucleating agents in the polymer matrix (Pavlidou and Papaspyrides 2008), which can also improve spherulites growth (Fragiadakis et al. 2005). At 4% wt., a high number of crystallites are formed, and the addition of more clay did not change the crystal structure of the polymer since no more space for the polymer molecules to reorganize is available inside the matrix. But nucleation and growth efficiency depend on clay dispersion.

4.4 Rheological Properties

- ***Melt rheological test***

Melt rheology tests were performed on an MCR 500 (Physica) Rheometer equipped with a CTD600 device. The measurements were carried out at 240 °C under small amplitude oscillatory shear mode using 25 mm parallel plate–plate geometry with 2-mm-thick samples. Frequency sweeps between 500 and 0.05 Hz were performed at a strain of 5%, for which the materials exhibit a linear viscoelastic behavior as verified by previous strain sweeps.

The rheological properties of the nano-composites were also investigated in the melt state, and the results are plotted in Fig. 11. It can be seen in Fig. 11a that the

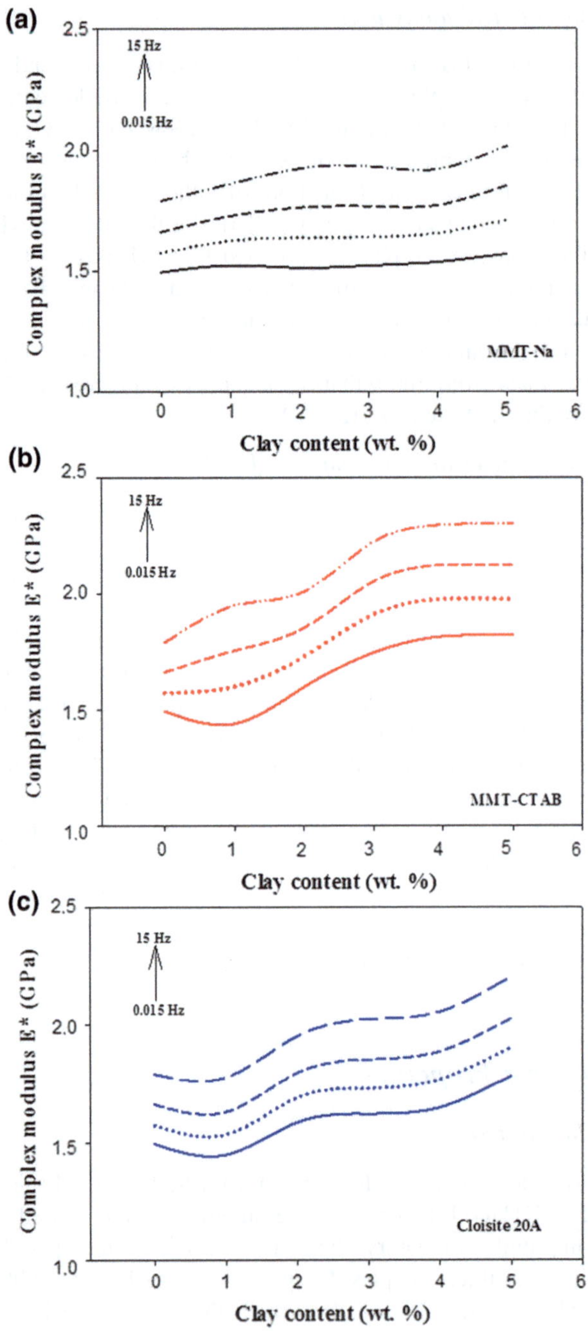

Fig. 9 Complex modulus as a function of nanoparticle content for different frequency: **a** MMT-Na, **b** MMT-CTAB, and **c** Cloisite 20A

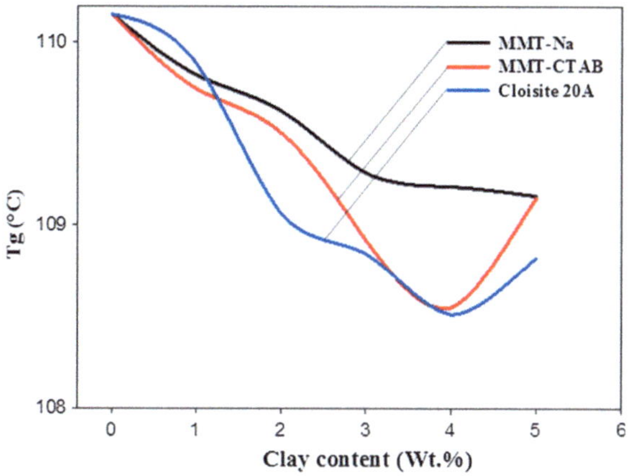

Fig. 10 Glass transition temperature (T_g) for the different nano-composites as a function of clay content

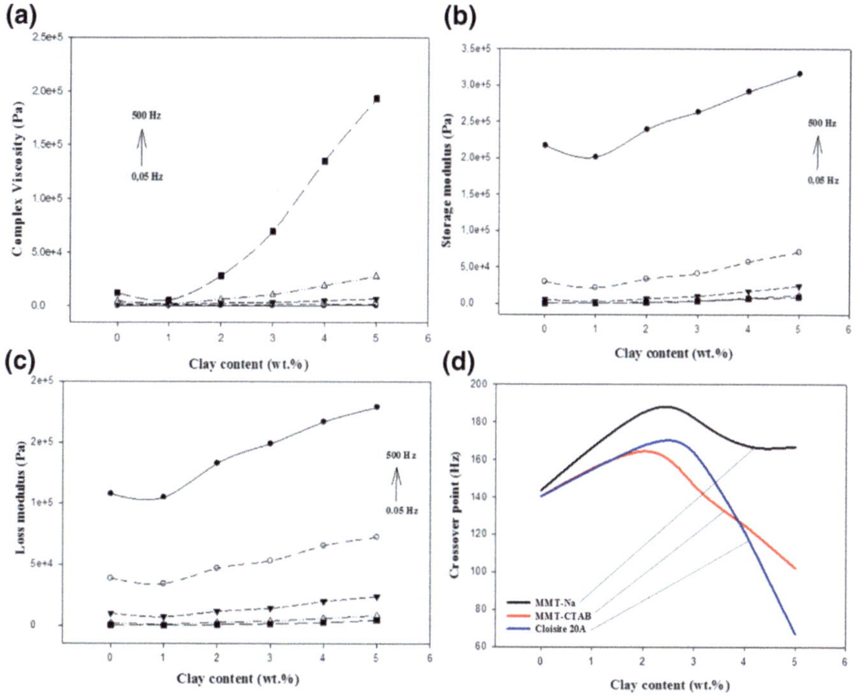

Fig. 11 Dynamic rheological properties of the nano-composites (240 °C) as a function of clay content: **a** complex viscosity, **b** storage modulus (G′), **c** loss modulus (G″), and **d** crossover point

complex viscosities of the clay composites have a plateau at low frequency or the Newtonian behavior, indicating a transition from a liquid-like to a solid-like viscoelastic behavior. Then, Fig. 11b shows that the nano-clay particles increase the storage moduli, indicating that the rheological properties of the reinforced copolymer were mainly influenced by the addition of nano-clay, and at higher clay content, the elastic character of the material prevails over its viscous behavior (Bensalah et al. 2017).

Figure 11c shows the loss modulus against frequency for various nano-clay contents. An increase in loss modulus with increasing MMT-CTAB content can clearly be observed when the nano-clay content increases over the whole frequency range. Due to good dispersion, distribution, and affinity of the rigid nano-clay inside the polymer blend, all these phenomena can prevent the melt to flow which is reflected by an increase in loss modulus/viscosity (Yürüdü et al. 2005).

Finally, Fig. 11d shows the variation of the crossover point (frequency where $G' = G''$) with clay content. The results show a decrease in the crossover point values until a maximum presented at 5 wt% of MMT-CTAB content. This reduction in crossover frequency indicates that G'' becomes smaller than G' or the materials have lower elasticity and lower energy dissipation. This behavior can be associated with insufficient time to allow for polymer chains to relax with increasing rigid particle content, which is a contribution from the increased elastic nature of the polymer melt (Raji et al. 2016b).

5 Conclusion

In summary, the montmorillonite can be successfully modified using hexadecyltrimethylammonium bromide (CTAB). The XRD and FTIR results clearly showed that intercalation of the organic cationic surfactant between the clay mineral layers occurred. The modified clay was then used as reinforcement inside a thermoplastic copolymer matrix based on polyamide 6/acrylonitrile butadiene styrene. The results were compared with the unmodified clay (MMT-Na) and a commercial organoclay (Cloisite 20A). The compounds were prepared by twin-screw extrusion, and the samples were molded via injection molding. For the nano-composites produced under the range of conditions tested (0–5% wt.), it was found that exfoliated clay platelets were selectively localized in the PA6 phase and also was found that 4% wt. is the optimum clay content. Finally, the results showed that thermal stability was significantly improved (20 °C) by the incorporation of these nano-clays, and similar improvements were observed for mechanical and rheological properties.

References

Abdellaoui H et al (2017) Laminated epoxy biocomposites based on clay and jute fibers. J Bionic Eng 14(2):379–389

Agag T, Koga T, Takeichi T (2001) Studies on thermal and mechanical properties of polyimide-clay nanocomposites. Polymer 42(8):3399–3408

Akin O, Tihminlioglu F (2018) Effect of organo-modified clay addition on properties of polyhydroxy butyrate homo and copolymers nanocomposite films for packaging applications. 26(3):1121–1132

Albano C, González J, Ichazo M, Kaiser D (1999) Thermal stability of blends of polyolefins and sisal fiber. Polym Degrad Stab 66(2):179–190

Amendola E et al (2012) Epoxy nanocomposites based on silylated montmorillonite: effect of the coupling agents structure on the mechanical properties. Macromolecules 6:33–36

Aowda SA, Jaffar Al-mulla EA, Baqir SJ (2011) Modification of montmorillonite, using different phosphonium salts, study their effect upon the structure. J Al-padisiyah Pure Sci 16:1–10

ASTM D 4092 – 01 (2013) Standard terminology for plastics: dynamic mechanical properties. ASTM Int 1–4

Bensalah H et al (2017) Mechanical, thermal, and rheological properties of polypropylene hybrid composites based clay and graphite. J Compos Mater 51(25):3563–3576

Bergaya F, Lagaly G (2013) General introduction: clays, clay minerals, and clay science. In: Handbook of clay science. Elsevier, Netherlands, pp 1–19

Bhat G, Hegde RR, Kamath MG, Deshpande B (2008) Nanoclay reinforced fibers and nonwovens. J Eng Fibers Fabr 3(3):22–34

Bhattacharya SS, Aadhar M (2014) Studies on preparation and analysis of organoclay nano particles. Res J Eng Sci 3(3):10–16

Bidsorkhi HC et al (2014) Mechanical, thermal and flammability properties of ethylene-vinyl acetate (EVA)/sepiolite nanocomposites. Polym Test 37:117–122

Boujmal R et al (2017) Alfa fibers/clay hybrid composites based on polypropylene. J Thermoplast Compos Mater 31(7):974–991

Carastan DJ et al (2013) Morphological evolution of oriented clay-containing block co-polymer nanocomposites under elongational flow. Eur Polym J 49(6):1391–1405

Carrado KA (2000) Synthetic organo- and polymer-clays: preparation, characterization, and materials applications. Appl Clay Sci 17(1–2):1–23

de Paiva LB, Morales AR, Díaz FRV (2008) Organoclays: properties, preparation and applications. Appl Clay Sci 42(1–2):8–24

Diagne M et al (2006) The effect of photo-oxidation on thermal and fire retardancy of polypropylene nanocomposites. J Mater Sci 41(21):7005–7010

Donescu D, Vuluga Z, Radovici C, Serban S (2008) Modification of organosilicate with silane coupling agents for polymer nanocomposites. Mater Plastice 45(4):305–309

Erden S, Ho K (2017) Composites science and engineering. Fiber technology for fiber-reinforced composites. Woodhead Publishing is an imprint of Elsevier

Essabir H et al (2016a) Structural, mechanical and thermal properties of bio-based hybrid composites from waste coir residues: fibers and shell particles. Mech Mater 93:134–144

Essabir H, Raji M, Bouh R (2016b) Nanoclay and natural fibers based hybrid composites: mechanical, morphological, thermal and rheological properties. In Nanoclay reinforced polymer composites, pp 29–49

Essabir H et al (2017) Morphological, thermal, mechanical, electrical and magnetic properties of ABS/PA6/SBR blends with Fe_3O_4 nano-particles. J Mater Sci Mater Electron 28(22):17120–17130

Fragiadakis D, Pissis P, Bokobza L (2005) Glass transition and molecular dynamics in poly (dimethylsiloxane)/silica nanocomposites. Polymer 46(16):6001–6008

Fujimori A, Kusaka J, Nomura R (2010) Formation and structure of organized molecular films for organo-modified montmorillonite and mixed monolayer behavior with poly(L-lactide). Polym Eng Sci 51(6):1–9

Gacitua EW, Aldo Ballerini A, Zhang J (2005) Polymer nanocomposites : synthetic and natural fillers. Maderas Ciencia y tecnología 7(3):159–178

Gao F (2004) Clay/polymer composites: the story. Mater Today 7(November):50–55

Gavrilko TA et al (2013) Molecular dynamics and phase transitions behavior of binary mixtures of fatty acids and cetyltrimethylammonium bromide as studied via davydov splitting of molecular vibrational modes. Ukrainian J Phys 58(7):636–645

González JA, Del M (2006) Bleaching of kaolins and clays by chlorination of iron and titanium. Appl Clay Sci 33:219–229

Guggenheim S et al (2006) Summary of recommendations of nomenclature committees relevant to clay mineralogy: report of Association Internationale Pour l'Etude Des Argiles (AIPEA) Nomenclature Committee for 2006. Clays Clay Miner 54:761–772

Hakeem KR, Jawaid M, Rashid U (2014) Fuel and energy abstracts. Biomass and bioenergy processing and properties. Springer, Berlin

He A et al (2010) Structural design of imidazolium and its application in PP/montmorillonite nanocomposites. Polym Degrad Stab 95(4):651–655. https://doi.org/10.1016/j.polymdegradstab.2009.12.003

Hoidy WH, Al-mulla EAJ (2013) Study of preparation for co-polymer nanocomposites using PLA/LDPE/CTAB modified clay. Iraqi Nat J Chem 49:61–72

Jeong E, Lim JW, Seo KW, Lee YS (2011) Effects of physicochemical treatments of illite on the thermo-mechanical properties and thermal stability of illite/epoxy composites. J Ind Eng Chem 17(1):77–82

Kokal I, Somer M, Notten PHL, Hintzen HT (2011) Sol–gel synthesis and lithium ion conductivity of $Li_7La_3Zr_2O_{12}$ with garnet-related type structure. Solid State Ionics 185(1):42–46

Kumar R, Yakabu MK, Anandjiwala RD (2010) Effect of montmorillonite clay on flax fabric reinforced poly lactic acid composites with amphiphilic additives. Compos A Appl Sci Manuf 41(11):1620–1627

Laaziz SA et al (2017) Bio-composites based on polylactic acid and argan nut shell: production and properties. Int J Biol Macromol 104:30–42

Laoutid F, Persenaire O, Bonnaud L, Dubois P (2013) Flame retardant polypropylene through the joint action of sepiolite and polyamide 6. Polym Degrad Stab 98(10):1972–1980

Leszczy A, Njuguna J, Pielichowski K, Banerjee JR (2007) Polymer/montmorillonite nanocomposites with improved thermal part II : thermal stability of montmorillonite nanocomposites based on different polymeric matrixes. Thermochim Acta 454(1):1–22

Li Y, Shimizu H (2005) Co-continuous polyamide 6 (PA6)/Acrylonitrile-Butadiene-Styrene (ABS) nanocomposites. Macromol Rapid Commun 26(9):710–715

Majeed K et al (2013) Potential materials for food packaging from nanoclay/natural fibres filled hybrid composites. Mater Des 46:391–410

Mansoori Y, Roojaei K, Zamanloo MR, Imanzadeh G (2012) Polymer-clay nanocomposites via chemical grafting of polyacrylonitrile onto cloisite 20A. Bull Mater Sci 35(7):1063–1070

Mejía A, García N, Guzmán J, Tiemblo P (2014) Surface modification of sepiolite nanofibers with PEG based compounds to prepare polymer electrolytes. Appl Clay Sci 95:265–274

Navrátilová Z, Wojtowicz P, Vaculíková L, Šugárková V (2007) Sorption of alkylammonium cations on montmorillonite. Acta Geodyn Geomater 4(3):59–65

Nazir MS et al (2016) Characteristic properties of nanoclays and characterization of nanoparticulates and nanocomposites. In Nanoclay reinforced polymer composites, pp 29–49. http://link.springer.com/10.1007/978-981-10-0950-1

Nekhlaoui S et al (2014) Fracture study of the composite using essential work of fracture method: PP–SEBS–g–MA/E1 clay. Mater Des 53:741–748

Nuzzo A et al (2014) Nanoparticle-induced co-continuity in immiscible polymer blends—a comparative study on bio-based PLA-PA11 blends filled with organoclay, sepiolite, and carbon nanotubes. Polymer 55(19):4908–4919

Oliveira M, Machado AV (2013) Preparation of polymer-based nanocomposites by different routes. In: Nanocomposites: synthesis, characterization and applications. NOVA Publishers, New York, pp 1–22
Pavlidou S, Papaspyrides CD (2008) A review on polymer–layered silicate nanocomposites. Prog Polym Sci 33:1119–1198
Pretschuh C, Schwarzinger C, Abdala AA, Vukusic S (2014) Characterization of conductive nanographite melamine composites. Open J Compos Mater 4:61–71
Raji M, Essabir H et al (2016a) Morphological, thermal, mechanical, and rheological properties of high density polyethylene reinforced with illite clay. Polym Polym Compos 16(2):101–113
Raji M, Mekhzoum MEM et al (2016b) Nanoclay modification and functionalization for nanocomposites development: effect on the structural, morphological, mechanical and rheological properties. In: Nanoclay reinforced polymer composites, pp 1–34
Raji M, Essabir H, Rodrigue D et al (2017a) Influence of graphene oxide and graphene nanosheet on the properties of polyvinylidene fluoride nanocomposites. https://doi.org/10.1002/pc.24292
Raji M, Essabir H, Bouhfid R, el kacem Qaiss A (2017b) Impact of chemical treatment and the manufacturing process on mechanical, thermal, and rheological properties of natural fibers-based composites. In: Handbook of composites from renewable materials. Wiley, Hoboken, pp 225–252
Santos KS et al (2013) The influence of screw configurations and feed mode on the dispersion of organoclay on PP. Políymer 23(2):175–181
Singla P, Mehta R, Upadhyay SN (2012) Clay modification by the use of organic cations. Green Sustain Chem 2:21–25
Šupová M, Martynková GS, Barabaszová K (2011) Effect of nanofillers dispersion in polymer matrices: a review. Sci Adv Mater 3(1):1–25
Tamura K, Yokoyama S, Pascua CS, Yamada H (2008) New age of polymer nanocomposites containing dispersed high-aspect-ratio silicate nanolayers. Chem Mater 20(6):2242–2246
Tartaglione G, Tabuani D, Camino G, Moisio M (2008) PP and PBT composites filled with sepiolite: morphology and thermal behaviour. Compos Sci Technol 68(2):451–460
Tian H, Tagaya H (2007) Preparation, characterization and mechanical properties of the polylactide/perlite and the polylactide/montmorillonite composites. J Mater Sci 42(9): 3244–3250
Tjong SC (2006) Structural and mechanical properties of polymer nanocomposites. Mater Sci Eng R Rep 53(3–4):73–197
Usuki A (2002) Preparation and properties of EPDM—clay hybrids. Polymer 43(8):2185–2189
Vuluga Z et al (2014) The effect of polystyrene blocks content and of type of elastomer blocks on the properties of block copolymer/layered silicate nanocomposites. J Alloy Compd 616:569–576
Wang K et al (2013) Effect of talc content on the degradation of re-extruded polypropylene/talc composites. Polym Degrad Stab 98(7):1275–1286
Wu SH, Chen DH (2004) Synthesis of high-concentration Cu nanoparticles in aqueous CTAB solutions. J Colloid Interface Sci 273:165–169
Xie A et al (2011) Microstructure and antibacterial activity of phosphonium montmorillonites. Bull Korean Chem Soc 32(6):1936–1938
Xue W, He H, Zhu J, Yuan P (2007) FTIR investigation of CTAB-Al-Montmorillonite complexes. Spectrochim Acta Part A Mol Biomol Spectrosc 67(3–4):1030–1036
Yan W et al (2012) Morphology and mechanical properties of Acrylonitrile-Butadiene-Styrene (ABS)/polyamide 6 (PA6) nanocomposites prepared via melt mixing. J Macromol Sci Part B 51(1):70–82
Yürüdü C et al (2005) Synthesis and characterization of HDA/NaMMT organoclay. Bull Mater Sci 28(6):623–628
Zari N, Raji M, El Mghari H, Bouhfid R (2018) Nanoclay and polymer-based nanocomposites: materials for energy efficiency. In: Polymer-based nanocomposites for energy and environmental applications. Woodhead, UK, pp 75–103
Zeng QH, Yu AB, Lu GQ, Paul DR (2005) Clay-based polymer nanocomposites: research and commercial development. J Nanosci Nanotechnol 5:1574–1592

Zhang Z, Zhang J, Liao L, Xia Z (2013) Synergistic effect of cationic and anionic surfactants for the modification of Ca-montmorillonite. Mater Res Bull 48(5):1811–1816

Zhao Y, Abdullayev E, Vasiliev Ae, Lvov Y (2013) Halloysite nanotubule clay for efficient water purification. J Colloid Interface Sci 406(June):121–129

Zotti A et al (2014) Effect of sepiolite filler on mechanical behaviour of a bisphenol A-based epoxy system. Compos B Eng 67:400–409

Zýková J, Kalendová A, Kovářová L, Maláč J (2009) Influence of intercalation agents on the thermal stability of Pvc/clay. Nanocon 20(10):1–4

Nanotube/Biopolymer Nanocomposites

Nilesh Kumar Shrivastava, Muhammad Akmal Ahmad Saidi, M. S. Z. Mat Desa, Mohamad Zurina, Norhayani Othman, Azman Hassan, A. K. M. Moshiul Alam, M. D. H. Beg and R. M. Yunus

Abstract Nanotubes are one of the most important classes of 1D nanomaterial which can be used as reinforcing filler for the polymers and biopolymers. Out of several organic/inorganic nanotubes, carbon nanotubes (CNTs) and halloysite nanotubes (HNTs) were most studied due to their high aspect ratio, outstanding mechanical and thermal properties. Nanotubes have potential to increase the physical properties of the biopolymer even at very low loading. The properties of nanotube-filled nanocomposite depend on various parameters like the aspect ratio, dispersion, and interaction between filler and polymer. To improve the utilization and maximum potential of these fillers, it is important to understand the mechanisms of reinforcement. It will help the future researchers to find out the limitations and issues with existent processing method of nanotube-based nanocomposite which is required for further improvement. In this regard, this chapter will help the researchers to fully understand the current progress in processing method, issues, and possible methods to prepare an ideal nanotube and biopolymer-based nanocomposites. This chapter will discuss only biopolymer-based nanocomposites filled by the two most important 1D tubular nanofillers, which are CNT and HNT. This study has been divided into three subtopics, namely (i) thermoplastic nanocomposites, (ii) thermoset nanocomposites, and (iii) natural rubber

N. K. Shrivastava · M. A. A. Saidi · M. Zurina · N. Othman · A. Hassan (✉)
Department of Bioprocess and Polymer Engineering, School of Chemical and Energy Engineering, Faculty of Engineering, Universiti Teknologi Malaysia, 81310 Skudai, Johor, Malaysia
e-mail: azmanh@cheme.utm.my

A. Hassan
Centre for Advanced Composite Materials, Universiti Teknologi Malaysia, 81310 Skudai, Johor, Malaysia

M. S. Z. Mat Desa · A. K. M. Moshiul Alam · M. D. H. Beg · R. M. Yunus
Faculty of Chemical and Natural Resources Engineering, Universiti Malaysia Pahang, Lebuhraya Tun Razak, 26350 Gambang, Kuantan, Pahang, Malaysia

M. L. Sanyang and M. Jawaid (eds.), *Bio-based Polymers and Nanocomposites*,
https://doi.org/10.1007/978-3-030-05825-8_7

(NR) nanocomposites. It focuses on the fabrication processes, properties, and potential applications of CNT and HNT biopolymer nanocomposites.

Keywords CNT · HNT · Nanotube · Biopolymer · Nanocomposites

1 Introduction

Nanotubes are one-dimensional (1D) cylindrical nanostructure which is hollow in the center. This unique structure of nanotube gives them outstanding electrical, mechanical, and thermal properties. An important class of nanotubes, carbon nanotube (CNT) which is an allotrope of carbon, was discovered in 1991. These cylindrical carbon molecules have unusual properties, which are valuable for nanotechnology, electronics, optics, and other fields of materials science and technology. Later, a significant research has been conducted for synthesizing nanoscale 1D nanotubes which are either organic (Shao et al. 2011; Yang et al. 2012) or transition metal sulfides, selenides, halogenides, or oxides such as CO_3S_4 nanotubes (Chen et al. 2007), CdTe nanotubes (Niu and Gao 2006), MnO_2 nanotubes (Wang et al. 2013), $Y(OH)_3$ single-crystalline nanotubes (Fang et al. 2003), TiO_2 nanotubes (Banerjee et al. 2008), and SnO_2 nanotubes (Ye et al. 2010). Another type of nanotube which is naturally occurring silica is halloysite nanotube (HNT). These nanotubes have several application areas such as in power storage, electromagnetic interference (EMI) shielding, piezoelectric, solid catalyst, sensors, drug delivery, nanoelectronics, and reinforcing fillers for polymers. Out of these nanotubes, CNT and HNT were the most studied as reinforcing fillers for polymers that is because the properties of nanofiller filled nanocomposite depend on various parameters like the shape of nanofillers, dispersion, and interaction between filler and polymer. The shape of nanofiller is most crucial parameter because rest of two parameters can be altered by different processing techniques.

In recent years, biopolymers have gained much popularity due to increasing environmental issues from plastic wastes. Therefore, many biopolymers were discovered and synthesized such as polylactic acid (PLA), polyhydroxyalkanoate (PHA), poly(ε-caprolactone), poly(butylene succinate/adipate), and poly(butylene adipate-co-terephthalate). However, biopolymer still holds a tiny fraction of the total global plastic market mainly due to high cost. Besides that, biopolymers have relatively poor mechanical properties, gas permeability, flame retardancy, and thermal properties. Many reinforcing agents were used to improve the physical and mechanical properties of biopolymers which include nanosized fillers (nanofillers) and reinforcements.

2 Carbon Nanotubes/Biopolymer Nanocomposites

2.1 Carbon Nanotubes

The revolution of super-strong carbon allotropes began with the discovery of fullerene in 1985 by R.E Smalley et al. from Rice University. They successfully discovered C_{60} buckminsterfullerene, a geometric cage-like structure of carbon atoms comprised of pentagonal and hexagonal arrangements (Kroto et al. 1985). It is then followed by the discovery of CNT by Iijima (1991) and graphene in 2004 by Andre Geim and Kostya Novoselov from University of Manchester, UK (Novoselov et al. 2004, 2005). All of the newly discovered synthetic carbon allotropes exhibit almost similar crystal arrangement with the natural graphite and diamond carbon allotropes with outstanding mechanical and electrical properties.

Studies have shown that CNT is a nanotube of long, slender fullerenes with walls of the tubes comprised of hexagonal carbon similar to graphite structure and often has a pentagonal hemispherical capped on at least at one end of the tubes (Thostenson et al. 2001) as illustrated in Fig. 1. The presence of the hemispherical cap made CNT a nanotube which chemically comprises of two major parts: the end caps and the sidewall (Niyogi et al. 2002). It was reported that due to the processing for CNT fabrications, nanotubes can exist as either single-walled carbon nanotube (SWCNT),

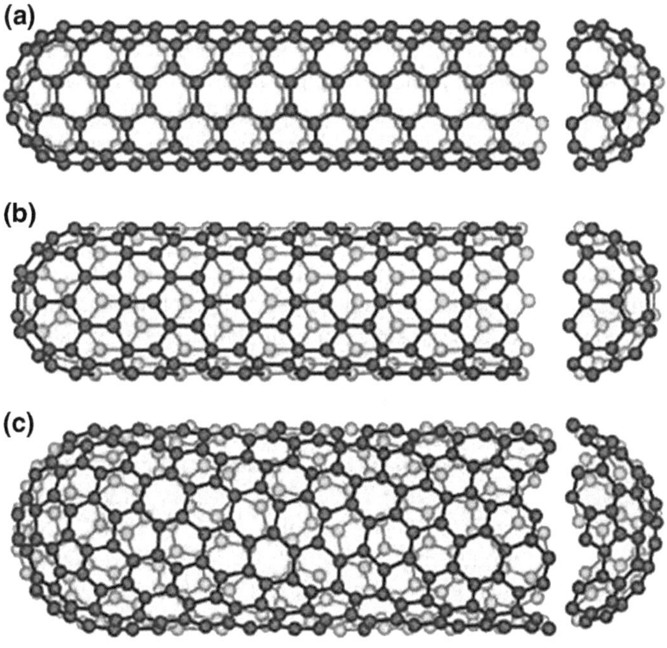

Fig. 1 Illustration of hemispherical capped carbon nanotubes with different chiralities (**a**) armchair, (**b**) zigzag, (**c**) chiral (Dresselhaus et al. 1995)

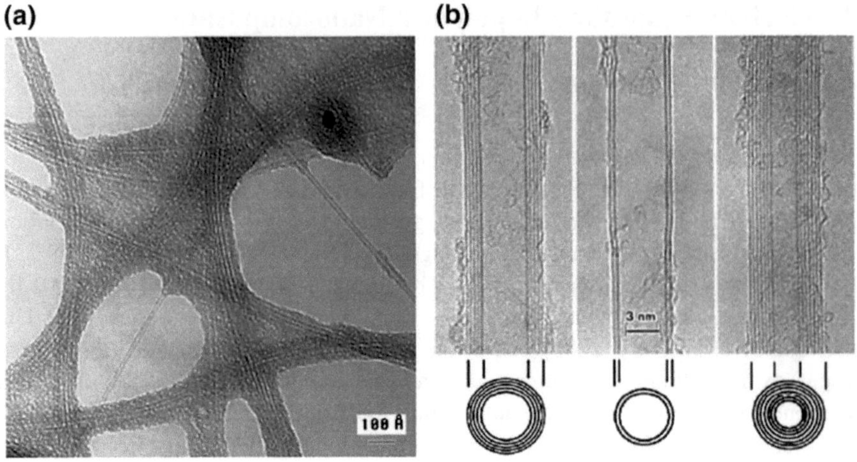

Fig. 2 TEM images of the different types of CNT (**a**) SWCNT (**b**) MWCNT with different layers (from left to right: 5, 2, and 7 layers respectively) (Iijima 1991; Bethune et al. 1993)

double-walled carbon nanotube (DWCNT) or multi-walled carbon nanotube (MWCNT) (Iijima 1991; Thostenson et al. 2001; Sahoo et al. 2010; Ma et al. 2010). Early discovery of CNT was in the form of MWCNT which is actually a collection of concentric graphene cylinder (Iijima 1991), while synthesis of SWCNT was first discovered in 1993 by Iijima (1993) and Bethune et al. (1993) with nanotube that consists of a rolled seamless cylinder of single graphene sheet as illustrated in Fig. 2.

SWCNT typically has diameters ranging from 0.4 to 2–3 nm and several micrometer lengths, and some researchers reported SWCNT length of up to centimeters. Initially, it was produced through arc discharge generators (Iijima 1993; Bethune et al. 1993), but recently the most common method of SWCNT production is by laser ablation and chemical vapor decomposition (CVD). The fabrication of SWCNT may produce various chiralities due to changes in chiral vector and chiral angle during rolling action which will produce bundles of SWCNT comprised of metallic, semimetallic, or semiconducting SWCNT. Even though isolated pristine SWCNT is more favored, experimentally it is difficult to produce such material due to their flexibility and high surface energy which made SWCNT tend to agglomerate into large bundles containing mixtures of various chirality SWCNTs (Coleman et al. 2006).

Fabrication of MWCNT through CVD is the most common method and the most promising fabrication technique due to the ability of mass production at relatively low cost for industrial-scale production (Coleman et al. 2006). Other than that, MWCNT can also be produced by the arc discharge method with better-quality MWCNT of less defect and well-graphitized wall (Cadek et al. 2002). The typical size of MWCNT may range from an inside diameter of 0.4 nm to an outside diameter of about 5 nm (Iijima 2002). Early production of MWCNT by Iijima (1991) was able to fabricate MWCNT with up to seven layers of the graphitic sheet with a diameter of 6.5 nm and five layers of the graphitic sheet with 6.7 nm in diameter. Transmission electron microscope (TEM) image in Fig. 2 clearly shows the cylindrical tube image of MWCNT with several graphitic carbon layers. As discussed earlier, the layers are essentially concentric single-walled tubes held together by secondary van der Waals bonding; thus in regard to the presence of multilayers in MWCNT, it creates a complicated structural property due to intra-tube interactions and different chiralities displayed by each individual tube. This unique nature has made MWCNT differ greatly from the ideal SWCNT and need to be further investigated in order to tap their full potential in real-life applications such as reinforcing polymer nanocomposites (Thostenson et al. 2001).

CNT is one of the most significant candidates for reinforcing polymer nanocomposites due to their outstanding mechanical strength, high surface area, and superior conductivity (Terrones et al. 1998). The CNT-based polymer nanocomposites find extensive applications in nanoelectronics, advanced composites for space and aircraft and gas storage devices (Lau et al. 2004). Unlike the microscopic particulate composites, the mechanical properties of nanocomposites depend strongly on the dispersion state of nanofillers, apart from the properties of fillers and matrix material. Usually, CNT-reinforced nanocomposite properties are dependent on the degree of dispersion, interfacial adhesion with the matrix in the nanocomposite system (Singh et al. 2013).

The dispersion can be defined as a system in which one type of particles is intermixed with a continuous phase of another composition or state (Wang et al. 2009). When the dispersed domains are nanometers and/or micrometers in size, the solution is known as colloidal solution, with the subcategories of suspensions; in case of CNT, the degree of dispersion in a colloidal solution cannot easily be defined or measured because the dispersion status of CNT changes over time (Yu et al. 2007).

Favorable mechanical and physical properties for the composites can only be achieved when the nanofillers are very well dispersed in the matrix (Subramaniam et al. 2013). Most common problems associated with CNT dispersion are a strong aggregation tendency and poor dissolvability in most of the organic solvents and water during the composite fabrication (Xiao and Zhang 2004). Such problems due to aggregation of CNT are often minimized by functionalization technique, which tailors the structure and property of CNT (Lou et al. 2004; Pirlot et al. 2003). This technique enhances solubility and the compatibility of CNT with the rubbers and thus helps in the fabrication of novel nanodevices and composites (Frogley et al. 2003).

The two common functionalization methods of CNT are grafting of functional groups by covalent bonding (Hill et al. 2002; Velasco-Santos et al. 2002; Vast et al. 2004) and wrapping of surfactants using noncovalent interaction method (Kokai et al. 2005). Noncovalent methods are most preferred to the covalent methods since the chemical bonds between carbon atoms on graphene sheets remain undisturbed and properties of the nanotubes are preserved (Lau 2003). In noncovalent approach, a number of surfactants have been reported for obtaining stable aqueous dispersions of carbon nanotubes (George et al. 2017; Rastogi et al. 2008; Vaisman et al. 2006). Addition to the surfactant in aqueous suspensions of CNT followed by sonication can very effectively de-bundle the CNT aggregates resulting in a homogeneous dispersion. But the sonication energy and CNT-to-surfactant ratio need to be optimized to prevent any detrimental effect on the inherent properties of the nanotubes (Vichchulada et al. 2010). Furthermore, high surfactant concentration can lead to a reduction in electrical properties of the nanocomposite since it forms an insulating layer around the nanotubes (Le et al. 2013).

2.2 Carbon Nanotubes/Thermoplastic Nanocomposites

Thermoplastic polymers are an important class of materials with a wide range of properties, some of which are unattainable from any other materials such as low density, flexibility, easy processing, recyclability, and low cost. At present, reinforced thermoplastics are receiving increased interest in engineering applications due to their attractive price/performance ratio with apparent enhancement in physical and thermal properties (Offringa 1996). Therefore, continuous research is being carried out to develop new thermoplastic composite materials with enhanced properties to meet various engineering applications. One of the strategies is by fabrication of thermoplastic composites through the incorporation of stiff organic and inorganic fillers as reinforcement agent such as glass fibers and carbon fibers. Although the use of organic/inorganic reinforcement agent is known to significantly improve the mechanical properties of thermoplastics, these conventional thermoplastic composites require high content of fillers, thus increasing the cost and density of thermoplastic composites, and make processing more difficult and expensive. The emergence of nanotechnology saw the birth of thermoplastic nanocomposites, where the use of nanofillers as reinforcement agent in thermoplastic polymers was shown to exhibit a significant improvement of mechanical strength even at low content of nanofiller.

Development of thermoplastic-based nanocomposites has been one of the main subjects of interest among researchers in both industries and academia. Apart from the improvement in mechanical properties, thermoplastic nanocomposites were also reported to exhibit a significant improvement of thermostability, barrier properties, electrical conductivity, and flame retardancy (Moniruzzaman and Winey 2006). One of the promising nanofillers to be used as nanoreinforcement for thermoplastics is CNT. The fascinating mechanical and physical properties of CNT have stimulated intensive research in the design and preparation of various novel CNT/polymer nanocomposites (Liu et al. 2004; McNally et al. 2005; Kanagaraj et al. 2007; Kulkarni et al. 2013). The incorporation of low CNT content (1–5 wt%) into various polymer matrixes has resulted in property enhancement comparable to conventional content (15–40 wt%) of traditional microfillers (Yuan and Misra 2006). One of the potential applications of CNT/polymer nanocomposite is in the automotive and aeronautical sector, where the parts made of CNT/polymer nanocomposite are expected to reduce weight compared to polymer microcomposites and hence decrease fuel consumption.

Properties of CNT/thermoplastic nanocomposites depend on several factors. In general, filler distribution and dispersion are two crucial factors that will largely determine a good polymer nanocomposite blend. Distribution is described as the randomness of filler in the polymer mixture, while dispersion is rather the breakup of discrete domains into smaller domains. The improvement in filler dispersion, especially in immiscible polymer blend, is proven to produce better thermal and mechanical properties of the blends. Decades after the discovery of CNT, various researches are still being carried out on utilizing CNT as nanofiller in the polymer nanocomposite system with promising outcome enabling real-life application of it in the near future. However, most of the works on incorporating CNT as reinforcement nanofiller are hindered by obstacles associated with achieving optimum dispersion of the nanotube in the polymer matrix and poor interfacial interaction between nanotube and polymer matrix.

The problem of poor dispersion originated from the entangle bundles of raw commercial CNT due to van der Waals interaction between the sidewall of each individual tube. Failure to overcome the bundling of the nanotube as illustrated in Fig. 3 will undermine the true potential of CNT as mechanical reinforcement in the polymer matrix. In addition to that, poor interfacial interactions between CNT and the polymer matrix are another major concern which attributed from the chemically stable aromatic carbon bonding of CNT sidewalls which are inert in nature and can only interact with van der Waals bonding with the surrounding matrix. Thus without the presence of stronger chemical bonding, the relatively weaker van der Waals bonding is unable to provide an efficient interfacial interaction between CNT and the polymer matrix (Ma et al. 2010).

Therefore, to unveil the true potential of CNT as reinforcement agent in a polymer matrix, improvement in CNT dispersion is the main priority. In order to improve dispersion of CNT in a polymer matrix, researchers employed various mechanical and chemical processing approaches and quite often both techniques were combined together in order to get better dispersion.

To date, various mechanical dispersion techniques are available that can effectively disperse CNT with some advantage and disadvantages between one

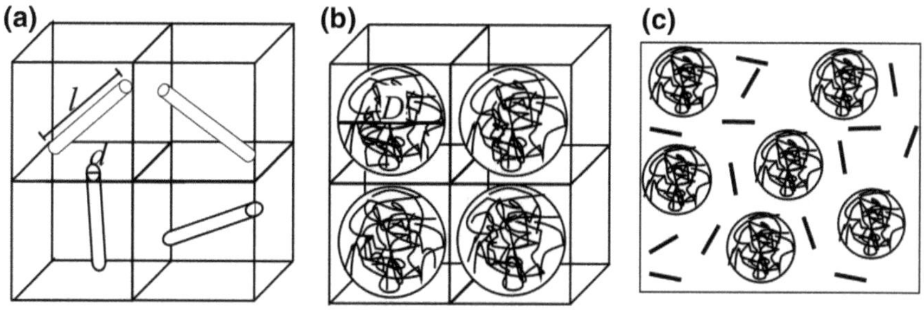

Fig. 3 Illustration of CNT-reinforced polymer nanocomposites containing, (**a**) perfectly dispersed cylindrical CNTs, (**b**) CNTs in the form of agglomerates, and (**c**) a mixture of individual CNTs and agglomerates (Li et al. 2007)

technique and the other. One of the employed techniques is a calendaring process which is also known as three-roll mills. This technique involves dispersion of CNT in a polymer matrix through shear force created by milling of the CNT/polymer mixture over three adjacent cylindrical rollers. The technique was employed by few researchers to fabricate CNT/thermoplastic nanocomposite and was shown to achieve promising CNT dispersion (Thostenson et al. 2001, 2009). However, the employment of high shear force during milling might not be suitable for heat-sensitive thermoplastics.

Another available mechanical method employed to disperse CNT in the polymer matrix is ball milling process. This technique is normally used to grind solid materials into a fine particle which not only increase the surface area of the material but can also enhance incorporation of the particle into the polymer matrix. In this technique, the materials are literally smashed into fine particles due to collision with the tiny, rigid ball in a concealed container. The ball is normally made of a tough material such as ceramic and stainless steel and has been successfully employed by several researchers to improve dispersion ability of CNT in a polymer matrix (Kim et al. 2002). In addition to that, by employing ball milling technique, Ma et al. successfully introduced functional group such as amine and amide group onto CNT sidewall which is useful for the enhancement of dispersion and interfacial bonding between CNT and polymer matrix (Ma et al. 2008, 2009). However, one of the major drawbacks of ball milling process is that it will produce CNT with a shorter length and might severely affect the aspect ratio of the nanotube.

One of the most widely used techniques in the fabrication of CNT/thermoplastic nanocomposite is ultrasonication process. It involves the application of ultrasound energy to untangle the CNT agglomerates into individually separated CNT during solution casting method. For example, in the first study on preparation of PLA reinforced with MWCNT nanocomposite films by Moon et al. (2005), they employed ultrasonication of MWCNT with PLA in chloroform solvent for 6 h prior to film casting of the nanocomposite films. Song et al. (2008) also performed ultrasonication of MWCNT at 100 W and the nominal frequency of 50 kHz at ambient temperature for 2 h to disperse and untangle MWCNT bundles. Similar sonication method employed by various researchers with different degrees of sonication strength and duration was proven to result in good dispersion of MWCNT in the PLA matrix as revealed by morphological analysis and hence resulted in better mechanical and thermal properties of MWCNT/PLA nanocomposite. However, too aggressive and/or too long sonication treatment may result in severe damage to CNT surface and may cut the brittle CNT into shorter nanotubes, thus decreasing the aspect ratio which in turn may seriously deteriorate both electrical and mechanical properties of the pristine CNT.

Shear mixing is another form of mechanical dispersion being widely employed by researchers to disperse CNT in the polymer matrix. This technique requires the intensive stirring action of CNT particles in the liquid system, typically a solution of the polymer matrix. The dispersion ability of CNT using this technique is very much attributed to the propeller's design and mixing speed control. Sandler et al. (1999) dispersed CNT in the polymer matrix by using shear mixing at

2000 rpm for 1 h. This intense mixing has resulted in a fine dispersion of CNT in the nanocomposite. Subsequent works on CNT/PLA nanocomposite also employed solution blending which involves dissolving PLA in a solvent such as chloroform and tetrahydrofuran (THF) followed by addition of CNTs. The mixture is then undergone several hours mixing aided with continuous stirring (Yang and Pan 2008; Yoon et al. 2009). The fabricated MWCNT/PLA nanocomposite using this method is normally in the form of nanocomposite film and was proven to be able to significantly disperse CNT in the PLA matrix. Studies reported by Wu et al. (2008, 2009) proved a better dispersion of CNT in PLA matrix by applying higher shear force through intensive shear mixing by using rheometer to compound PLA with functionalized MWCNT.

Recently, melt blending technique has become the most popular method being employed to improve dispersion of CNTs in the polymer matrix. In polymer processing technology, the extruder is the most ideal technique to perform melt blending process. Extrusion process involves multiple controllable steps comprising of melting, mixing, de-volatilization, and pumping actions. This technique is also suitable to perform reactive melt blending to enhance the compatibility of fillers into the polymer matrix. The efficiency of blending in an extruder is very much dependent on the processing condition, and there are various researches being carried in the quest to find the optimum processing condition to manufacture polymer composites (Shearer and Tzoganakis 1999, 2001; Modesti et al. 2009). The melt blending for PLA-MWCNT was first reported by Wu and Liao (2007) on the incorporation of chemically modified MWCNT into acrylic-grafted PLA. It was then followed by Kuan et al. (2008a) and Kuan et al. (2008b) reported investigations on MWCNT/PLA nanocomposites which have successfully compounded functionalized MWCNT into PLA system by using corotating-type twin screw extruder. Villmow et al. (2008) went on to further scrutinize the effect of twin screw extruder processing condition for producing better MWCNT dispersion in MWCNT/PLA nanocomposite. Through various works on utilization of extrusion in compounding PLA with CNT, better understanding on the optimization of processing condition was obtained, thus making melt blending technique particularly by using corotating twin screw extruder as promising method to produce CNT/PLA with better dispersion (Ramontja et al. 2009; Bourbigot et al. 2011; Zou et al. 2009; Ko et al. 2009; Shi et al. 2011; Mat Desa et al. 2014; Desa et al. 2016).

2.3 Carbon Nanotubes/Thermosetting Nanocomposites

A small amount of SWCNT and MWCNT have been utilized for reinforcing thermosetting polymers, such as epoxy, polyester, polyurethane, and phenol formaldehyde resins (Kota et al. 2007; Mina et al. 2014). Thermoset polymers are formed through condensation polymerization. Polymer molecules are joined together by the cross-linking process and raise the molecular weight. They are cured with a different process, for instance, chemical cross-linking, heat and pressure curing, and ionizing radiations such as electron beam, gamma, or X-ray curing (Alam et al. 2011; Shubhra and Alam 2011; Mina et al. 2013). It can be different according to the type of polymer. In all cases, the chemical structure of the polymer is altered through this process.

Curing is an important process because it can improve the mechanical, thermal and chemical properties of the cured product (Mina et al. 2013). The cross-linking of unsaturated polyester (UP) resins is characterized by a complex mechanism involving copolymerization of the polyester and styrene molecules induced by the decomposition of the initiator (Rouison et al. 2004). The thermoset polymers become hard and brittle due to the curing process. As a result, thermoset plastics show increased resistance to heat but are typically hard and brittle which limit their potential for load-bearing applications (Raquez et al. 2010). These shortfalls can be overcome by using a macro-, micro-, and nanofiller to reinforce thermoset plastics and fabricate composite materials.

Usually, CNT-reinforced nanocomposite properties are dependent on the degree of dispersion, interfacial adhesion with the matrix in the nanocomposite system (Singh et al. 2013). Among CNT, the van der Waals forces and π–π electrostatic interactions increase the agglomeration and twist them as rope (Gryshchuk et al. 2006). The entanglement and agglomeration of CNT are not only caused by the intermolecular forces but also the aspect ratio and flexibility of CNT (Breton et al. 2004; Li et al. 2014). Thus, homogeneous dispersion of CNT is challenging in the polymer matrix. Besides that, CNT delays the curing process of thermoset matrix through the free radical scavenging behavior of nanotubes. The free radical initiator is absorbed on the surface of CNT through π–π interaction in a state of initiation of the polymer chain, as a result, delaying the curing process (Monti et al. 2011). Different approaches are applied to break the nanotube bundles and disentangle the nanotubes.

The disentanglement processes are carried out by two methods: (i) physical method and (ii) chemical method (Chen et al. 2001; Britz and Khlobystov 2006; Hilding et al. 2003; Sahoo et al. 2010). The purpose of these techniques is to reduce nanotube agglomeration and to develop their wetting or adhesion characteristics with matrix materials (Stephen and Thomas 2010; Gkikas et al. 2012). Moreover, this method involves shear mixing of CNT in the matrix at room temperature with different nonhydrogen-bonded Lewis base solvents which reduce van der Waals force in nanotubes (Ausman et al. 2000; Lau et al. 2005; Liu and Choi 2012).

Beg et al. (2015) fabricated MWCNT-reinforced thermoset UP nanocomposites. The microstructure and thermomechanical properties of nanocomposites enhanced due to the application of pre-dispersion as a physical approach. In that case, the premixing of MWCNT was carried out in the nonhydrogen-bonded THF Lewis base solvent. On the other hand, the same properties of nanocomposites where CNT was dispersed directly in UP matrix had not improved the properties adequately due to poor dispersion of CNT. The shear thinning behavior of nanosuspension revealed well interaction in pre-dispersed MWCNT and UP matrix compared to directly dispersed MWCNT in UP matrix. The pre-dispersion technique improves the dispersion quality of CNT in thermoset polymer matrix. The MWCNT agglomeration tendency in the composite system is reduced, and most of MWCNTs are rooted in the matrix as compared to straight incorporation in the polymer matrix. Pre-dispersed MWCNT improves the nucleation effect of UP matrix in the nanocomposite. Therefore, the mechanical properties as well as the load-sharing capacity of pre-dispersed MWCNT are improved and most of the nanotubes are broken down in the state of pullout from the matrix.

Nanotube content in thermoset polymer matrix greatly influences the morphology and thermomechanical properties of nanocomposite system. Alam et al. (2017) studied the microstructure and fractography of CNT-reinforced thermoset nanocomposites which fabricated with different concentrations of pre-dispersed MWCNT. The shear thinning behaviors of nanosuspensions were evaluated at the elevated shear rate. The viscosity of nanosuspension is like complex fluid, and an optimum amount of MWCNT-blended nanosuspension demonstrates the highest viscosity at low shear rate due to well interaction and sound dispersion in the matrix. The greater concentration of MWCNT reduces the viscosity of nanosuspension at low shear rate due to agglomeration of nanotubes, lack of wettability by the matrix, and poor interaction between nanotubes and matrix.

Besides that, the optimum content of nanotubes reduces the crack propagation as well as shields the crack through the crack bridge of nanotubes in the nanocomposite system. The optimal nanotube content in nanocomposites can increase the mechanical properties including tensile strength and tensile modulus and reduce the breaking strain. It is interesting that excess amount of nanotubes enhances the flexibility of nanocomposites due to slipping the agglomerated or twisted nanotubes. Therefore, excess nanotubes are pulled out at the breaking stress, as a consequence excess nanotube-incorporated thermoset nanocomposites exhibit poor load-bearing capacity.

Usually, the partial crystalline thermoset polyester resin contains a lot of molecular disorder. Nanotubes reduce the amorphousness of thermoset polymer matrix. It extends the molecular arrangement in the nanocomposite system. The nanotube content greatly influences microstructure of thermoset nanocomposites. The nucleation efficiency of optimal MWCNT is greater as compared to the excess amount of MWCNT in nanocomposites. Therefore, optimal nanotube improves the crystal size and lattice parameter of nanocomposites. Thermal transitions as well as the thermal stability of thermoset polymer matrix influence by the carbon nanotubes. The optimal amount of MWCNT increase the glass transition and melt transition temperature of UP nanocomposites.

The chemical methods are carried out by the action of surfactants, surface modification, and polymer wrapping technology (Kim et al. 2012a, b). This modification can be carried out with synthetic and natural polymers which contain functional end groups (Pan et al. 2009; Soradech et al. 2013). For example, hyperbranched polymers are highly branched macromolecules used in a wide range of application. They have a huge number of reactive end groups which are effective for rapid cross-linking in the thermoset network (Wang et al. 2008). Considering these advantages, the hyperbranched polymer has been used for nondestructive functionalization of MWCNT to remove entanglement as well as enhance the dispersion of nanotubes. The functionalization was carried out via solvent evaporation technique to functionalize MWCNT. The functional MWCNT was incorporated in UP thermoset matrix to fabricate functional nanotube-reinforced nanocomposites. It reduces the cross-linking time of UP matrix as compared to pristine MWCNT in nanocomposite system. Alam et al. (2016) stated that hyperbranched polymer-coated MWCNT not only well dispersed but also performed as anti-scavenger during the curing process of UP matrix.

2.4 Carbon Nanotubes/Natural Rubber Nanocomposite

NR is one of the most important industrially significant elastomeric polymers which harvests from tree *Hevea brasiliensis*. It has an extensive renege of application ranging from household to industrial production. Rubber technologies have developed rapidly after the discovery of vulcanization process by Charles Goodyear. NR is a combination of many different components mainly *cis*-1,4-polyisoprene but also contains nonrubber components such as lipids, proteins, and carbohydrates (Hwee and Tanaka 1993). The long-chain *cis*-isoprene chain of NR contents *trans*-2-isoprene units and two terminal groups. These terminal groups are generally mono- and/or di-phosphate and dimethylallyl group (Tarachiwin et al. 2005). However, unvulcanized and unmodified NR has low strength and modulus and is not suitable for engineering application such as tires, hose, and gaskets. Therefore, it is required to reinforce the rubber which is necessary to enhance the properties of rubber (Bokobza 2007; Sui et al. 2008).

For improving the modulus and strength of NR, generally conventional microfillers such as carbon black (CB), calcium carbonate, zinc oxide, magnesium oxide, talc, mica, and silicates are used (Ismail and Chia 1998; Ismail and Freakley 1996). The reinforcing effect of this filler material depends on the size of the particles, aspect ratios, dispersion, orientation in the matrix, and the adhesion with the polymer chains (Sui et al. 2008). However, to improve the properties of NR, high amount of conventional microfiller is required which leads to agglomeration and part failure (Job et al. 2003). To overcome this problem, many researchers worked on nanomaterial-filled NR nanocomposite which exhibits exceptional enhancement in mechanical, thermal, and electromagnetic properties with very low content of nanofiller (Sadasivuni et al. 2014). The physical, electrical, and thermal properties of NR may be improved by a low amount of nanofillers than microscale fillers due to the high strength, surface area with preserving the elasticity of the rubber (Berahman et al. 2016). For reinforcing of NR, several nanomaterials were used such as nanosilica, nano-CB, HNT, CNT, graphene, and nanoclay. Nanofiller-reinforced NR was firstly reported by using nanoclay as filler, where an exponential increment in modulus was observed by adding a small amount of nanoclay (Joly et al. 2002; Varghese and Karger-Kocsis 2003). The results suggest much better reinforcement effect may be possible by using CNT in the NR matrix due to their superior properties.

The CNT/NR nanocomposites have a great impact on the field of flexible electronics because of their mechanical strength and conductivity. Additionally, CNT/NR nanocomposite application benefits a lot in industrial materials such as rubber hoses, tire components, sensing devices, electrical shielding, and electrical heating devices (Baskaran et al. 2004; Lee et al. 2011). However, some of the main challenges in the processing of CNT are to overcome the strong van der Waals attraction energy of ca. 500 eV/mm of tube–tube contact, which makes them aggregate into bundles or ropes in the polymer nanocomposites (Girifalco et al. 2000). Moreover, the second most important which governs the final properties of the nanocomposite is good interfacial adhesion between CNT and the NR (Ponnamma et al. 2013).

Le et al. (2013) reported that the use of ionic liquid, as a noncovalent surfactant for surface modification of CNT, was found very effective as a dispersing agent for CNT in rubber matrix. On the contrary, covalent modification is permanent, stronger, and more efficient than the noncovalent approach. It gives strong interfacial adhesion between polymer and filler but reduces the conductivity of CNT itself by creating surface defects.

Apart from surface modification of CNT, fabrication methods also influence the dispersion of filler particles and play a very important role in the final properties of the nanocomposite. There are mainly three methods for fabricating CNT/NR nanocomposites, namely mechanical method, solution method, and latex method. The mechanical method involves batch mixture, extruder, or two-roll mill for mixing of NR and CNT. This method is most acceptable for industrial use because it is easy and environment-friendly. However, in this method, dispersion of CNT is the main concern. Many researchers developed different roots to increase the dispersion level of CNT in NR by using mechanical mixing method because of the high viscosity of NR. For making CNT/NR nanocomposites by the mechanical method, most of the researchers used functionalized CNT. For example, Sui et al. (2007) prepared MWCNT/NR nanocomposite by using a batch mixture. In this work, they used a three-component CNT modification method. In that, the acid-treated CNT was further treated with hydrated silica, resorcinol, and hexamethylenetetramine with a weight ratio of 15:10:6 bonding systems. The prepared CNT/NR nanocomposite by modified CNT was compared with CB and found that temperature of CNT/NR mixing system was lower compared to CB/NR system. This result indicates that CNT/NR system required less energy consumption. However, there was no effect on T_g of CNT/NR nanocomposite with increasing CNT content. They also observed improved interfacial bonding between NR and CNT by the surface treatment process.

The second most important method for preparing NR/CNT nanocomposite is solution casting method. This method involved a solvent as a medium for dispersing the nanomaterial in NR. In the first step of solution casting, NR is dissolved in a proper solvent such as toluene or cyclohexene and in two separate vessels; CNT will disperse in the same solvent by sonication. In the second step, CNT dispersion will be added slowly in the NR solution with continuous stirring. After drying the solvent, a high dispersed CNT/NR nanocomposite is formed. Some researchers for an instant, Ismail et al. (2010), prepared CNT/NR nanocomposite by solution casting method by using unmodified MWCNT. In their experiments, they used toluene as a solvent for unvulcanized NR. They also prepared CNT/NR nanocomposite with mechanical mixing method and compared the results. In their study, they found that tensile properties of solution mixing method are better than composite that is prepared by mechanical mixing method. This is due to high dispersion of CNT in NR in case of solution mixing method. The advantage of this method is it is possible to get high dispersion of CNT. On the other hand, the major disadvantage of this method is the remaining trace of solvent in the final nanocomposite and it is not environment- and industrial friendly.

Latex technology is the third most important preparation method for the preparation of CNT/NR nanocomposite. In this method, nanomaterial like CNT was dispersed in the rubber latex and after drying a highly homogeneous nanocomposite was formed (Pang et al. 2014). The main challenge in this method is hydrophobicity of CNT as NR latex is an aqueous emulation of rubber; it is difficult to disperse unmodified CNT in latex without any surfactant. Using of surfactant is one of the best methods to dispersed CNT in an aqueous medium. Ponnamma et al. (2014) prepared CNT/NR nanocomposite by latex technology and used four different kinds of surfactant for dispersing the CNT. The surfactant systems involved anionic sodium dodecyl sulfate (SDS), cationic cetyltrimethylammonium bromide (CTAB), nonionic Tween 20, and a mixture of anionic and cationic (SDS+CTAB). The efficacies of surfactant were studied by UV visible spectroscopy and sedimentation index analysis. From these experiments, they concluded that SDS/CNT system is most efficient for NR as the contact angle between SDS/CNT and NR latex is lowermost compared to another surfactant system. Moreover, they found, respectively, high storage modules of CNT/NR nanocomposite which is prepared by SDS/

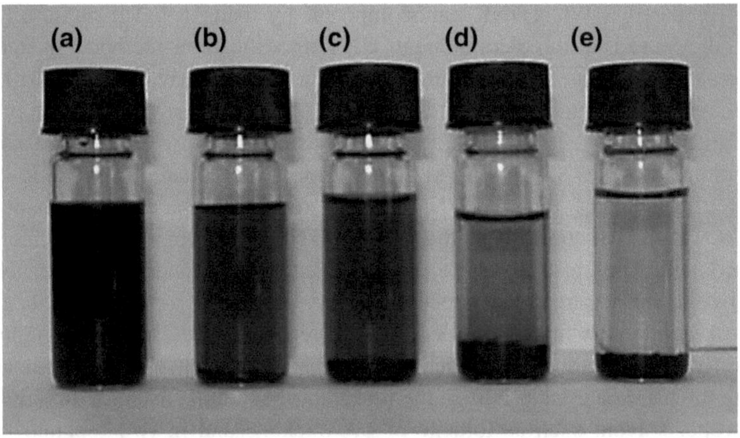

Fig. 4 Vials containing aqueous dispersions of SWCNT in (**a**) NaDDBS, (**b**) SDS, (**c**) PVA, (**d**) Criton X-100, and (**e**) sodium benzoate (photograph was taken after six weeks of the sample preparation) (Anand et al. 2009)

CNT that indicate good dispersion and interfacial adhesion between filler and polymer. In another report, Anand et al. (2009) used five different types of surfactant for dispersing of SWCNT in the NR latex as presented in Fig. 4. These five surfactants are, namely sodium dodecylbenzene sulfonate (NaDDBS), SDS, polyvinyl alcohol (PVA), isooctylphenoxy polyethoxyethanol (Criton X-100), and sodium benzoate. In their study, they found NaDDBS as an ideal surfactant for the aqueous dispersion of SWCNT; in this case, the dispersion was stable even after six weeks. They reported significantly improved mechanical properties of SWCNT/NR nanocomposite prepared by using NaDDBS as a surfactant. They also found that by adding even a small amount of SWCNT (2.0 phr) to the NR latex, the tensile strength and modulus were increased by 56 and 63% respectively, compared to pure NR. Moreover, only in 1.5 phr SWCNT content is enough for achieving electrical percolation threshold. Conversely, the thermal stability of NR remains unaffected by low SWCNT content.

3 Halloysite Nanotubes/Biopolymer Nanocomposites

3.1 Halloysite Nanotubes

HNT is natural silicates which have 1D nanotube shape. Chemical composition and structure of HNT are similar to kaolinite (a form of layered silicate mineral). In fact, nanotube structure of HNT was evolved by rolling up the layers of kaolinite under natural conditions. The average outer diameter of HNT varies from 10 to 50 nm, the length about 0.5–2 μm, and the inside diameter about 5–20 nm. The surface area of HNT is moderate compared to another nanomaterial with a similar shape (Brunauer–Emmett–Teller (BET) surface area value 22.10–81.59 m^2/g), but it has 10.7–39% lumen space (Pasbakhsh et al. 2013). The lumen space between HNT structures leads to a low density of the material which varies from 2.14 to 2.59 g/cm^3. The density of HNT is much lower than conventional filler such as talc and calcite ($CaCO_3$) (Alhuthali and Low 2013b). As a result of unique chemical–physical properties and rich reserve in nature, in a very short time, HNT has become an important nanofiller for the polymers. It also used in pesticides, anti-bacterial coatings, nanoscale reactors, catalyst carriers, cosmetics, etc. In the field of polymer nanocomposite, HNT is extensively used to reinforce various polymers such as polypropylene (PP) (Prashantha et al. 2011), epoxy resin (Ye et al. 2007), PLA (Touny et al. 2010), nylon (Marney et al. 2008), polyethylene (PE) (Singh et al. 2016), NR (Rooj et al. 2010), and different synthetic rubbers (Ismail et al. 2008). These results indicate that HNT has a significant reinforcement, flame retardant, and modifying effects on polymer materials and exhibit promising application prospect (Fig. 5).

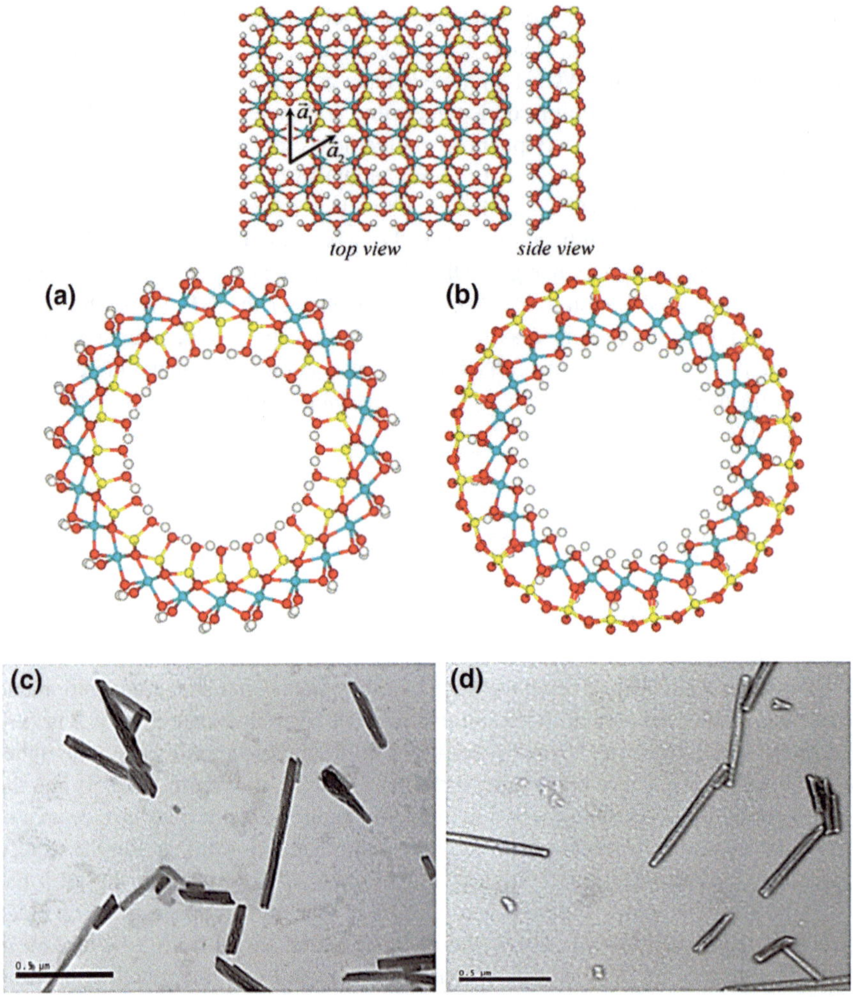

Fig. 5 Structure of halloysite monolayer and cross-sectional views of **(a)** a(12,0) imogolite nanotube and **(b)** a(12,0) halloysite nanotube. White atoms are hydrogen, red oxygen, gray aluminum, and yellow silicon (Guimaraes et al. 2010). **(c)** and **(d)** TEM micrograph of HNT (Jia et al. 2014)

3.2 Halloysite Nanotubes/Thermoplastic Nanocomposite

Many intensive reports and reviews have recorded the performance of HNT in polymer nanocomposites (Liu et al. 2014; Szpilska et al. 2015; Gaaz et al. 2017; Kausar 2018). HNT is reported to be a promising nanofiller for the thermoplastic polymer nanocomposites due to its high specific surface area, high aspect ratio, good dispersion, and excellent mechanical properties. Many studies investigated the effect of HNT as nanofiller on the mechanical properties (Qiao et al. 2017;

Sahnoune et al. 2017; Wan et al. 2017), thermal stability (Wang and Huang 2014; Qiao et al. 2017), and flame retardancy (Lecouvet et al. 2012) of thermoplastic polymers. These polymers include PP (Prashantha et al. 2011; Wang and Huang 2013), PE (Qiao et al. 2017), PLA (Therias et al. 2017; Krishnaiah et al. 2017), polyamide (Sahnoune et al. 2017), PHA, poly(vinylidene fluoride) (Wu et al. 2016a; Wang and Huang 2014), and poly(arylene ether nitrile) (Wan et al. 2017).

HNT thermoplastic polymer nanocomposite exhibited better reinforcing ability in terms of the mechanical properties. The excellent reinforcing effect of HNT on thermoplastic polymer could be attributed to the uniformly distributed rigid nanotubes in the thermoplastic polymer matrix and the specific bonding interactions between treated HNT and thermoplastic polymer. Prashantha et al. (2011) reported the ability of quaternary ammonium salt-treated halloysite nanotubes to reduce the agglomeration of HNT in PP matrix. This was attributed to the fact that functional groups brought by quaternary ammonium salt present on the surface of halloysites decrease their surface free energy and hinder nanotube/nanotube interaction, thereby breaking up the aggregates and improving the interface between HNT and the PP matrix. Pedrazzoli et al. (2015) confirmed the treated HNT that they produced can be finely dispersed within the LLDPE matrix even at high filler contents (8 wt%) and without using a polymeric compatibilizer.

Table 1 records the mechanical property enhancement of a few thermoplastic polymer nanocomposites at their optimum HNT content. It can be concluded that there are two main factors determining the performance of HNT nanocomposites;

Table 1 Mechanical properties of selected HNT thermoplastic polymer nanocomposites at their optimum HNT content

Thermoplastic polymer nanocomposite	Optimum HNT content (wt%)	Tensile strength (MPa)	Young's modulus (MPa)	Impact strength (kJ/m^2)	References
HNT/PP	6	36.0	1700	3.7	Prashantha et al. (2011)
Treated HNT/PP	6	38.0	1800	3.9	Prashantha et al. (2011)
HNT/UHMWPE	2	150.0 longitudinal 70.0 lateral	–	–	Qiao et al. (2017)
HNT/PA 11	5	44.6	1511	3.6	Sahnoune et al. (2017)
HNT/PVDF	5	48.5	1450	–	Wu et al. (2016a)
HNT/PVDF	5	50.0	1400	–	Wang and Huang (2014)
Treated HNT/LDPE	1	21.8	220	–	Pedrazzoli et al. (2015)

UHMWPE Ultra-high-molecular-weight PE; *LDPE* low-density PE; *PA 11* polyamide 11; and *PVDF* polyvinylidene fluoride dispersion of HNT in the polymer matrix and interfacial affinity between HNT and the polymer. Among the performances of HNT thermoplastic polymer nanocomposite that have been investigated are the mechanical, thermal, flammability (Bartolucci et al. 2013), crystallization (Liu et al. 2009; Qiao et al. 2017), dielectric (Wan et al. 2017), and anti-bacterial (Meira et al. 2017; Gorrasi 2015) properties. The common processes for the preparation of the HNT thermoplastic polymer nanocomposites are melt mixing and solution casting.

3.3 Halloysite Nanotubes/Thermoset Nanocomposites

HNT has been incorporated into the various types of thermosets such as epoxy (Saif et al. 2016; Vahedi et al. 2015), unsaturated polyester (UP) (Albdiry and Yousif 2014), polyester (Saharudin et al. 2016), and vinyl ester (Alhuthali and Low 2013a). Mechanical mixing method was mainly used to prepare the thermoset HNT nanocomposites. Ultrasonic mixing was an effective method to improve the dispersion of HNT in epoxy resin. Viscosity and impact properties of ultrasonically mixed nanocomposites were considerably higher than those prepared by mechanical mixing. The effect was clearly seen at high HNT content and for modified HNT where there are more tendencies for agglomeration. Moreover, the performance of HNT as reinforcing filler in epoxy resin was also affected by curing agent and HNT type used. DETA-cured epoxy deteriorated the impact resistance, and using HNT in MDA-cured HNT led to enhancement of impact resistance. Longer and more uniform HNT showed significantly higher impact strength (Vahedi et al. 2015). Saif et al. (2016) prepared epoxy/HNT nanocomposites by slow mechanical mixing followed by the ultrasonic bath for 30 min. γ-irradiated HNT was used for structural modification of natural pristine HNT under air-sealed environment at different absorbed doses. The irradiated HNT improved modulus, flexural, and tensile strength along with better thermal performance and crack resistance.

Uniform dispersion of vinyltrimethoxysilane-treated HNT (s-HNT) is observed in the s-HNT/UP versus skewed-like clusters in the HNT/UP nanocomposite. The toughening mechanisms of highly brittle fracture of neat UP have been shifted to shear yielding and zone shielding mechanisms with the presence of halloysite particles in the nanocomposite. Tensile strength and modulus of UP nanocomposites steadily increased with increasing HNT up to 5 wt%. Either 3 wt% HNT or 3 wt% s-HNT demonstrated the highest increase in mechanical properties by 7 and 10% in tensile strength and 7 and 12% in Young's modulus, respectively (Albdiry and Yousif 2014). Enhancements in toughness are attributed to crack bridging, deflection, and localized plastic deformation, while strength improvements can be attributed to the large aspect ratio of HNT, favorable interfacial adhesion and dispersion, and inter-tubular interaction.

Addition of 0.5 wt% modified HNTs significantly improved the tensile strength, Young's modulus, and elongation at the break of the waterborne polyurethane, by approximately 200, 200, and 30%, respectively. An excess amount of HNT added into the system could weaken the reinforcing effect and stability of the composite emulsion (Wu et al. 2016b). In the presence of HNT in polyester matrix, it was found that seawater uptake increased and reduced mechanical properties compared to monolithic polyester. Young's modulus decreased from 0.6 to 0.4 GPa (33% decrease), flexural modulus decreased from 0.6 to 0.34 GPa (43% decrease), and impact toughness dropped from 0.71 to 0.48 kJ/m^2 (32% decrease). However, the fracture toughness increased with the addition of HNT due to the plasticization effect. The reduction in mechanical properties was attributed to the degradation of the nanoclay–matrix interface influenced by seawater absorption and agglomeration of halloysite nanoclay (Saharudin et al. 2016).

3.4 Halloysite Nanotubes/Natural Rubber Nanocomposites

As discussed above in the HNT/thermoset and thermoplastic sections, it is known that HNT has a great potential to reinforce polymers. Similarly, HNT/NR nanocomposites received special attention from researchers. This section of the chapter will discuss HNT/NR nanocomposites as NR comes from natural resources and is used in various rubber products. Thereby, it is necessary to reinforce NR to improve the properties of vulcanizates. There are mainly three methods (as similar to CNT/NR nanocomposites) to prepare HNT/NR nanocomposites: (i) mechanical mixing method, (ii) latex method, and (iii) solution method.

The most common preparation method of HNT/NR nanocomposites is mechanical mixing method. In this method, all ingredients with NR and HNT were mixed together in the mechanical mixtures such as a two-roll mill, melt mixer, or extruder. In this method, the dispersion of HNT is a main concern because of the high viscosity of NR. To overcome this problem of mechanical mixing method, many researchers have worked in latex method. In the latex method, HNT was mixed with NR in the latex stage itself. In this stage, the viscosity of the medium is relatively low, and due to hydrophilic surface of HNT, it dispersed in the latex very well. After drying the latex, it forms a nanocomposite with highly dispersed HNT. Baochun et al. (2009) prepared HNT/NR by latex method and found a very uniform dispersion of HNT in NR matrix which leads comparatively high mechanical properties. In solution method, NR is dissolved in a suitable solvent such as toluene or cyclohexane and HNT is dispersed in it. After removing the solvent, the HNT/NR nanocomposite was formed. In this method, it is possible to control the viscosity of the system very precisely by adding or removing the solvent which is important to increase the HNT content in the nanocomposite (Fig. 6).

Apart from processing, surface properties of the nanocomposite also play an important role in the properties of final nanocomposite. Jia et al. (2014) used unmodified HNT to prepare unmodified HNT/NR nanocomposite by mechanical mixing method and investigated the morphology and mechanical properties. They found that by using unmodified HNT it is possible to achieve dispersion up to 20 phr content of HNT. Increasing HNT content beyond that limit, agglomeration and

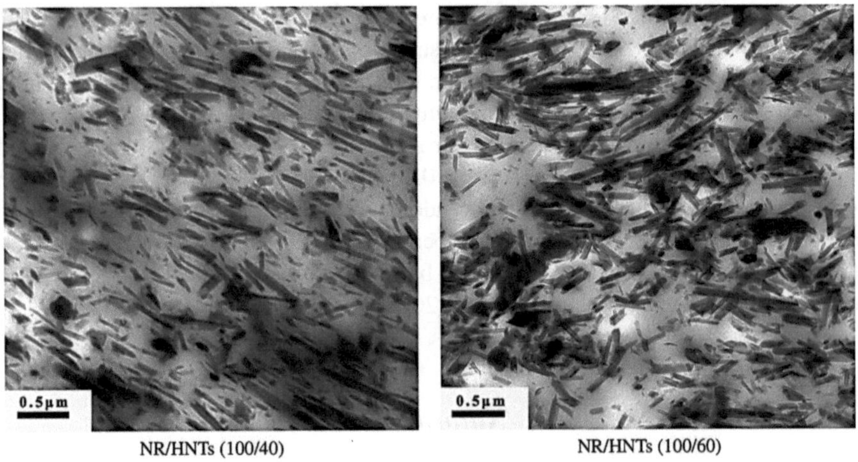

Fig. 6 TEM photographs of HNT/NR composites prepared by mechanical mixing method (Jia et al. 2014)

un-uniformity in nanocomposite increase. They also prepared similar nanocomposite by latex mixing method using unmodified HNT. In this case, also, they found agglomeration beyond 20 phr content. However, the dispersion of HNT in case of latex mixing method is better than mechanical mixing. Although unmodified HNT has some reinforcing and modifying effects on rubbers, the compatibility of unmodified HNT with the majority of rubbers is unfavorable. Generally, unmodified HNT is not easily dispersed in rubber matrices. Furthermore, the interfacial combination between unmodified HNT and rubber is generally weak.

In order to improve the interfacial combination between HNT and rubber, some surface modification of HNT is required since HNT has a polar, hydrophilic surface because of Si–O or Si–OH group which is not compatible with most of the organic nonpolar rubber. Consequently, further property improvement of the HNT/NR is restricted, and it is necessary to modify HNT and the nanocomposites with the modifiers. The simplest and most effectual modification method of HNT/NR nanocomposites is in situ mechanical mixing method with different modifiers. The interfacial interactions between HNT and rubber matrixes mainly involve covalent bonds, hydrogen bonds, ion bonds, charge transfer, and multiple interactions.

4 Conclusions

This chapter has discussed the fabrication processes, properties, and potential applications of CNT and HNT biopolymer nanocomposites. From the previous research works, it can be concluded that CNT and HNT are ideal fillers for reinforcing biopolymers as they can increase the properties of the biopolymer at very low content. However, the dispersion of such nanotubes into biopolymer materials is a problem due to the high aspect ratio of nanotubes and high viscosity of the polymer. Therefore, the interaction between the nanotubes and the biopolymer matrix still needs to be improved to fully utilize the maximum potential of these fillers.

CNTs are most studied nanotube by the scientific community for its potential application as fillers for biopolymer nanocomposites. Many researches have been reported to improve processing methods of CNT/polymer nanocomposites for better dispersion and also on surface modification of CNT for interfacial adhesion between polymer and CNT. Even though the full potential of CNT cannot be used for several applications of nanocomposites, several researchers have shown the outstanding results of CNT/biopolymer nanocomposites in numerous applications such as biosensors, EMI-shielded biopackaging, and high-performance nanocomposites.

On the other hand, HNT as an environmental-friendly natural inorganic tubular nanomaterial has a reinforcing effect on biopolymer even in unmodified condition. The modification with various modifiers such as ionic liquid and surfactants could effectively improve the dispersion of HNT in biopolymer matrixes, strengthen the interfacial bond between HNT and biopolymer matrixes, and thus remarkably improve the mechanical properties of the HNT/biopolymer composites. It is worth mentioning that HNT/NR nanocomposites have unique dynamic mechanical properties, suggesting that the nanocomposites are appropriate for use in high-performance tires or green tires.

References

Alam A, Shubhra QT, Al-Imran G et al (2011) Preparation and characterization of natural silk fiber-reinforced polypropylene and synthetic E-glass fiber-reinforced polypropylene composites: a comparative study. J Compos Mater 45(22):2301–2308

Alam AM, Beg M, Yunus R et al (2016) Evolution of functionalized multi-walled carbon nanotubes by dendritic polymer coating and their anti-scavenging behavior during curing process. Mater Lett 167:58–60

Alam A, Beg M, Yunus R (2017) Microstructure and fractography of multiwalled carbon nanotube reinforced unsaturated polyester nanocomposites. Polym Compos 38(S1):E462–E471

Albdiry M, Yousif B (2014) Role of silanized halloysite nanotubes on structural, mechanical properties and fracture toughness of thermoset nanocomposites. Mater Des 57:279–288

Alhuthali A, Low IM (2013a) Mechanical and fracture properties of halloysite nanotube reinforced vinyl-ester nanocomposites. J Appl Polym Sci 130(3):1716–1725

Alhuthali A, Low IM (2013b) Water absorption, mechanical, and thermal properties of halloysite nanotube reinforced vinyl-ester nanocomposites. J Mater Sci 48(12):4260–4273

Anand KA, Jose TS, Alex R et al (2009) Natural rubber-carbon nanotube composites through latex compounding. Int J Polym Mater 59(1):33–44

Ausman KD, Piner R, Lourie O et al (2000) Organic solvent dispersions of single-walled carbon nanotubes: toward solutions of pristine nanotubes. J Phys Chem B 104(38):8911–8915

Banerjee S, Mohapatra SK, Das PP et al (2008) Synthesis of coupled semiconductor by filling 1D TiO_2 nanotubes with CdS. Chem Mater 20(21):6784–6791

Baochun G, Yanda L, Quanliang Z et al (2009) Structure and properties of natural rubber/halloysite nanotubes composites prepared by latex-suspension coagulation. Chin Synth Rubber Ind 32(2):131–134

Bartolucci SF, Supan KE, Wiggins JS et al (2013) Thermal stability of polypropylene-clay nanocomposites subjected to laser pulse heating. Polym Degrad Stab 98(12):2497–2502

Baskaran D, Mays JW, Bratcher MS (2004) Polymer-grafted multiwalled carbon nanotubes through surface-initiated polymerization. Angew Chem Int Ed 43(16):2138–2142

Beg M, Alam AM, Yunus R et al (2015) Improvement of interaction between pre-dispersed multi-walled carbon nanotubes and unsaturated polyester resin. J Nanopart Res 17(1):53

Berahman R, Raiati M, Mazidi MM et al (2016) Preparation and characterization of vulcanized silicone rubber/halloysite nanotube nanocomposites: effect of matrix hardness and HNT content. Mater Des 104:333–345

Bethune D, Kiang CH, De Vries M et al (1993) Cobalt-catalysed growth of carbon nanotubes with single-atomic-layer walls. Nature 363(6430):605

Bokobza L (2007) Multiwall carbon nanotube elastomeric composites: a review. Polymer 48 (17):4907–4920

Bourbigot S, Fontaine G, Gallos A et al (2011) Reactive extrusion of PLA and of PLA/carbon nanotubes nanocomposite: processing, characterization and flame retardancy. Polym Adv Technol 22(1):30–37

Breton Y, Desarmot G, Salvetat J et al (2004) Mechanical properties of multiwall carbon nanotubes/epoxy composites: influence of network morphology. Carbon 42(5–6):1027–1030

Britz DA, Khlobystov AN (2006) Noncovalent interactions of molecules with single walled carbon nanotubes. Chem Soc Rev 35(7):637–659

Cadek M, Murphy R, McCarthy B et al (2002) Optimisation of the arc-discharge production of multi-walled carbon nanotubes. Carbon 40(6):923–928

Chen RJ, Zhang Y, Wang D et al (2001) Noncovalent sidewall functionalization of single-walled carbon nanotubes for protein immobilization. J Am Chem Soc 123(16):3838–3839

Chen X, Zhang Z, Qiu Z et al (2007) Hydrothermal fabrication and characterization of polycrystalline linneite (Co_3S_4) nanotubes based on the Kirkendall effect. J Colloid Interface Sci 308(1):271–275

Coleman JN, Khan U, Blau WJ et al (2006) Small but strong: a review of the mechanical properties of carbon nanotube-polymer composites. Carbon 44(9):1624–1652

Desa M, Zaidi MS, Hassan A et al (2016) Influence of rubber content on mechanical, thermal, and morphological behavior of natural rubber toughened poly(lactic acid)-multiwalled carbon nanotube nanocomposites. J Appl Polym Sci. https://doi.org/10.1002/app.44344

Dresselhaus MS, Dresselhaus G, Saito R (1995) Physics of carbon nanotubes. Carbon 33(7):883–891

Fang YP, Xu AW, You LP et al (2003) Hydrothermal synthesis of rare earth (Tb, Y) hydroxide and oxide nanotubes. Adv Funct Mater 13(12):955–960

Frogley MD, Ravich D, Wagner HD (2003) Mechanical properties of carbon nanoparticle-reinforced elastomers. Compos Sci Technol 63(11):1647–1654

Gaaz TS, Sulong AB, Kadhum AAH et al (2017) The impact of halloysite on the thermo-mechanical properties of polymer composites. Molecules. https://doi.org/10.3390/molecules22050838

George N, Bipinbal P, Bhadran B et al (2017) Segregated network formation of multiwalled carbon nanotubes in natural rubber through surfactant assisted latex compounding: a novel technique for multifunctional properties. Polymer 112:264–277

Girifalco L, Hodak M, Lee RS (2000) Carbon nanotubes, buckyballs, ropes, and a universal graphitic potential. Phys Rev B 62(19):13104

Gkikas G, Barkoula NM, Paipetis A (2012) Effect of dispersion conditions on the thermo-mechanical and toughness properties of multi walled carbon nanotubes-reinforced epoxy. Compos Part B Eng 43(6):2697–2705

Gorrasi G (2015) Dispersion of halloysite loaded with natural antimicrobials into pectins: characterization and controlled release analysis. Carbohydr Polym 127:47–53

Gryshchuk O, Karger-Kocsis J, Thomann R et al (2006) Multiwall carbon nanotube modified vinylester and vinylester-based hybrid resins. Compos Part A Appl Sci Manuf 37(9):1252–1259

Guimaraes L, Enyashin AN, Seifert G et al (2010) Structural, electronic, and mechanical properties of single-walled halloysite nanotube models. J Phys Chem C 114(26):11358–11363
Hilding J, Grulke EA, Zhang ZG et al (2003) Dispersion of carbon nanotubes in liquids. J Dispersion Sci Technol 24(1):1–41
Hill DE, Lin Y, Rao AM et al (2002) Functionalization of carbon nanotubes with polystyrene. Macromolecules 35(25):9466–9471
Hwee E, Tanaka Y (1993) Structure of natural rubber. Trends Polym Sci 3:493–513
Iijima S (1991) Helical microtubules of graphitic carbon. Nature 354:56–58
Iijima S (1993) Growth of carbon nanotubes. Mater Sci Eng B 19(1–2):172–180
Iijima S (2002) Carbon nanotubes: past, present, and future. Phys Rev B Condens Matter 323(1–4):1–5
Ismail H, Chia H (1998) The effects of multifunctional additive and vulcanization systems on silica filled epoxidized natural rubber compounds. Eur Polym J 34(12):1857–1863
Ismail H, Freakley P (1996) Determination of the modes of action of a cationic surfactant for property development in silica-filled natural rubber compounds. Eur Polym J 32(4):411–416
Ismail H, Pasbakhsh P, Fauzi MA et al (2008) Morphological, thermal and tensile properties of halloysite nanotubes filled ethylene propylene diene monomer (EPDM) nanocomposites. Polym Test 27(7):841–850
Ismail H, Ramly F, Othman N (2010) Multiwall carbon nanotube-filled natural rubber: the effects of filler loading and mixing method. Polym Plast Technol Eng 49(3):260–266
Jia Z, Guo B, Jia D (2014) Advances in rubber/halloysite nanotubes nanocomposites. J Nanosci Nanotechnol 14(2):1758–1771
Job A, Oliveira F, Alves N et al (2003) Conductive composites of natural rubber and carbon black for pressure sensors. Synth Met 135:99–100
Joly S, Garnaud G, Ollitrault R et al (2002) Organically modified layered silicates as reinforcing fillers for natural rubber. Chem Mater 14(10):4202–4208
Kanagaraj S, Varanda FR, Zhil'tsova TV et al (2007) Mechanical properties of high density polyethylene/carbon nanotube composites. Compos Sci Technol 67(15–16):3071–3077
Kausar A (2018) Review on polymer/halloysite nanotube nanocomposite. Polym Plast Technol Eng 57(6):548–564
Kim Y, Hayashi T, Fukai Y et al (2002) Effect of ball milling on morphology of cup-stacked carbon nanotubes. Chem Phys Lett 355(3–4):279–284
Kim MG, Moon JB, Kim CG (2012a) Effect of CNT functionalization on crack resistance of a carbon/epoxy composite at a cryogenic temperature. Compos Part A Appl Sci Manuf 43(9):1620–1627
Kim SW, Kim T, Kim YS et al (2012b) Surface modifications for the effective dispersion of carbon nanotubes in solvents and polymers. Carbon 50(1):3–33
Ko S, Hong M, Park B et al (2009) Morphological and rheological characterization of multi-walled carbon nanotube/PLA/PBAT blend nanocomposites. Polym Bull 63(1):125–134
Kokai F, Koshio A, Shiraishi M et al (2005) Modification of carbon nanotubes by laser ablation. Diamond Relat Mater 14(3–7):724–728
Kota AK, Cipriano BH, Duesterberg MK et al (2007) Electrical and rheological percolation in polystyrene/MWCNT nanocomposites. Macromolecules 40(20):7400–7406
Krishnaiah P, Ratnam CT, Manickam S (2017) Development of silane grafted halloysite nanotube reinforced polylactide nanocomposites for the enhancement of mechanical, thermal and dynamic-mechanical properties. Appl Clay Sci 135:583–595
Kroto H, Heath J, O'Brien S et al (1985) C_{60}: buckminsterfullerene. Nature 318:162–163
Kuan CF, Chen CH, Kuan HC et al (2008a) Multi-walled carbon nanotube reinforced poly(l-lactic acid) nanocomposites enhanced by water-crosslinking reaction. J Phys Chem Solids 69(5–6):1399–1402
Kuan CF, Kuan HC, Ma CCM et al (2008b) Mechanical and electrical properties of multi-wall carbon nanotube/poly(lactic acid) composites. J Phys Chem Solids 69(5–6):1395–1398

Kulkarni M, Charhate N, Bhavsar K et al (2013) Development of polyaniline-multiwalled carbon nanotube (PANI-MWCNT) nanocomposite for optical pH sensor. Mater Res Innovations 17 (4):238–243
Lau KT (2003) Interfacial bonding characteristics of nanotube/polymer composites. Chem Phys Lett 370(3–4):399–405
Lau KT, Chipara M, Ling HY et al (2004) On the effective elastic moduli of carbon nanotubes for nanocomposite structures. Compos Part B Eng 35(2):95–101
Lau KT, Lu M, Lam CK et al (2005) Thermal and mechanical properties of single-walled carbon nanotube bundle-reinforced epoxy nanocomposites: the role of solvent for nanotube dispersion. Compos Sci Technol 65(5):719–725
Le H, Oßwald K, Wießner S et al (2013) Location of dispersing agent in rubber nanocomposites during mixing process. Polymer 54(26):7009–7021
Lecouvet B, Bourbigot S, Sclavons M et al (2012) Kinetics of the thermal and thermo-oxidative degradation of polypropylene/halloysite nanocomposites. Polym Degrad Stab 97(9):1745–1754
Lee RS, Chen WH, Lin JH (2011) Polymer-grafted multi-walled carbon nanotubes through surface-initiated ring-opening polymerization and click reaction. Polymer 52(10):2180–2188
Li J, Ma PC, Chow WS et al (2007) Correlations between percolation threshold, dispersion state, and aspect ratio of carbon nanotubes. Adv Funct Mater 17(16):3207–3215
Li Q, Church JS, Kafi A et al (2014) An improved understanding of the dispersion of multi-walled carbon nanotubes in non- aqueous solvents. J Nanopart Res. https://doi.org/10.1007/s11051-014-2513-0
Liu CX, Choi JW (2012) Improved dispersion of carbon nanotubes in polymers at high concentrations. Nanomaterials 2(4):329–347
Liu T, Phang IY, Shen L et al (2004) Morphology and mechanical properties of multiwalled carbon nanotubes reinforced nylon-6 composites. Macromolecules 37(19):7214–7222
Liu M, Guo B, Du M et al (2009) Halloysite nanotubes as a novel β-nucleating agent for isotactic polypropylene. Polymer 50(13):3022–3030
Liu M, Jia Z, Jia D et al (2014) Recent advance in research on halloysite nanotubes-polymer nanocomposite. Prog Polym Sci 39(8):1498–1525
Lou X, Detrembleur C, Sciannamea V et al (2004) Grafting of alkoxyamine end-capped (co) polymers onto multi-walled carbon nanotubes. Polymer 45(18):6097–6102
Ma PC, Tang BZ, Kim JK (2008) Conversion of semiconducting behavior of carbon nanotubes using ball milling. Chem Phys Lett 458(1–3):166–169
Ma PC, Wang SQ, Kim JK et al (2009) In-situ amino functionalization of carbon nanotubes using ball milling. J Nanosci Nanotechnol 9(2):749–753
Ma PC, Siddiqui NA, Marom G et al (2010) Dispersion and functionalization of carbon nanotubes for polymer-based nanocomposites: a review. Compos A 41(10):1345–1367
Marney D, Russell L, Wu D et al (2008) The suitability of halloysite nanotubes as a fire retardant for nylon 6. Polym Degrad Stab 93(10):1971–1978
Mat Desa M, Hassan A, Arsad A et al (2014) Mechanical properties of poly(lactic acid)/ multiwalled carbon nanotubes nanocomposites. Mater Res Innovations 18(6):S6-14–S16-17
McNally T, Pötschke P, Halley P et al (2005) Polyethylene multiwalled carbon nanotube composites. Polymer 46(19):8222–8232
Meira SMM, Zehetmeyer G, Werner JO et al (2017) A novel active packaging material based on starch-halloysite nanocomposites incorporating antimicrobial peptides. Food Hydrocoll 63:561–570
Mina MF, Shohrawardy M, Khan MA et al (2013) Improved mechanical performances of triple super phosphate treated jute-fabric reinforced polypropylene composites irradiated by gamma rays. J Appl Polym Sci 130(1):470–478
Mina M, Beg M, Islam M et al (2014) Structures and properties of injection-molded biodegradable poly(lactic acid) nanocomposites prepared with untreated and treated multiwalled carbon nanotubes. Polym Eng Sci 54(2):317–326

Modesti M, Besco S, Lorenzetti A (2009) Effect of processing conditions on the morphology and properties of polymer nanocomposites. In: Mittal V (ed) Optimization of polymer nanocomposite properties, 1st edn. Wiley-VCH, Germany, pp 369–405
Moniruzzaman M, Winey KI (2006) Polymer nanocomposites containing carbon nanotubes. Macromolecules 39(16):5194–5205
Monti M, Puglia D, Natali M et al (2011) Effect of carbon nanofibers on the cure kinetics of unsaturated polyester resin: thermal and chemorheological modelling. Compos Sci Technol 71 (12):1507–1516
Moon SI, Jin F, Lee CJ et al (2005) Novel carbon nanotube/poly(l-lactic acid) nanocomposites; their modulus, thermal stability, and electrical conductivity. Macromol Symp 224(1):287–296
Niu H, Gao M (2006) Diameter-tunable CdTe nanotubes templated by 1D nanowires of cadmium thiolate polymer. Angew Chem Int Ed 45(39):6462–6466
Niyogi S, Hamon M, Hu H et al (2002) Chemistry of single-walled carbon nanotubes. Acc Chem Res 35(12):1105–1113
Novoselov KS, Geim AK, Morozov SV et al (2004) Electric field effect in atomically thin carbon films. Science 306(5696):666–669
Novoselov K, Jiang D, Schedin F et al (2005) Two-dimensional atomic crystals. Proc Natl Acad Sci U S A 102(30):10451–10453
Offringa AR (1996) Thermoplastic composites-rapid processing applications. Compos A 27 (4):329–336
Pan B, Cui D, Xu P et al (2009) Synthesis and characterization of polyamidoamine dendrimer-coated multi-walled carbon nanotubes and their application in gene delivery systems. Nanotechnology. https://doi.org/10.1088/0957-4484/20/12/125101
Pang H, Xu L, Yan DX et al (2014) Conductive polymer composites with segregated structures. Prog Polym Sci 39(11):1908–1933
Pasbakhsh P, Churchman GJ, Keeling JL (2013) Characterisation of properties of various halloysites relevant to their use as nanotubes and microfibre fillers. Appl Clay Sci 74:47–57
Pedrazzoli D, Pegoretti A, Thomann R et al (2015) Toughening linear low-density polyethylene with halloysite nanotubes. Polym Compos 36(5):869–883
Pirlot C, Mekhalif Z, Fonseca A et al (2003) Surface modifications of carbon nanotube/polyacrylonitrile composite films by proton beams. Chem Phys Lett 372(3–4):595–602
Ponnamma D, Sadasivuni KK, Strankowski M et al (2013) Synergistic effect of multi walled carbon nanotubes and reduced graphene oxides in natural rubber for sensing application. Soft Matter 9(43):10343–10353
Ponnamma D, Sung SH, Hong JS et al (2014) Influence of non-covalent functionalization of carbon nanotubes on the rheological behavior of natural rubber latex nanocomposites. Eur Polym J 53:147–159
Prashantha K, Lacrampe M, Krawczak P (2011) Processing and characterization of halloysite nanotubes filled polypropylene nanocomposites based on a masterbatch route: effect of halloysites treatment on structural and mechanical properties. Express Polym Lett 5(4):295–307
Qiao X, Na M, Gao P et al (2017) Halloysite nanotubes reinforced ultrahigh molecular weight polyethylene nanocomposite films with different filler concentration and modification. Polym Test 57:133–140
Ramontja J, Ray SS, Pillai SK et al (2009) High-performance carbon nanotube-reinforced bioplastic. Macromol Mater Eng 294(12):839–846
Raquez JM, Deléglise M, Lacrampe MF et al (2010) Thermosetting (bio) materials derived from renewable resources: a critical review. Prog Polym Sci 35(4):487–509
Rastogi R, Kaushal R, Tripathi S et al (2008) Comparative study of carbon nanotube dispersion using surfactants. J Colloid Interface Sci 328(2):421–428
Rooj S, Das A, Thakur V et al (2010) Preparation and properties of natural nanocomposites based on natural rubber and naturally occurring halloysite nanotubes. Mater Des 31(4):2151–2156
Rouison D, Sain M, Couturier M (2004) Resin transfer molding of natural fiber reinforced composites: cure simulation. Compos Sci Technol 64(5):629–644

Sadasivuni KK, Ponnamma D, Thomas S et al (2014) Evolution from graphite to graphene elastomer composites. Prog Polym Sci 39(4):749–780
Saharudin MS, Wei J, Shyha I et al (2016) The degradation of mechanical properties in halloysite nanoclay-polyester nanocomposites exposed in seawater environment. J Nanomater. https://doi.org/10.1155/2016/2604631
Sahnoune M, Taguet A, Otazaghine B et al (2017) Effects of functionalized halloysite on morphology and properties of polyamide-11/SEBS-g-MA blends. Eur Polym J 90:418–430
Sahoo NG, Rana S, Cho JW et al (2010) Polymer nanocomposites based on functionalized carbon nanotubes. Prog Polym Sci 35(7):837–867
Saif MJ, Naveed M, Zia KM et al (2016) Pristine and γ-irradiated halloysite reinforced epoxy nanocomposites-insight study. Radiat Phys Chem 127:115–121
Sandler J, Shaffer M, Prasse T et al (1999) Development of a dispersion process for carbon nanotubes in an epoxy matrix and the resulting electrical properties. Polymer 40(21):5967–5971
Shao H, Gao M, Kim SH et al (2011) Aqueous self-assembly of L-lysine-based amphiphiles into 1D n-type nanotubes. Chem A Eur J 17(46):12882–12885
Shearer G, Tzoganakis C (1999) Analysis of mixing during melt-melt blending in twin screw extruders using reactive polymer tracers. Polym Eng Sci 39(9):1584–1596
Shearer G, Tzoganakis C (2001) Distributive mixing profiles for co-rotating twin-screw extruders. Adv Polym Technol 20(3):169–190
Shi Y, Li Y, Wu J et al (2011) Toughening of poly (L-lactide)/multiwalled carbon nanotubes nanocomposite with ethylene-co-vinyl acetate. J Polym Sci Part B Polym Phys 49(4):267–276
Shubhra QT, Alam A (2011) Effect of gamma radiation on the mechanical properties of natural silk fiber and synthetic E- glass fiber reinforced polypropylene composites: a comparative study. Radiat Phys Chem 80(11):1228–1232
Singh RP, Jain S, Ramarao P (2013) Surfactant-assisted dispersion of carbon nanotubes: mechanism of stabilization and biocompatibility of the surfactant. J Nanopart Res. https://doi.org/10.1007/s11051-013-1985-7
Singh VP, Vimal K, Kapur G et al (2016) High-density polyethylene/halloysite nanocomposites: morphology and rheological behaviour under extensional and shear flow. J Polym Res. https://doi.org/10.1007/s10965-016-0937-1
Song W, Zheng Z, Lu H et al (2008) Incorporation of multi-walled carbon nanotubes into biodegradable telechelic prepolymers. Macromol Chem Phys 209(3):315–321
Soradech S, Limatvapirat S, Luangtana-anan M (2013) Stability enhancement of shellac by formation of composite film: effect of gelatin and plasticizers. J Food Eng 116(2):572–580
Stephen R, Thomas S (2010) Nanocomposites: state of the art, new challenges and opportunities. In: Thomas S, Stephen R (eds) Rubber nanocomposites: preparation, properties, and applications, 1st edn. Wiley, Singapore, pp 1–20
Subramaniam K, Das A, Simon F et al (2013) Networking of ionic liquid modified CNTs in SSBR. Eur Polym J 49(2):345–352
Sui G, Zhong W, Yang X et al (2007) Processing and material characteristics of a carbon-nanotube-reinforced natural rubber. Macromol Mater Eng 292(9):1020–1026
Sui G, Zhong W, Yang X et al (2008) Curing kinetics and mechanical behavior of natural rubber reinforced with pretreated carbon nanotubes. Mater Sci Eng A 485(1–2):524–531
Szpilska K, Czaja K, Kudła S (2015) Halloysite nanotubes as polyolefin fillers. Polimery 60:359–371
Tarachiwin L, Sakdapipanich J, Ute K et al (2005) Structural characterization of α-terminal group of natural rubber. 2. Decomposition of branch-points by phospholipase and chemical treatments. Biomacromolecules 6(4):1858–1863
Terrones M, Hsu W, Schilder A et al (1998) Novel nanotubes and encapsulated nanowires. Appl Phys A Mater Sci Process 66(3):307–317
Therias S, Murariu M, Dubois P (2017) Bionanocomposites based on PLA and halloysite nanotubes: from key properties to photooxidative degradation. Polym Degrad Stab 145:60–69

Thostenson ET, Ren Z, Chou TW (2001) Advances in the science and technology of carbon nanotubes and their composites: a review. Compos Sci Technol 61(13):1899–1912
Thostenson ET, Ziaee S, Chou TW (2009) Processing and electrical properties of carbon nanotube/vinyl ester nanocomposites. Compos Sci Technol 69(6):801–804
Touny AH, Lawrence JG, Jones AD (2010) Effect of electrospinning parameters on the characterization of PLA/HNT nanocomposite fibers. J Mater Res 25(5):857–865
Vahedi V, Pasbakhsh P, Chai SP (2015) Toward high performance epoxy/halloysite nanocomposites: new insights based on rheological, curing, and impact properties. Mater Des 68:42–53
Vaisman L, Wagner HD, Marom G (2006) The role of surfactants in dispersion of carbon nanotubes. Adv Colloid Interface Sci 128:37–46
Varghese S, Karger-Kocsis J (2003) Natural rubber-based nanocomposites by latex compounding with layered silicates. Polymer 44(17):4921–4927
Vast L, Philippin G, Destree A et al (2004) Chemical functionalization by a fluorinated trichlorosilane of multi-walled carbon nanotubes. Nanotechnology 15(7):781
Velasco-Santos C, Martinez-Hernandez A, Lozada-Cassou M et al (2002) Chemical functionalization of carbon nanotubes through an organosilane. Nanotechnology 13(4):495–498
Vichchulada P, Cauble MA, Abdi EA et al (2010) Sonication power for length control of single-walled carbon nanotubes in aqueous suspensions used for 2-dimensional network formation. J Phys Chem C 114(29):12490–12495
Villmow T, Pötschke P, Pegel S et al (2008) Influence of twin-screw extrusion conditions on the dispersion of multi-walled carbon nanotubes in a poly(lactic acid) matrix. Polymer 49 (16):3500–3509
Wan X, Zhan Y, Zeng G et al (2017) Nitrile functionalized halloysite nanotubes/poly(arylene ether nitrile) nanocomposites: Interface control, characterization, and improved properties. Appl Surf Sci 393:1–10
Wang B, Huang HX (2013) Effects of halloysite nanotube orientation on crystallization and thermal stability of polypropylene nanocomposites. Polym Degrad Stab 98(9):1601–1608
Wang B, Huang HX (2014) Incorporation of halloysite nanotubes into PVDF matrix: nucleation of electroactive phase accompany with significant reinforcement and dimensional stability improvement. Compos A 66:16–24
Wang J, Chu H, Li Y (2008) Why single-walled carbon nanotubes can be dispersed in imidazolium-based ionic liquids. ACS Nano 2(12):2540–2546
Wang S, Chang K, Yuan C (2009) Enhancement of electrochemical properties of screen-printed carbon electrodes by oxygen plasma treatment. Electrochim Acta 54(21):4937–4943
Wang J, Liu J, Zhou Y et al (2013) One-pot facile synthesis of hierarchical hollow microspheres constructed with MnO_2 nanotubes and their application in lithium storage and water treatment. RSC Adv 3(48):25937–25943
Wu CS, Liao HT (2007) Study on the preparation and characterization of biodegradable polylactide/multi-walled carbon nanotubes nanocomposites. Polymer 48(15):4449–4458
Wu D, Wu L, Zhang M et al (2008) Viscoelasticity and thermal stability of polylactide composites with various functionalized carbon nanotubes. Polym Degrad Stab 93(8):1577–1584
Wu D, Zhang Y, Zhang M et al (2009) Selective localization of multiwalled carbon nanotubes in poly(ε-caprolactone)/ polylactide blend. Biomacromol 10(2):417–424
Wu M, Huang HX, Tong J (2016a) Enhancing β-phase content and tensile properties in poly (vinylidene fluoride) by adding halloysite nanotubes and injecting water during extrusion. Mater Des 108:761–768
Wu Y, Du Z, Wang H et al (2016b) Preparation of waterborne polyurethane nanocomposite reinforced with halloysite nanotubes for coating applications. J Appl Polym Sci. https://doi.org/10.1002/app.43949
Xiao K, Zhang L (2004) The stress transfer efficiency of a single-walled carbon nanotube in epoxy matrix. J Mater Sci 39(14):4481–4486
Yang LP, Pan CY (2008) A non-covalent method to functionalize multi-walled carbon nanotubes using six-armed star poly(L-lactic acid) with a triphenylene core. Macromol Chem Phys 209 (8):783–793

Yang Z, Huang X, Zhao Q et al (2012) Hydrogen-bonded 1D nanotubes and 2D layers of group 12 metal complexes with a pyridylurea ligand. Cryst Eng Comm 14(17):5446–5453

Ye Y, Chen H, Wu J et al (2007) High impact strength epoxy nanocomposites with natural nanotubes. Polymer 48(21):6426–6433

Ye J, Zhang H, Yang R et al (2010) Morphology-controlled synthesis of SnO_2 nanotubes by using 1D silica mesostructures as sacrificial templates and their applications in lithium-ion batteries. Small 6(2):296–306

Yoon JT, Jeong YG, Lee SC et al (2009) Influences of poly(lactic acid)-grafted carbon nanotube on thermal, mechanical, and electrical properties of poly(lactic acid). Polym Adv Technol 20 (7):631–638

Yu J, Grossiord N, Koning CE et al (2007) Controlling the dispersion of multi-wall carbon nanotubes in aqueous surfactant solution. Carbon 45(3):618–623

Yuan Q, Misra R (2006) Polymer nanocomposites: current understanding and issues. Mater Sci Technol 22(7):742–755

Zou J, Zhang YC, Huang J et al (2009) Preparation and properties of PP/PLA/multiwall carbon nanotube composites filaments obtained by melt compounding. Mater Sci Forum 620–622:465–468

Starch-Based Nanocomposites: Types and Industrial Applications

Faruq Mohammad, Tanvir Arfin, Ibrahim B. Bwatanglang and Hamad A. Al-lohedan

Abstract The extraordinary physicochemical and functional features offered by the starch material segregated from various sources of plants, such as rice, corn and wheat, are put into use for a large extent of applications. The physicochemical features of starch namely lipids content, a ratio of amylose to amylopectin, the size distribution of granule play a significant role to grasp the concept related to the mechanism on the functionality of starch in various systems. The starch-modified chemistry along with a large number of reactive sites carries the biologically active compounds as biocompatible carriers and are metabolized in the human body quickly and comfortably. The current chapter focusses on the different composites made up of starch along with polymers like polylactic acid, polycaprolactone, polyhydroxy alkaloid where the synthesis, chemistry and application part are greatly discussed. Further, the physicochemical stability of the nanocomposites relating the specific structure is compared in addition to their deployment in various industrial applications.

1 Introduction

In the present era, the everyday human life is entirely governed by the automated systems supported by the advanced materials with novel functionalities where a variety of synthesis approaches are adopted so as to form the human-made materials for the

F. Mohammad (✉) · H. A. Al-lohedan
Surfactant Research Chair, Department of Chemistry, College of Science, King Saud University, Riyadh 11451, Kingdom of Saudi Arabia
e-mail: fmohammad@ksu.edu.sa

T. Arfin
Environmental Materials Division, CSIR-National Environmental and Engineering Research Institute (NEERI), Nehru Marg, Nagpur, Maharashtra, India

I. B. Bwatanglang
Department of Pure and Applied Chemistry, Faculty of Science, Adamawa State University, Mubi, Nigeria

M. L. Sanyang and M. Jawaid (eds.), *Bio-based Polymers and Nanocomposites*,
https://doi.org/10.1007/978-3-030-05825-8_8

betterment of life quality (Arfin and Athar 2018). In a similar way, the biomaterials from the renewable resources are the user-friendly substances and also considered to be the ideal ones for many different applications in several scientific fields like biotechnology, biomedicine, biofuels and bioenergy (Arfin et al. 2014; Arfin 2015). The naturally obtained materials are regarded as the sustainable ones with not much lessening to the efficiency and can be produced in high quantities with minimal reaction processes that can be employed in a variety of applications (Mohammad et al. 2015). Within the class of nature derived biomaterials, the starch-derived materials have attracted many advanced applications by taking advantage of the physiochemical and functional characteristics supported by the reactive oxygen containing groups. The products derived from starch-based materials have generated high level of interest in the design and formulation of green-based technology considering the raising treat posed by fossil-based product. Molecularly, the unique signature of starch as a heterogeneous material makes it the most essential and useful biocompatible material compared to all other polymers of natural origin (Arfin and Tarannum 2017). It occurs naturally and is the second extensive polysaccharide after cellulose. The various sources of starch are shown in Fig. 1.

Starch is available in the granules form and is composed by joining different glucose molecules through the glycosidic bonds (Mograkar and Arfin 2017). The features of starch granules such as chemical composition, size and structure are dependent on the original state from where it is derived. Starch contains two macromolecules namely the amylopectin and the amylose. In the architecture of starch, the amylopectin and the amylose constitute the basic signature, structurally showing the amylose appearing as a linear structure of 1,4 linked glucose units, while the highly branched structure of short 1,4 chains linked by 1,6 bonds represents amylopectin (Nasseri and Mohammadi 2014). The structures of the respective macromolecules are represented in Fig. 2.

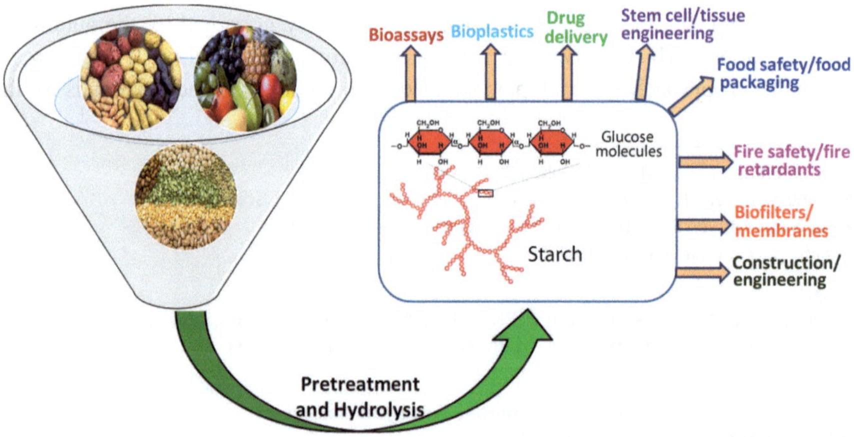

Fig. 1 Different sources of starch, its extraction and applications into different sectors

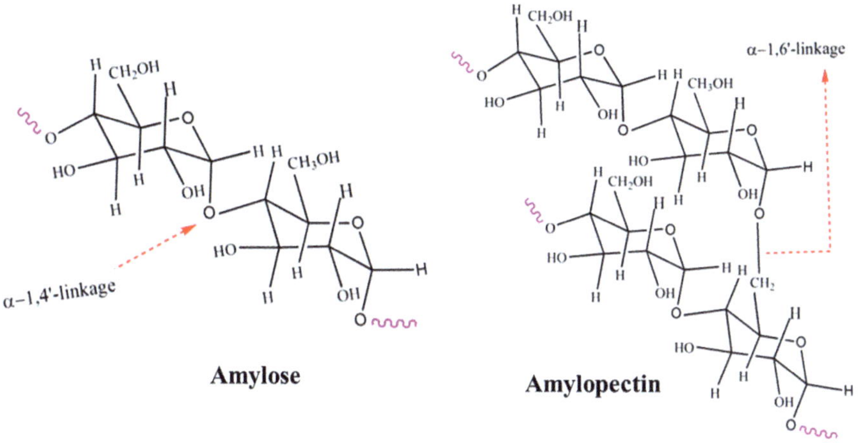

Fig. 2 Structures of amylose and amylopectin

As mentioned earlier, the structural architecture of the starch granules is directly influenced by the orientation in space of both the amylose and amylopectin units. The amylose molecules are oriented parallel between each unit and between adjacent chains held by hydrogen bonding. The orientation is also implicated in defining the chemistry of starch, influencing its water absorption capacity, degradation rate, gelatinization, etc (Mograkar and Arfin 2017; Nasseri and Mohammadi 2014).

As interesting as these properties entail starch in its natural form suffers from a number of limitations. Poor mechanical properties and high permeation are some of the few limitations of natural starch materials. However, process improvement following the doping of starch granules with filler materials or blending of the starch granules with other polymer-based materials to form starch-based nanocomposites has significantly ameliorated the aforementioned limitations. The chemistry was made possible owing to the interfacial molecular bridges between the starch matrix and the incorporated nanofiller (Ghanbarzadeh et al. 2011; Zuraida et al. 2012; Olsson et al. 2013; Gutiérrez et al. 2014; Famá et al. 2012).

1.1 Biodegradable Packaging Materials

Lorcks in 1997 proposed the properties and applications of plastics (Lörcks 1998). The biodegradable plastics stand out as a rare invention and substitute for the pollution creating plastics. Typically owed to its high biodegradability and less half-life, they can be disposed or degraded following environmental-friendly manner. Plastics are the polymers made up of monomers through chemicals reactions. Condensation polymerization and polyaddition (chain and step reaction) are the necessary processes for plastic making.

- In poly-addition, the chemical reactions occur which forms polymers from monomers. For this process, energy and catalyst are mostly required. In some cases where the availability of insufficient catalyst/energy, the addition of monomers results in the loss of hydrogen movement that further restricts the separation of by-products.
- In case of poly-addition step reactions, the sufficient movement of hydrogen bonds results in the prevention of double bonded activities.
- Also, the poly-condensation reaction in which the actual plastic formation takes place and is a step reaction.

According to the properties and structure, they are classified into different classes. They may be thermoplastics, thermosets, polymer blends, elastomers and semi-conductor thermoplastics. This quest for biodegradable plastics leads to a number of engineering efforts that involved blending of either synthetic polymer or biopolymers with starch-based materials. Among many different applications, the important application of starch-based materials being the environmental safe biodegradable plastics production. In addition for being relatively cheap, spread widely in nature, the starch-based materials also possessed good oxygen barrier and biodegradability (Fabunmi et al. 2007). Starch has been used as a reinforcement/fillers polymer matrix, in the production of thermoplastic starch prepared following the plasticization of starch in the presence of plasticizers (e.g. glycerol), and it is also used in the production of synthetic polymer films like polylactic acid (PLA), polycaprolactone (PCL) and polyhydroxyalkanoate (PHA) (Avérous and Pollet 2012).

2 Biodegradable Starch-Based Nanocomposites

As mentioned earlier, the limitations associated with nature-based starch materials are readily overcome by converting the starch matrix into nanocomposites while retaining their active green chemistry. The search for biodegradable materials further reinforces the drive towards contextual application of starch-based materials reinforced with either natural or synthetic polymers specifically to meets the industrial demand for food packaging, surgery, pharmaceutical, biomedical applications (Bouyer et al. 2012). Of greater interest, cellulose, clays and number of synthetic materials are recently used as fillers in the development of biodegradable starch-based compounds (Thakur et al. 2012, 2014; Hassani and Nafchi 2014; Thakur and Thakur 2014). Preceding sections will cover development into the chemistry of starch-based nanocomposites.

2.1 Starch–Clay Nanocomposites

The formation of starch nanocomposites using various fillers has significantly improved the quality of starch-based materials. In comparison to the conventional clays, the chart of nano-fillers to improve the efficiency of starch was found to rank the top (Mondragón et al. 2008). Avella et al. (2005) discovered that the starch–clay nanocomposite could be used for the food packaging purpose. The use of the nanoparticle in the different starch-based clays was applied to produce the starch-based film for packaging. Use of such materials for preparing films increases the properties of the packaging films and also tensile strength and modulus. The central aspect of this product is that it is the novel biodegradable material used especially for food packaging purpose.

Clay is a polymer composite of silicate consisting of a central positive charge silicon ion sandwiched by a negatively charged oxygen ion linking another silicon ion in a progressive manner. This arrangement can assume a different combination consisting of sheets arranged either in tetrahydral $[SiO_4]_4$ or octahedral $[AlO_3(OH)_3]_6$ shape. Among these structural silicates, the clay composites consisting of two-dimensional layered silicates (montmorillonite) are widely used owing to its higher percentage of water molecules in the layer available for cation exchange processes (Huang et al. 2006; Ikeo et al. 2006). The cation exchange processes and the ease of intercalation make this clay types flexible and compatible with other polymer like starch. Park et al. (2002) introduced the preparation of hybrid composite of starch–clay. The process used for the preparation was melt intercalation. Natural and the starch–clays which are organically modified were only used for hybrid preparation. The storage modulus of this hybrid was higher in comparison with other composites.

The widely reported processing techniques used in the intercalation starch/clay are in situ intercalative polymerization, intercalation of polymer and melt intercalation (Schlemmer et al. 2010; Barzegar et al. 2014). In addition to the ion exchange processes, the polymerization processes that required the application of heat or passing of irradiation or use of a catalyst to facilitate the swollen up of the clay in the presence of a liquid monomer is called in situ intercalative polymerization. Conversably, in melt intercalation processes, the solvent molecules are reportedly desorbed from the silicate layer during the polymerization process allowing easy flow of the incoming polymer. This method is considered a greener approach and highly compatible with environmental best practices. In the intercalation of polymer, the polymer is dissolved in a suitable solvent in order to exfoliate (uniformly dispersed silicate layers in a continuous polymeric matrix) layered clays into a single platelet. This method is influenced by the choice of polymer/solvent pairs and is considered environmentally unfriendly owed to the use of unfriendly solvents (Cui et al. 2015).

In order to overcome the poor water resistance, poor tensile strength and high brittleness of starch materials, a number of efforts in addition to the use of nanoclay as discussed earlier are reported (Zabihzadeh 2010; Jamshidian et al. 2010; Takegawa et al. 2010). Early study shows that the reinforcement of potato starch with 5% clay significant increases $\%E$ and tensile strength by >20 and 25%, respectively, and further shows a reduction in water vapour transmission rate by 35% (Park et al. 2003). Avella et al. (2005) and Pandey and Singh (2005) in a separate study reported increase in mechanical properties potato starch–clay nanocomposite films and cornstarch–clay films, respectively. The cornstarch/clay nanocomposites following a study conducted by Huang et al (2006) reported about 450 and 20% increase in the tensile strength and strain, respectively, with the addition of 5% clay. In another study, reinforcement of Cará root starch/hectorite films with 30% clay level increases the tensile strength of the nanocomposites 70% and elongation at break (*%E*) decreased by 50% (Wilhelm et al. 2003). This observed improvement in both thermal stability and water absorbance capacity of the starch–clay composites makes it an attractive candidate for packaging materials (Chiou et al. 2007). Table 1 is a list of few process development made in engineering starch–clay nanocomposites.

Noble efforts were also made by using biopolymer extracted from cellulose, gelatin, chitosan and plant-based extracts to prepare starch-based nanocomposites. This quest for biodegradable polymers leads to the discovery that microorganism

Table 1 Engineering starch–clay nanocomposites

Preparation/methods	Result	References
Green starch/clay nanocomposites were prepared by solution-induced intercalation method using starch, jute, glutaraldehyde, nanoclay and glycerol	In the study, the thermal stability and mechanical strength of the prepared composite were significantly improved with the addition of glutaraldehyde and nanoclay	Iman and Maji (2012)
Bio-nanocomposite films were prepared from potato starch by casting method using halloysite nanoclay as the reinforcing materials	The incorporation of halloysite nanoclay into the polymer matrix decreased the permeability of the material to gaseous molecules and improved the mechanical properties. Tensile strength increased from 7.33 to 9.82 MPa and elongation at break decreased from 68.0 to 44.0% with the filler addition	Hassani and Nafchi (2014)
The biodegradable trays of cassava starch and organically modified montmorillonite were prepared using a baking process	In the study, the stress at break of the samples was observed to be strongly affected on incorporation of the nanoclay	Matsuda et al. (2013)
In this study, biodegradable starch–clay nanocomposites were prepared by incorporating a dilute clay dispersion to a starch solution that was followed by co-precipitation in ethanol	Well-dispersed starch–clay nanocomposites were obtained	Chung et al. (2010)
In this study, a film blowing technique was used to prepare the starch–clay films using hydroxypropyl distarch phosphate and five different kinds of clays as materials	High tensile strength and improved barrier properties were observed in the prepared starch–clay film nanocomposites. Furthermore, a drop in glass transition temperature (T_g) and better heat endurance were equally observed	Gao et al. (2012)

can synthesize biopolyesters from both natural and synthetic monomers. The biopolyesters synthesized through this process include but not limited to thermoplastic starch (TPS), polylactic acid (PLA), polycaprolactone (PCL), and polyhydroxyalkanoate (PHA) (Avérous and Pollet 2012).

2.2 *Starch–Polylactic Acid (PLA) Composites*

Muller et al. (2017) proposed in their recent work about the use of PLA-starch composite for making biodegradable food packaging plastics. The replacement of pollution creating plastics was done by using the plastic made from PLA and starch. Interestingly, PLA is nature-based materials derived from sugar stock, rice and corn by fermentation and chemical conversion of the sources to dextrose and then to lactic acid using poly-condensation processes (Rydz et al. 2015).

The suitable liquids for PLA in which it is soluble are acetonitrile, chloroform, benzene, ethyl acetate, acetone etc and is not soluble in ethanol, methanol and water. The biochemical aspects of this material being the biocompatibility, biodegradability, environmental friendly, renewable, and readily hydrolyzes to natural, nontoxic products on disposal. Most of food packaging industries require PLA for making biodegradable plastics. It has hydrophobic nature. Thus, before using PLA the drying of the pallets should be done at about 60–100 °C. This process modifies it physically (Lim et al. 2008). It has application in the tissue engineering field for wound healing (Armentano et al. 2013). It is resistant to elasticity, which enables it to be rigid plastic and brittle.

The properties of PLA and starch match each other. On the combination of both these materials, the product form will be more efficient and without creating any issue whatever it may be economical or pollution. As described PLA is hydrophobic whereas starch is hydrophilic, this makes the processing of the material difficult. The product obtained from PLA and starch is used in the food packaging industries. The films and packaging materials formed by the combination of PLA and starch are found to have enhanced properties as compared to the ones which are formed from the individual materials. By varying the ratios of starch and PLA, the blend can be made. It serves as the best alternative for reducing pollution caused due to plastics. The obtained plastic will have the properties of antioxidant and antimicrobial. The plastic or film has excellent flexibility and mechanical properties. However, the PLA material when applied alone, suffers from the required indices of efficient biodegradability which is a must so as to use in the industrial application due to its brittleness. Hence, the combination of PLA with other thermoplastic materials like polyethylene glycol (PEG) or dispersed nanoclay into the matrix found to significantly improved its biodegradability and associated marketability (Rydz et al. 2015).

Sheth et al. (1997) proved about the use of PLA and PEG for making the plastic whereby varying the concentrations of both materials, and the composite can be prepared. Depending on the concentrations, the mixture becomes partial miscible and miscible. If PLA is more than 50% of a mixture, then PEG crystallinity increases, and if PEG is more than 30%, then the weight loss occurs and it will be mainly due to PEG as it gets dissolved. Also, the tensile strength decreases if PEG is high in concentration. Ayana et al. (2014) prepared the composite from thermoplastic starch (TPS), PLA and dispersed nanoclay. Potato starch was in situ gelatinized. The thermo-mechanical properties and tensile strength of the composite were increased by the use of PLA and clay. The synthesis of novel material was done in the environmental fever, so that it can be used as a packaging material also.

In another study, solvent casting methods using *N*,*N*-dimethylacetamide was used to prepare uniformly dispersed cellulose nanofibrils using bleached wood pulp/PLA with PEG to improve the interfacial bonding/adhesion between the matrix and the fibre. In the study, the tensile strength increases by 28.2% and similar increases in the percentage elongation by 25% following the addition of PEG to the blend (Qu et al. 2010). One-step extrusion process was used to prepare binary and ternary TPS/PLA-PCL nanocomposites. The result from the study shows that at 36% plasticizer (glycerol), a transition temperature for the TPS/PLA-PCL nanocomposites was recorded at −57 °C. It was further observed that varying up the TPS concentration in the blend has a direct impact towards increasing the strain at break (Sarazin et al. 2008).

2.3 Cellulose Fibre

Cellulose in addition to its natural responsibility as shielding material in maintaining the rigidity of plant also presents remarkable properties for various applications. Cellulose-based fibres derived from nature are widely used to reinforce polymer composites due to its compatibility with many polymer matrixes. Essentially, the chemistry between the cellulose nanofibre and the polymer matrix enables excellent dispersion of the reinforcement species within the matrix. The molecular structure and interfacial similarity of the cellulose material further favours effective chemistry among many different matrix species (Rydz et al. 2015). This organic polymer is widely spread in nature and consisting of *D*-anhydroglucopyranose units linked together with β-(1-4) glycosidic bonds (Thakur et al. 2010, 2012). Despite being nature derived, is a poor material to dissolve in aqueous media. However, further modification following simple esterification or etherification of the hydroxyl groups was reported to confer a solubility characteristic in water (Mischnick and Momcilovic 2010).

Wan et al. (2009) used a biodegradable reinforcement in the form of bacterial cellulose nanofibres. They discussed the entire phenomenon occurring during the formation of nanocomposite. The method used was solution impregnation. The comparison was made between the nanocomposite of cellulose and starch and the unreinforced starch. Sorption diffusion process was used for the kinetic analysis, and then the different parameters were determined. The study of tensile strength and microbial attacks was performed. The use of bacterial cellulose in making of nanocomposites increases the strength and ability to resist the attack of microbes.

This can have broad application in the protection of the environment. In another study, Khan et al. (2010) improved the properties of cellulose by synthesizing a water-soluble methylcellulose (MC)-based biodegradable nanocomposite films. In the study, simple mechanical processes involving stirring and sonication in the presence of plasticizer (glycerol), vegetable oil and Tween 80 were followed under room temperature. The outcome of the study revealed that the addition of CN enhanced the mechanical and barrier properties of MC-based films. Similarly, a decrease in water vapour permeability was observed indicating the moisture barrier properties of MC-based films. Zhou et al. (2009) follow a low energy pathway to synthesis cellulose-based nanocomposite from bacterial cellulose nanofibrils coated with hydroxyethyl cellulose (HEC). The experiment was conducted by press drying of the water suspension of the cellulose nanofibres. The characterization revealed a material with encouraging tensile strength and optical transparency.

The potential of hydroxypropyl methylcellulose (HPMC)/cellulose whiskers nanocomposite films as suitable material for packaging applications was investigated and reported. The addition of cellulose nanowhiskers was observed to have a direct relationship with increase in tensile strength, Young's modulus and the water barrier properties of the composite films (Sáinz et al. 2010, 2011). Furthermore, carboxymethyl cellulose (CMC) is another water-soluble derivative of cellulose often combines with carbon nanotubes (CNTs) as fillers to reinforce polymer nanocomposites. The CNTs in the presence of plasticizer (glycerine) were reported to have improved the strength and stiffness of the CMC polymer nanocomposite (Choi and Simonsen 2006).

2.4 *Starch-Polyhydroxy Alkaloid (PHA) Composites*

Molyneux (1993) gave their review on the isolation, characterization and application of PHA. It is the class of phytochemicals. It is a newly discovered class, and therefore, the analysis and detection are ongoing. Polyhydroxy is the compounds which have the active inhibitory property of glycosidase. The activities depend on the stereochemistry and the hydroxyl groups. For the processing of glycoproteins in organisms, PHA is needed so that the biological mechanisms get regulated. It is proved that glycoproteins are indirectly the enhancer of the immune response (Molyneux 1993). The source of PHA is ground plant material. It can be extracted by using ethanol or methanol or water by just varying the proportion of solvent. Since few compounds isolated at each time, chemical determination is limited. Mass spectroscopy, UV-Vis spectroscopy, nuclear magnetic resonance spectroscopy and X-ray crystallography are the characterization technique for PHA.

In the class of starch-based biodegradable polymers, PHA has the highest degradation rate of about 100% and thus stands out as a remarkable eco-friendly biosynthetic polymer of the century. The microorganisms synthesize this material intracellularly by means of a carbon-based chemical approach and can be stored as an energy reservoir (granules) (Jendrossek and Handrick 2002; Shan et al. 2011).

The storage granules in the bacteria following anaerobic respiration converts the stored granule into a hydroxyl acid homopolyester, derivatives of PHA called or poly(3-hydroxybutyrate) (PHB). The PHB and its copolymers consisting of few units of hydroxyvalerate (HV), and significant amount of hydroxybutyrate (HB) units (PHBV) is the product of the anaerobic degradation of the granules (El-Hadi et al. 2002; Lenz and Marchessault 2005).

Study revealed that the amount of HV units in PHBV significantly improved the mechanical and physical properties compared to its homopolymer (PHB) (Choi and Park 2004). Similarly, the HV units provide a platform for the properties to be modified to suits specific objectives (Hu et al. 2004; Shang et al. 2011; Ferreira et al. 2002). Thus, this unique disposition allowed modification such as the addition of plasticizer, formation of blends with other biopolymers such as PCL, polylactides, starch and cellulose (El-Hadi et al. 2002; Ferreira et al. 2002).

Reis et al. (2008) studied the characteristics of the composite film formed by the combination of PHB-HV (polyhydrixybutyrate-hydroxyvalerate) maize starch prepared by means of a casting method. The use of maize makes the blend more economical as compared to the original PHB-HV blend. As the concentration of starch varies, the properties vary including the parameters like Young's modulus and tensile strength. In a separate study, similar solution casting methods were used to PHB/HV and organo-modified mica composites. The preparation was achieved by dispersing the modified clay in chloroform to produce a material with intercalated structures (Garcia and Lagaron 2010; D'Amico et al. 2012). Other study also reported the preparation of organo-modified clay/PHB, PHB/HV and PHB/PCL (Bordes et al. 2008; Botana et al. 2010; Bruzaud and Bourmaud 2007).

2.5 Starch–Polycaprolactone (PCL) Composites

Polycaprolactone) (PCL) is a highly versatile nature-based polyester with low melting point and viscosity derived essentially from oil-producing plants. There are supporting materials widely recruited to reinforce the mechanical properties of polymer like collagen or gelatin to form PCL/collagen or PCL/gelatin nanofibrous, a material suitable for tissue engineering purposes (Sohier et al. 2014; Choi et al. 2008). The beauty of PCL-based material is their simple method of preparation, using simple conventional melt blending technologies to synthesize same. The PCL material can be blended with other biopolymers for improving the fundamental properties to suit the specific objectives that range from food packaging to biomedical applications (Ludueña et al. 2011).

Lo et al. (2010) studied the morphology and properties of the composites made from PCL-starch and PCL-pine composite in the presence of silane as coupling agent. The formed composites were observed to be non-toxic and highly biodegradable, thus showing good prospect in the design of food packaging materials. The mechanical properties of layered silicate/starch PCL blend nanocomposites were investigated by Pérez et al. (2007). In the study, melt intercalation method followed by compression moulding was used to prepare the nanocomposites. The characterization using various instruments shows a strong intercalated silicate or starch PCL blend nanocomposites. In another study, the efficacy of modified PCL-based nanocomposites was investigated using melt intercalation on a twin-screw extruder. The study shows that the modification processes improve the stability and chemical compatibility of the PCL with the clay, leading to the formation of a dispersed material with strong mechanical stability (Ludueña et al. 2011). In the milt intercalation processes, it was observed that shear forces are responsible for further breaking down of the PCL/clay to nanoform. The same author further reported that the exfoliation of the PCL/layered is facilitated by the chemical compatibility between the PCL matrix and clay layers (Homminga et al. 2005). Processing technique such as the melt intercalation methods is reported by a number of researchers to produce largely an exfoliated and well-intercalated material (Pantoustier et al. 2002; Gain et al. 2005). Similar intercalation/exfoliation structure was reported by solvent casting (Wu et al. 2000), by swelling effect under ultrasonic agitation (Ludueña et al. 2011). Other process improvement in the synthesis of PCL-based nanocomposites was also reported (Chang et al. 2009; DeKesel et al. 1997; Pérez et al. 2008).

Fibre-based PCL nanocomposites were also synthesized and reported. Azimi et al. (2014) synthesize different forms of fibre-based PCL nanocomposites, and according to their content, they were used in the different areas of biomedical field. The choice of using PCL in the preparation being aliphatic polyester was highly biodegradable. Furthermore, the fibre was selected because it is environmentally friendly and biocompatible. The study reported that the formed PCL fibres are a good candidate for tissue engineering and drug delivery applications.

2.6 Starch-Polyethylene Terephthalate (PET) Composites

Siracusa and co-workers have given the review on the use of PET as a biodegradable polymer (Siracusa et al. 2008). They have presented the complete review on the use of plastics. It is not possible to replace the plastic, but to a certain extent it is possible. The consumption of plastics is more than about 200 million, and hence, it is the critical issue. The substitute for the petrochemical-based plastics is an important point of the discussion.

The United States-based chemist E.I. du Pont developed the polymer PET when he was trying to make textile fibres. It is a thermoplastic polymer, and the main components or monomers required for the synthesis of PET are ethylene glycol and terephthalic acid. The esterification process converts two monomers mainly into alcohol and an organic acid with ester and water. But PET is a polymer; hence, it can also be called as "polyester", where it is resistant to moisture and so it can be used for the packaging-related applications in food, textile and agriculture industry. It can also be used for preparing fibre for clothes as it does not have colour, clearly transparent and semi-crystalline in nature. In order to increase the strength and other associated properties, it can be used by combining with the other materials like

CNTS and glass. It is straightforward to use PET for everyday applications as compared to many other synthetic fibres where the PET is considered to be cheap, non-breakable unlike glass, and the most interesting thing with this material being its recycling property. It is a lightweight material but very strong. Since it is lightweighted, less amount of compound is required and transport also becomes easy with less consumption of fuel. It can be easily used in food packaging industries because it has the property which does not affect food and water when gets in contact with it (Xie et al. 2015).

2.7 *Starch–Gelatin Nanocomposites*

Gelatin is a soluble protein-based biopolymer consisting of α-amino acids linked by peptide bonds (Guillen et al. 2011). They form the basic component in the design of drug capsules and in food industry as additive used in preserving the composition of food active ingredient. The biochemical properties of gelatin such as its high hydrophilicity were exploited in the design of nanofibrous systems with the ability to affect cell–tissue regeneration or engineering (Lee et al. 2012; Jafari et al. 2011; Zhuang et al. 2010).

The property of gelatin was made better by blending it with compatible materials such as PCL, polyaniline (PAN)–CNTs to form engineered nanofibrous scaffolds. Li et al. (2013a) introduced the use of gelatine starch octenyl succinic anhydride (OSA) to enhance the stability and properties of orange oil in water emulsion. The cornstarch was modified by using OSA before blending it with gelatine. This OSA starch absorbs the interface of oil and water and acts as a stabilizer. Another cited example is blending of PCL with gelatin to form gelatin/PCL-blended nanofibres. The blending was reported significantly to enhance the proliferation of tissue cells (Gupta et al. 2009). Other study by Li et al. shows a significant enhancement in the myocardial proliferation following the blending of gelatin with PAN (Li et al. 2006). Also, some similar end results in terms of tissue-cell proliferation and engineering were obtained following the blending of gelatin with chitosan to obtain chitosan/gelatin hybrid nanofibers (Jafari et al. 2011; Dhandayuthapani et al. 2011; Qian et al. 2011).

In addition to biomedical applications as discussed above, George and Siddaramaiah (2012) in a separate study observed that blending of gelatin to CNTs results in the formation of percolated networks of CNTs within the gelatin matrix. The obtained gelatin/CNT matrix was observed to have significantly improved mechanical properties of the films, thus providing adequate properties for the design of packaging materials.

2.8 *Chitosan-Based Films*

Structurally, the unique signature of chitosan lies on the solution sensitivity of the positive charged NH_2 groups in its molecular chains (Poverenov et al. 2014).

This structural characteristic increases the advantages of chitosan as a matrix material and nanoenforcement in composite formation (Rydz et al. 2015; Yu et al. 2014). These cationic properties increase the adsorption capacity of the polymer, its biocompatibility and biodegradability (Yu et al. 2014; Shawky et al. 2012; Nitayaphat and Jintakosol 2014; Popuri et al. 2014).

Chitosan is a product of de-acetylated chitin made up of about greater than 5000 units of glucosamine and readily found in nature in whole body of fungi, sea animals like crabs, insects and shrimp shells. The chitosan from the crustacean is extracted by the de-acetylation of the parent chitin while enzymatic extraction is used to extract same from fungi (Arfin and Mohammad 2016). The de-acylated chitosan consists of β-1,4 linkage of D-glucosamine and *N*-acetyl-D-glucosamine (Mohammad et al. 2015) and possesses excellent properties with the wide range of application, as excipient in drug formulation and reinforcement materials in the design of non-toxic biodegradable packaging.

As highlighted above, chitosan-based films related to its physiochemistry play a key role in the formulation of biodegradable packaging materials with properties that could increase the shelf life of food (Wang et al. 2018). Aider (2010) in an effort to develop a biodegradable material for packaging purposes introduced the application of chitosan bio-based films as a substitute for fossil-based packaging material. This effort attracted considerable interest owing to the rising cases of food spoilage relating to food storage and security. The most common method of preserving food using chitosan-based materials is in vacuum packaging where pure chitosan film is used in the vacuum preparation (Saiz et al. 2013). Studies further show that chitosan within nanorange compared to microparticle has a better chemistry in forming packaging materials. Similarly, reinforcement or blending chitosan with other polymer fillers or plasticizers was reported to have improved the mechanical properties and chain mobility of the composites (Gol et al. 2013; Garza et al. 2015; Vimaladevi et al. 2015).

An excellent chitosan/gelatin composite based on emulsion of oil in water was prepared and reported by Rui et al. (2017). The reaction between the polymers was favoured by hydrogen bonding (Rui et al. 2017) leading to the formation of biocompatible film and micro-carriers used for animal cell culture on a large scale (Taravel and Domard 1993). Other effort reported also involved in the union of chitin and starch. In a study, Chang et al. (2010a, b) generated the starch from glycerol plasticized potato using casting/evaporation method to make the starch–chitin nanoparticles. The formed nanoparticles when tested are found to improve the tensile strength, water vapour bearing, and increased glass transition temperature, i.e. all the characteristic features are mostly required to check the suitability for a material to be good or not towards packaging-related applications. Both acetylated chitin and de-acetylated chitin were blended to produce chitin/chitosan nanofibres as material for tissue engineering and other biomedical applications (Jayakumar et al. 2010).

An interesting effort was presented by Jantanasakulwong et al. (2016), introduced the blending of thermoplastic starch (TPS) with chitosan and rubber. In the study, the thermoplastic plastic was prepared by using plasticizer (glycerol), cassava starch and chitosan. Due to the NH_2 group, the reaction of chitosan with any other group becomes easy and fast. In a study, the blend leads to increase mechanical strength and tensile strength forming a low weight composite. Lopez et al. (2014) in a separate experiment used chitin, chitosan and thermoplastic cornstarch to formulate a biodegradable film by thermo-compression method. The modifications were done in the film by the addition of chitosan. Increase in crystallinity was obtained. The obtained film was of very efficient quality, i.e. with a smooth surface and homogeneous nature. There were no cracks or pores in the film. The combination of chitin and chitosan increases the elastic modulus and strength.

The chemistry exerted between chitosan and cellulose leads to the formation of a composite film that exhibited improved water vapour barrier property and reduced bacterial adhesion on the packaging material (Xiao et al. 2013; Bansal et al. 2016; Li et al. 2013a, b; Sundaram et al. 2016). A cited example shows and increases in shelf life of cheese and wheat bread preserved using a packaging material made from chitosan/carboxymethyl cellulose film (Sundaram et al. 2016). Similar chemistry was also exploited to develop different chitosan/cellulose films (Gol et al. 2013; Noshirvani et al. 2017; Chen et al. 2016; Liu et al. 2013). Table 2 is a summary of selected events on efforts made in using chitosan-based composite in food industries and other related applications.

Table 2 Different types of starch–chitosan-based composites

Preparation/methods	Result	References
The study reported on the reinforcing effect of chitin whiskers (ChW) in PVA matrix	The result shows that the tensile strength and the mechanical properties of the polymer increase to 1880 MPa and 535 toughness (68 J/g) at 5 wt% ChW loading. Furthermore, tensile modulus was observed to reach its peak value of 50 GPa at 30 wt% ChW loading	Uddin et al. (2012)
The study covered the preparation of bio-based nanocomposite films using cellulose whiskers as the reinforcing phase and chitosan as the matrix	The results showed that the whisker content enhances the mechanical properties of the composites. The tensile strength of the composite films in the dry state increased from 85 to 120 MPa with increasing filler content from 0 to 20 wt%	Li et al. (2009)
In this study, chitosan was used to reinforced nanocrystalline cellulose and reported the mechanical and barrier properties nanocomposite film	The nanocomposites showed a 25% increase in tensile strength and n 87% increase in tensile modulus at this CNC content reported an optimum content of 5 wt% of CNC in chitosan matrix	Khan et al. (2012)
In this study, a packaging material was fabricated using glycerol plasticized starch matrix films reinforced with chitosan nanoparticles by physical crosslinking	The nanocomposites show a good degree of dispersion, an increase in tensile strength and storage modulus	Chang et al. (2010a, b)

(continued)

Table 2 (continued)

Preparation/methods	Result	References
The researchers prepared bio-based plastics for packaging using eugenol-loaded chitosan nanoparticles by extrusion method. The base materials used in the study consist of mixture of cassava, rice and waxy rice flours	The incorporation of the base materials with eugenol introduced some changes in the physicochemical properties of the composites. The elongation at break behaved as the matrix. The incorporation of these nanoparticles caused a reduction of WVP, indicating that this property was enhanced by the incorporation of the nanofillers	Woranucha and Yoksana (2013)
The study reported on porous CNTs/chitosan composite with lamellar structure prepared by ice templating	The incorporation of chitosan leads to a well-defined microchannel porous structure that is biodegradable and biocompatible with promising adsorption properties	Wu and Yan (2013)
The study reported on novel chitosan/PVA thin adsorptive membranes modified with amino functionalized multi-walled carbon nanotubes	The formed nanocomposites effectively remove Cu(II) from water. Thus, the composite systems could offer exclusive properties as a composite in the removal of heavy metal ions and treatment of wastewater	Salehi et al. (2012)
The study talked about the synthesis of multi-walled CNTs/chitosan polymer composite modified glassy carbon electrode for sensitive simultaneous determination of levodopa and morphine	The CNTs-filled chitosan demonstrated a modulated release of dexamethasone	Babaei and Babazadeh (2011)
The researchers prepared polymeric composites containing chitosan/CNTs for improved blood biocompatibility and bone tissue engineering	The study shows that the manipulation of CNTs/chitosan composites leads to the formation of scaffold with promising properties for cell–tissue engineering applications	Sahithi et al. (2010)
In this study, chitosan/alginate was developed via layer-by-layer electrostatic deposition edible coating on fresh-cut melon model	The obtained film showed excellent gas-exchange performance and water vapour permeability property. The chitosan/alginate film displayed great potential for food packaging	Da Silva et al. (2013)
The researchers incorporated photoactive of chlorophyll, essential oil and carvacrol in chitosan/cyclodextrin films to prepare a bioactive food packaging material	The incorporation of the cyclodextrin increased the tensile strength of chitosan film and also enhanced the antimicrobial activities of the film	Sun et al. (2014), Higueras et al. (2014)

(continued)

Table 2 (continued)

Preparation/methods	Result	References
Caseinate, with excellent thermoplastic and film-forming properties, was combined with chitosan to prepare chitosan/caseinate film, through ionic interaction to form a coating material for paper packaging	Improvement like water vapour permeability	Khwaldia et al. (2014)
In the study, controlled release of nisin from HPMC, sodium caseinate, PLA and chitosan was investigated for active packaging applications	The material was observed to be favourably used in extending the shelf life of packaged foods	Iman and Maji (2012)
Combination of nisin and ε-polylysine with chitosan coating inhibits the white blush of fresh-cut carrots	When the chitosan/ε-polylysine/nisin film was applied to fresh-cut carrots, the carrots showed the inhibited respiration rate, the declined ascorbic acid, the decreased growth of microorganism and the suppressed synthesis of white blush and lignin	Song et al. (2017)
The extent of Maillard reaction on antioxidant properties was investigated using chitosan–glucose conjugates	The chitosan/glucose films, with enhanced antioxidant property, were observed to effectively delayed the declines of total soluble solids, decreased decay and weight loss, suppressed respiration rate and ensured better berry texture and higher sensory scores of food	Kosaraju et al. (2010)
The antimicrobial activity of lysozyme–chitosan was investigating and its effects in prolonging the shelf life of chicken eggs during storage	Lysozyme–chitosan films were reported to enhance the freshness of the egg during storage, improve shell strength and maintain the internal quality	Yuceer and Caner (2014)
A film-based gelatin reinforce with chitosan was reported to demonstrate improved properties for the fabrication of fish packaging	Films based on chitosan/gelatin were reported to show improved properties, like mechanical properties and barrier properties, against water vapour and light	Hosseini et al. (2016), Kowalczyk et al. (2015)
The mechanical/antimicrobial properties of PVA reinforced chitosan and chitosan/PVA/clay nanocomposites were investigated for its application in food packaging	The PVA was observed to contribute to the plasticization and improves the water and oxygen barrier properties of the obtained films, affirming its suitability as food packaging material	Wang et al. (2018), Giannakas et al. (2016)
The study reported the synthesis and characterization of nanoamphiphilic chitosan dispersed PLA–bionanocomposite films for packaging application	The formed films demonstrated improved thermal, mechanical and gas barrier properties	Pal and Katiyar (2016)

(continued)

Table 2 (continued)

Preparation/methods	Result	References
The study reported on the antimicrobial properties of composite films derived from PLA/starch/chitosan blended matrix	The increase in the tensile and thermal properties of the nanocomposite film further verified the promising use of the obtained material in packaging food	Bie et al. (2013)
The application of chitosan-g-salicylic acid towards arresting postharvest losses was investigated.	The formed nanocomposites were reported to alleviate the chilling injury and were able to preserve the quality of cucumber	Zhang et al. (2015)
The study reported on the effect of chitosan/nanosilica coating on the physicochemical characteristics of longan fruit under ambient temperature	The chitosan/nanosilica films were observed to lower the decay of food by preventing the membrane structure from peroxidation, thus preserving the quality and elongating the storage shelf life	Shi et al. (2013)
The study investigated the barrier properties of nanosilicon carbide/chitosan nanocomposites for food packaging application	In the study, the nanosilicon carbide incorporated chitosan nanocomposite films was reported to substantially decrease the oxygen barrier properties and enhances the thermal stability and strengthen the chemical resistance of the film	Pradhan et al. (2015)
The molecular dynamics simulations of hydration effects on solvation, diffusivity and permeability properties were investigated on chitosan/chitin films	The film-forming ability of the chitin blended with chitosan was reported to introduce special properties on the films such as lower oxygen permeability. The chitosan/silver nanoparticle bionanocomposite films on the other hand were also reported to have improved the mechanical and barrier properties of the obtained films	McDonnell et al. (2016)

2.9 Starch–Propolis-Based Composites

Propolis is a natural resinous blend extracted from plants buds, honey, wax and vanilla naturally utilized as anti-predator and sealant in maintaining beehives (Hashemi 2016). The material composition of raw propolis consists mainly of 50% resins, 30% waxes, 10% essential oils, 5% pollen and 5% of various organic compounds (Park et al. 2002; Burdock 1998; Pietta et al. 2002; Huang et al. 2014). The antioxidant, antimicrobial and antifungal activities of propolis are recently utilized by food industries as a preservative in food packing material in addition to other uses, especially in pharmaceutical (Wagh 2013).

The assessment of propolis as a suitable material for food packaging was also reported by da Silva et al. (2013). The purpose of the study was to prepare a powdered form of propolis with a good antioxidant property, storage stability and dispensability in both aqueous and alcohol medium. The antimicrobial properties of propolis were used as a bio-additive in coatings to control fruit quality losses. For this purpose, chitosan containing propolis-ethanolic extract was used as the coating material. The results showed a reduced deterioration index and infection diameter of the fungus Colletotrichum gloeosporioides, as compared to the control papayas, thus elongating the shelve life of the fruit by two days (Barrera et al. 2015). In a similar development, a patent filed by Mizuno (1989), further confirmed the active role played by propolis as a preservative in food packing material. Tosi et al. (Tosi et al. 2007) used the bactericidal and bacteriostatic properties of propolis by agar diffusion and plate culture methods. By relating the zone of cellular growth inhibition with the propolis extracts concentration, a linear inhibition response was obtained, thus confirming the suitability of propolis as component for food preservation.

Other study showed the effectiveness of blending propolis with biopolymers. Chitosan film containing beeswax was prepared as a coating material to prolong the shelf life of strawberries on storage following layer-by-layer coating method. In the study, the composite was observed to modify the respiration rates, slowed down the metabolism of the fruit and at the same time preserved the physical architecture and taste of strawberries (Velickova et al. 2013). In another study, chitosan film containing propolis extract high in polyphenols was developed. In another study, chitosan film containing propolis extract high in polyphenols, were developed, where the characteristics of the resultant films reported for high tensile strength, elongation at break, and antioxidant activity due to the increased addition of propolis to the chitosan polymer. The blending of chitosan with the propolis further reduces the water vapour/oxygen permeability index, thus improving the potentials of employing the chitosan-based propolis films as active materials for the food packaging (Siripatrawan and Vitchayakitti 2016).

Acknowledgements Drs. FM and HAAL are grateful to the Deanship of Scientific Research, King Saud University for funding through Vice Deanship of Scientific Research Chairs.

References

Aider M (2010) Chitosan application for active bio-based films production and potential in the food industry. LWT Food Sci Technol 43:837–842

Arfin T (2015) Chitosan and its derivatives: overlook of commercial application in diverse field. In: Ahmed S, Ikram S (eds) Chitosan: derivatives, composites and applications. Scrivener Publishing LLC, USA

Arfin T, Athar S (2018) Graphene for advanced organic photovoltaics. In: Kanchi S, Ahmed S, Sabela MI, Hussian CM (eds) Nanomaterials: biomedical and environmental applications. Scrivener Publishing LLC, USA

Arfin T, Mohammad F (2016) Chemistry and structural aspects of chitosan towards biomedical applications. In: Ikram S, Ahmed S (eds) Natural polymers: derivatives, blends and composites. Nova Science Publishers, New York

Arfin T, Tarannum A (2017) Polymer materials: from the past to the future. In: Ahmed S, Annu S, Ikram S (eds) Green polymeric materials: advanced and sustainable development. Nova Science Publishers, New York

Arfin T, Mohammad F, Yusof NA (2014) Biomass resources in environmental and socio-economic analysis of fuel-wood consumption. In: Hakeem KR, Jawaid M, Rashid U (eds) Biomass and bioenergy: processing and properties. Springer International Publishing, Switzerland

Armentano I, Bitinis N, Fortunati E et al (2013) Multifunctional nanostructured PLA materials for packaging and tissue engineering. Prog Polym Sci 38:1720–1747

Avella M, De Vlieger JJ, Errico ME, Fischer S, Vacca P, Volpe MG (2005) Biodegradable starch/clay nanocomposite films for food packaging applications. Food Chem 93:467–474

Avérous L, Pollet E (2012) Environmental silicate nano-biocomposites. Springer International Publishing, Switzerland

Ayana B, Suin S, Khatua BB (2014) Highly exfoliated eco-friendly thermoplastic starch (TPS)/poly(lactic acid) (PLA)/clay nanocomposites using unmodified nanoclay. Carbohydr Polym 110:430–439

Azimi B, Nourpanah P, Rabiee M, Arbab S (2014) Poly(ε-caprolactone) fiber: an overview. J Eng Fibers Fabr 9:74–90

Babaei A, Babazadeh M (2011) Multi-walled carbon nanotubes/chitosan polymer composite modified glassy carbon electrode for sensitive simultaneous determination of levodopa and morphine. Anal Methods 3:2400–2405

Bansal M, Chauhan GS, Kaushik A, Sharma A (2016) Extraction and functionalization of bagasse cellulose nanofibres to Schiff-base based antimicrobial membranes. Int J Biol Macromol 91:887–894

Barrera E, Gil J, Restrepo A, Mosquera K, Durango D (2015) A coating of chitosan and propolis extract for the postharvest treatment of papaya (Carica papaya L. cv. Hawaiiana). Rev Fac Nac Agron Medellín 68:7667–7678

Barzegar H, Azizi MH, Barzegar M, Hamidi-Esfahani Z (2014) Effect of potassium sorbate on antimicrobial and physical properties of starch–clay nanocomposite films. Carbohydr Polym 110:26–31

Bie P, Liu P, Yu L, Li X, Chen L, Xie F (2013) The properties of antimicrobial films derived from poly(lactic acid)/starch/chitosan blended matrix. Carbohydr Polym 98:959–966

Bordes P, Pollet E, Bourbigot S, Avérous L (2008) Structure and properties of PHA/clay nano-biocomposites prepared by melt intercalation. Macromol Chem Phys 209:1473–1484

Botana A, Mollo M, Eisenberg P, Torres Sanchez RM (2010) Effect of modified montmorillonite on biodegradable PHB nanocomposites. Appl Clay Sci 47:263–270

Bouyer E, Mekhloufi G, Rosilio V, Grossiord JL, Agnely F (2012) Proteins, polysaccharides, and their complexes used as stabilizers for emulsions: alternatives to synthetic surfactants in the pharmaceutical field? Int J Pharm 436:359–378

Bruzaud S, Bourmaud A (2007) Thermal degradation and (nano) mechanical behavior of layered silicate reinforced poly(3-hydroxybutyrate-co-3-hydroxyvalerate) nanocomposites. Polym Test 26:652–659

Burdock GA (1998) Review of the biological properties and toxicity of bee propolis (propolis). Food Chem Toxicol 36:347–363

Chang PR, Ai F, Chen Y, Dufresne A, Huang J (2009) Effects of starch nanaocrystals-graftpolycaprolactone on mechanical properties of waterbone polyurethane-based nanocomposites. J Appl Polym Sci 111:619–627

Chang PR, Jian R, Yu J, Ma X (2010a) Starch-based composites reinforced with novel chitin nanoparticles. Carbohydr Polym 80:420–425

Chang PR, Jian R, Yu J, Ma X (2010b) Fabrication and characterisation of chitosan nanoparticles/plasticised-starch composites. Food Chem 120:736–740

Chen GG, Qi XM, Guan Y, Peng F, Yao CL, Sun RC (2016) High strength hemicellulose-based nanocomposite film for food packaging applications. ACS Sustain Chem Eng 4:1985–1993
Chiou BS, Wood D, Yee E, Imam SH, Glenn GM, Orts WJ (2007) Extruded starch-nanoclay nancomposites: effects of glycerol and nanoclay concentration. Polym Eng Sci 47:1898–1904
Choi JS, Park WH (2004) Effect of biodegradable plasticizers on thermal and mechanical properties of poly(3-hydroxybutyrate). Polym Test 23:455–460
Choi Y, Simonsen J (2006) Cellulose nanocrystal-filled carboxymethyl cellulose nanocomposites. J Nanosci Nanotechnol 6:633–639
Choi JS, Lee SJ, Christ GJ, Atala A, Yoo JJ (2008) The influence of electrospun aligned poly (epsilon-caprolactone)/collagen nanofiber meshes on the formation of self-aligned skeletal muscle myotubes. Biomaterials 29:2899–2906
Chung YL, Ansari S, Estevez L et al (2010) Preparation and properties of biodegradable starch–clay nanocomposites. Carbohydr Polym 79:391–396
Cui Y, Kumar S, Kona BR, van Houcke D (2015) Gas barrier properties of polymer/clay nanocomposites. RSC Adv 5:63669–63690
D'Amico DA, Manfredi LB, Cyras VP (2012) Relationship between thermal properties, morphology, and crystallinity of nanocomposites based on polyhydroxybutyrate. J Appl Polym Sci 123:200–208
Da Silva MA, Iamanaka B, Taniwaki MH, Kieckbusch TG (2013) Evaluation of the antimicrobial potential of alginate and alginate/chitosan films containing potassium sorbate and natamycin. Packag Technol Sci 26:479–492
DeKesel C, Wauven CV, David C (1997) Biodegradation of polycaprolactone and its blends with poly(vinylalcohol) by microorganisms from a compost of house-hold refuse. Polym Degrad Stab 55:107–113
Dhandayuthapani B, Yoshida Y, Maekawa T, Kumar DS (2011) Polymeric scaffolds in tissue engineering application: a review. Int J Polym Sci 2011:290602 (19 pp)
El-Hadi A, Schnabel R, Straube E, Müller G, Henning S (2002) Correlation between degree of crystallinity, morphology, glass temperature, mechanical properties and biodegradation of poly (3-hydroxyalkanoate) PHAs and their blends. Polym Test 21:665–674
Fabunmi OO, Lope G. Tabil LG, Panigrahi S, Chang PR (2007) Developing biodegradable plastics from starch. ASABE section meeting paper no. RRV-07130. St. Joseph, Mich
Famá LM, Gañan P, Bernal CR, Goyanes S (2012) Biodegradable starch nanocomposites with low water vapour permeability and high storage modulus. Carbohydr Polym 87:1989–1993
Ferreira BMP, Zavaglia CAC, Duek EAR (2002) Films of PLLA/PHBV: thermal, morphological, and mechanical characterization. J Appl Polym Sci 86:2898–2906
Gain O, Espuche E, Pollet E, Alexandre M, Dubois P (2005) Gas barrier properties of poly (ε-caprolactone)/clay nanocomposites: influence of the morphology and polymer/clay interactions. J Polym Sci Polym Phys 43:205–214
Gao W, Dong H, Hou H, Zhang H (2012) Effects of clays with various hydrophilicities on properties of starch–clay nanocomposites by film blowing. Carbohydr Polym 88:321–328
Garcia MDS, Lagaron JM (2010) Novel clay-based nanobiocomposites of biopolyesters with synergistic barrier to UV light, gas, and vapour. J Appl Polym Sci 118:188–199
Garza MZT, Garcia S, Gonzalez MDSF, Nino KAL (2015) Edible active coatings based on pectin, pullulan, and chitosan increase quality and shelf life of strawberries (Fragaria ananassa). J Food Sci 80:M1823–M1830
George J, Siddaramaiah (2012) High performance edible nanocomposite films containing bacterial cellulose nanocrystals. Carbohydr Polym 87:2031–2037. https://www.sciencedirect.com/science/article/pii/S0144861711009258.
Ghanbarzadeh B, Almasi H, Entezami A (2011) Improving the barrier and mechanical properties of corn starch-based edible films: effect of citric acid and carboxymethyl cellulose. Ind Crop Prod 33:229–235
Giannakas A, Vlacha M, Salmas C et al (2016) Preparation, characterization, mechanical, barrier and antimicrobial properties of chitosan/PVOH/clay nanocomposites. Carbohydr Polym 140:408–415

Gol NB, Patel PR, Rao TVR (2013) Improvement of quality and shelf-life of strawberries with edible coatings enriched with chitosan. Postharvest Biol Technol 85:185–195
Guillen MCG, Gimenez B, López-Caballero ME, Montero MP (2011) Functional and bioactive properties of collagen and gelatin from alternative sources: a review. Food Hydrocoll 25:1813–1827
Gupta D, Venugopal J, Prabhakaran MP et al (2009) Aligned and random nanofibrous substrate for the in vitro culture of Schwann cells for neural tissue engineering. Acta Biomater 5:2560–2569
Gutiérrez TJ, Pérez E, Guzmán R, Tapia MS, Famá L (2014) Physicochemical and functional properties of native and modified by crosslinking, dark cush-cush yam (*Dioscorea trifida*) and cassava (*Manihot esculenta*) starch. J Polym Biopolym Phys Chem 2:1–5
Hashemi JM (2016) Biological effect of bee propolis: a review. Eur J Appl Sci 8:311–318
Hassani FS, Nafchi AM (2014) Preparation and characterization of bionanocomposite films based on potato starch/halloysite nanoclay. Int J Biol Macromol 67:446–458
Higueras L, Carballo GL, Munoz PH, Catala R, Gavara R (2014) Antimicrobial packaging of chicken fillets based on the release of carvacrol from chitosan/cyclodextrin films. Int J Food Microbiol 188:53–59
Homminga D, Goderis B, Hoffman S, Reynaers H, Groeninckx G (2005) Influence of shear flow on the preparation of polymer layered silicate nanocomposites. Polymers 46:9941–9954
Hosseini SF, Rezaei M, Zandi M, Farahmandghavi F (2016) Development of bioactive fish gelatin/chitosan nanoparticles composite films with antimicrobial properties. Food Chem 194:1266–1274
Hu SG, Jou CH, Yang MC (2004) Biocompatibility and antibacterial activity of chitosan and collagen immobilized poly(3-hydroxybutyric acid–co–3-hydroxyvaleric acid). Carbohydr Polym 58:173–179
Huang M, Yu J, Ma X (2006) High mechanical performance MMT-urea and formamide-plasticized thermoplastic cornstarch biodegradable nanocomposites. Carbohydr Polym 63:393–399
Huang S, Zhang CP, Wang K, GLi GQ, Hu FU (2014). Recent ddvances in the chemical composition of propolis. Molecules 19:19610–19632
Ikeo Y, Aoki K, Kishi H et al (2006) Nano clay reinforced biodegradable plastics of PCL starch blends. Polym Adv Technol 17:940–944
Iman M, Maji TK (2012) Effect of crosslinker and nanoclay on starch and jute fabric based green nanocomposites. Carbohydr Polym 89:290–297
Jafari J, Emami SH, Samadikuchaksaraei A, Bahar MA, Gorjipour F (2011) Electrospun chitosangelatin nanofiberous scaffold: fabrication and in vitro evaluation. Biomed Mater Eng 21:99–112
Jamshidian M, Tehrany EA, Imran M, Jacquot M, Desobry S (2010) Polylactic acid: production, applications, nanocomposites, and release studies. Compr Rev Food Sci Food Saf 9:552–571
Jantanasakulwong K, Leksawasdi N, Seesuriyachan P et al (2016) Reactive blending of thermoplastic starch, epoxidized natural rubber and chitosan. Eur Polym J 84:292–299
Jayakumar R, Prabaharan M, Nair SV, Tamura H (2010) Novel chitin and chitosan nanofibers in biomedical applications. Biotechnol Adv 28:142–150
Jendrossek D, Handrick R (2002) Microbial degradation of polyhydroxyalkanoates. Annu Rev Microbiol 56:403–432
Khan RA, Salmieri S, Dussault D et al (2010) Production and properties of nanocellulose-reinforced methylcellulose based biodegradable films. J Agri Food Chem 58:7878–7885
Khan A, Khan RA, Salmieri S et al (2012) Mechanical and barrier properties of nanocrystalline cellulose reinforced chitosan based nanocomposite films. Carbohydr Polym 90:1601–1608
Khwaldia K, Basta AH, Aloui H, El-Saied H (2014) Chitosancaseinate bilayer coatings for paper packaging materials. Carbohydr Polym 99:508–516
Kosaraju SL, Weerakkody R, Augustin MA (2010) Chitosanglucose conjugates: influence of extent of Maillard reaction on antioxidant properties. J Agric Food Chem 58:12449–12455

Kowalczyk D, Wiater MK, Nowak J, Baraniak B (2015) Characterization of films based on chitosan lactate and its blends with oxidized starch and gelatin. Int J Biol Macromol 77:350–359

Lee EJ, Khan SA, Park JK, Lim KH (2012) Studies on the characteristics of drug-loaded gelatin nanoparticles prepared by nanoprecipitation. Bioprocess Biosyst Eng 35:297–307

Lenz RW, Marchessault RH (2005) Bacterial polyesters: biosynthesis, biodegradable plastics and biotechnology. Biomacromol 6:1–8

Li M, Guo Y, Wei Y, MacDiarmid AG, Lelkes PI (2006) Electrospinning polyaniline-contained gelatin nanofibers for tissue engineering applications. Biomaterials 27:2705–2715

Li Q, Zhou J, Zhang L (2009) Structure and properties of the nanocomposite films of chitosan reinforced with cellulose whiskers. J Polym Sci 47:1069–1077

Li C, Fu X, Luo F, Huang Q (2013a) Effects of maltose on stability and rheological properties of orange oil-in-water emulsion formed by OSA modified starch. Food Hydrocoll 32:79–86

Li F, Biagioni P, Finazzi M, Tavazzi S, Piergiovanni L (2013b) Tunable green oxygen barrier through layer-by-layer self-assembly of chitosan and cellulose nanocrystals. Carbohydr Polym 92:2128–2134

Lim LT, Auras R, Rubino M (2008) Processing technologies for poly(lactic acid). Prog Polym Sci 33:820–852

Liu K, Lin X, Chen L, Huang L, Cao S, Wang H (2013) Preparation of microfibrillated cellulose/chitosan-benzalkonium chloride biocomposite for enhancing antibacterium and strength of sodium alginate films. J Agric Food Chem 61:6562–6567

Lo HY, Kuo HT, Huang YY (2010) Application of polycaprolactone as an anti-adhesion biomaterial film. Artif Organs 34:648–653

Lopez O, Garcia MA, Villar MA, Gentili A, Rodriguez MS, Albertengo L (2014) Thermo-compression of biodegradable thermoplastic corn starch films containing chitin and chitosan. LWT Food Sci Technol 57:106–115

Lörcks J (1998) Properties and applications of compostable starch-based plastic material. Polym Degrad Stability 59:245–249

Ludueña LNN, Kenny JMM, Vázquez A, Alvarez VA (2011) Effect of clay organic modifier on the final performance of PCL/clay nanocomposites. Mater Sci Eng A 529:215–223

Matsuda DKM, Verceheze AES, Carvalho GM, Yamashita F, Mali S (2013) Baked foams of cassava starch and organically modified nanoclays. Ind Crops Prod 44:705–711

McDonnell MT, Greeley DA, Kit KM, Keffer DJ (2016) Molecular dynamics simulations of hydration effects on solvation, diffusivity, and permeability in chitosan/chitin films. J Phys Chem B 120:8997–9010

Mischnick P, Momcilovic D (2010) Chemical structure analysis of starch and cellulose derivatives. Adv Carbohydr Chem Biochem 64:117–210

Mizuno M (1989) Food packaging materials containing propolis as a preservative. Japanese patent no. JP Ol, 243(974), 89

Mograkar PR, Arfin T (2017) Chemical and structural importance of starch based derivatives and its applications. In: Ikram S, Ahmed S (eds) Natural polymers: derivatives, blends and composite. Nova Science Publishers, New York

Mohammad F, Arfin T, Yusof NA (2015) Chemical processes and reaction by-products involved in the biorefinery concept of biofuel production. In: Hakeem KR, Jawaid M, Alothman OY (eds) Agricultural biomass based potential materials. Springer International Publishing, Switzerland

Molyneux RJ (1993) Isolation, characterization and analysis of polyhydroxy alkaloids. Phytochem Anal 4:193–204

Mondragón M, Mancilla JE, Rodríguez-González FJ (2008) Nanocomposites from plasticized high-amylopectin, normal and high-amylose maize starches. Polym Eng Sci 48:1261–1267

Muller J, González-Martínez C, Chiralt A (2017) Combination of poly(lactic) acid and starch for biodegradable food packaging. Materials 10:952 (22 pp)

Nasseri R, Mohammadi N (2014) Starch-based nanocomposites: a comparative performance study of cellulose whiskers and starch nanoparticles. Carbohydr Polym 106:432–439

Nitayaphat W, Jintakosol T (2014) Removal of silver(I) from aqueous solutions by chitosan/carbon nanotube nanocomposite beads. Adv Mater Res 893:166–169

Noshirvani N, Ghanbarzadeh B, Mokarram RR, Hashemi M (2017) Novel active packaging based on carboxymethyl cellulose-chitosan-ZnO NPs nanocomposite for increasing the shelf life of bread. Food Pack Shelf Life 11:106–114

Olsson E, Hedenqvist M, Johansson C, Järnström L (2013) Influence of citric acid and curing on moisture sorption, diffusion and permeability of starch films. Carbohydr Polym 94:765–772

Pal AK, Katiyar V (2016) Nanoamphiphilic chitosan dispersed poly(lactic acid) bionanocomposite films with improved thermal, mechanical, and gas barrier properties. Biomacromol 17:2603–2618

Pandey JK, Sing RP (2005) Green nanocomposites from renewable resources: effect of plasticizer on the structure and material properties of clay-filled starch. Starch 57:8–15

Pantoustier N, Lepoittevin B, Alexandre M et al (2002) Biodegradable polyester layered silicate nanocomposites based on poly(ε-caprolactone). Polym Eng Sci 42:1928–1937

Park H, Li X, Jin C, Park C, Cho W, Ha C (2002) Preparation and properties of biodegradable thermoplastic starch/clay hybrids. Macromol Mater Eng 287:553–558

Park H, Lee W, Park C, Cho W, Ha C (2003) Enviromentally friendly polymer hybrids. Part 1 mechanical, thermal and barrier properties of thermoplastic starch/clay nanocomposites. J Mater Sci 38:909–915

Pérez CJ, Alvarez VA, Mondragón I, Vázquez A (2007) Mechanical properties of layered silicate/starch polycaprolactone blend nanocomposites. Polym Int 56:686–693

Pérez CJ, Alvarez VA, Vázquez A (2008) Creep behaviour of layered silicate/starch–polycaprolactone blends nanocomposites. Mater Sci Eng A 480:259–265

Pietta PG, Gardana C, Pietta AM (2002) Analytical methods for quality control of propolis. Fitoterapia 73:S7–S20

Popuri SR, Frederick R, Chang CY, Fang SS, Wang CC, Lee LC (2014) Removal of copper(II) ions from aqueous solutions onto chitosan/carbon nanotubes composite sorbent. Desalin Water Treat 52:691–701

Poverenov E, Danino S, Horev B, Granit R, Vinokur Y, Rodov V (2014) Layer-by-layer electrostatic deposition of edible coating on fresh cut melon model: Anticipated and unexpected effects of alginate-chitosan combination. Food Bioprocess Technol 7:1424–1432

Pradhan GC, Dash S, Swain SK (2015) Barrier properties of nano silicon carbide designed chitosan nanocomposites. Carbohydr Polym 134:60–65

Qian YF, Zhang KH, Chen F, Ke QF, Mo XM (2011) Cross-linking of gelatin and chitosan complex nanofibers for tissue-engineering scaffolds. J Biomater Sci Polym Ed 22:1099–1113

Qu P, Gao YA, Wu GF, Zhang LP (2010) Nanocomposites of poly(lactic acid) reinforced with cellulose nanofibrils. BioResources 5:1811–1823

Reis KC, Pereira J, Smith AC et al (2008) Characterization of polyhydroxybutyrate-hydroxyvalerate (PHB-HV)/maize starch blend films. J Food Eng 89:361–369

Rui L, Xie M, Hu B, Zhou L, Yin D, Zeng X (2017) A comparative study on chitosan/gelatin composite films with conjugated or incorporated gallic acid. Carbohydr Polym 173:473–481

Rydz J, Sikorska W, Kyulavska M, Christova D (2015) Polyester-based (bio)degradable polymers as environmentally friendly materials for sustainable development. Int J Mol Sci 16:564–596

Sahithi K, Swetha M, Ramasamy K, Srinivasan N, Selvamurugan N (2010) Polymeric composites containing carbon nanotubes for bone tissue engineering. Int J Biol Macromol 46:281–283

Sáinz CB, Bustillos RJA, Wood DF, Williams TG, McHugh TH (2010) Composite edible films based on hydroxypropyl methylcellulose reinforced with microcrystalline cellulose nanoparticles. J Agric Food Chem 58:3753–3760

Sáinz CB, Bras J, Williams T et al (2011) HPMC reinforced with different cellulose nano-particles. Carbohydr Polym 86:1549–1557

Saiz PF, Sánchez G, Soler C, Lagaron JM, Ocio MJ (2013) Chitosan films for the microbiological preservation of refrigerated sole and hake fillets. Food Control 34:61–68

Salehi E, Madaeni SS, Rajabi L et al (2012) Novel chitosan/poly(vinyl) alcohol thin adsorptive membranes modified with amino functionalized multi-walled carbon nanotubes for Cu(II) removal from water: preparation, characterization, adsorption kinetics and thermodynamics. Sep Purif Technol 89:309–319

Sarazin P, Li G, Orts WJ, Favis BD (2008) Binary and ternary blends of polylactide, polycaprolactone and thermoplastic starch. Polymer 49:599–609

Schlemmer D, Angélica RS, Sales MJA (2010) Morphological and thermomechanical characterization of thermoplastic starch/montmorillonite nanocomposites. Comp Struct 92:2066–2070

Shan GF, Gong X, Chen WP, Chen L, Zhu MF (2011) Effect of multi-walled carbon nanotubes on crystallization behavior of poly(3-hydroxybutyrate–co–3-hydroxyvalerate). Colloid Polym Sci 289:1005–1014

Shang L, Fei Q, Zhang YH, Wang XZ, Fan DD, Chang HN (2011) Thermal properties and biodegradability studies of poly(3-hydroxybutyrate–co–3-hydroxyvalerate). J Polym Environ 20:23–28

Shawky HA, El-Aassar AHM, Abo-Zeid DE (2012) Chitosan/carbon nanotube composite beads: preparation, characterization, and cost evaluation for mercury removal from wastewater of some industrial cities in Egypt. J Appl Polym Sci 125:E93–E101

Sheth M, Kumar RA, Davé V, Gross RA, McCarthy SP (1997) Biodegradable polymer blends of poly(lactic acid) and poly(ethylene glycol). J Appl Polym Sci 66:1495–1505

Shi S, Wang W, Liu L, Wu S, Wei Y, Li W (2013) Effect of chitosan/nano-silica coating on the physicochemical characteristics of longan fruit under ambient temperature. J Food Eng 118:125–131

Siracusa V, Rocculi P, Romani S, Rosa MD (2008) Biodegradable polymers for food packaging: a review. Trends Food Sci Technol 19:634–643

Siripatrawan U, Vitchayakitti W (2016) Improving functional properties of chitosan films as active food packaging by incorporating with propolis. Food Hydrocoll 61:695–702

Sohier J, Carubelli I, Sarathchandra P, Latif N, Chester AH, Yacoub MH (2014) The potential of anisotropic matrices as substrate for heart valve engineering. Biomaterials 35:1833–1844

Song Z, Li F, Guan H, Xu Y, Fu Q, Li D (2017) Combination of nisin and ε-polylysine with chitosan coating inhibits the white blush of fresh-cut carrots. Food Control 74:34–44

Sun X, Sui S, Ference C, Zhang Y, Sun S, Zhou N, Zhu W, Zhou K (2014) Antimicrobial and mechanical properties of betacyclodextrin inclusion with essential oils in chitosan films. J Agric Food Chem 62:8914–8918

Sundaram J, Pant J, Goudie MJ, Mani S, Handa H (2016) Antimicrobial and physicochemical characterization of biodegradable, nitric oxide-releasing nanocellulose-chitosan packaging membranes. J Agric Food Chem 64:5260–5266

Takegawa A, Murakami M, Kaneko Y, Kadokawa J (2010) Preparation of chitin/cellulose composite gels and films with ionic liquids. Carbohydr Polym 79:85–90

Taravel MN, Domard A (1993) Relation between the physicochemical characteristics of collagen and its interactions with chitosan: I. Biomaterials 14:930–938

Thakur VK, Thakur MK (2014) Processing and characterization of natural cellulose fibers/thermoset polymer composites. Carbohydr Polym 109:102–117

Thakur VK, Singha AS, Kaur I et al (2010) Silane functionalization of saccharum cilliare fibers: thermal, morphological, and physicochemical study. Int J Polym Anal Charact 15:397–414

Thakur VK, Yan J, Lin MF et al (2012) Novel polymer nanocomposites from bioinspired green aqueous functionalization of BNNTs. Polym Chem 3:962–969

Thakur VK, Thakur MK, Gupta RK (2014) Review: raw natural fiber–based polymer composites. Int J Polym Anal Charact 19:256–271

Tosi EA, Re E, Ortega ME, Cazzoli AF (2007) Food preservative based on propolis: bacteriostatic activity of propolis polyphenols and flavonoids upon *Escherichia coli*. Food Chem 104:1025–1029

Uddin AJ, Fujie M, Sembo S, Gotoh Y (2012) Outstanding reinforcing effect of highly oriented chitin whiskers in PVA nanocomposites. Carbohydr Polym 87:799–805

Velickova E, Winkelhausen E, Kuzmanova S, Alves VD, Martins MM (2013) Impact of chitosan-beeswax edible coatings on the quality of fresh strawberries (Fragaria ananassa cv Camarosa) under commercial storage conditions. LWT Food Sci Technol 52:80–92
Vimaladevi S, Panda SK, Xavier KA, Bindu J (2015) Packaging performance of organic acid incorporated chitosan films on dried anchovy (*Stolephorus indicus*). Carbohydr Polym 127:189–194
Wagh VD (2013) Propolis: a wonder bees product and its pharmacological potentials. Adv Pharm Sci 308249 (11 pp)
Wan YZ, Luo H, He F, Liang H, Huang Y, Li XL (2009) Mechanical, moisture absorption, and biodegradation behaviours of bacterial cellulose fibre-reinforced starch biocomposites. Comp Sci Technol 69:1212–1217
Wang H, Qian J, Ding F (2018) Emerging chitosan-based films for food packaging applications. J Agric Food Chem 66:395–413
Wilhelm HM, Sierakowski MR, Souza GP, Wypych F (2003) Starch films reinforced with mineral clay. Carbohydr Polym 52:101–110
Woranucha S, Yoksana R (2013) Eugenol-loaded chitosan nanoparticles: II. Application in biobased plastics for active packaging. Carbohydr Polym 96:586–592
Wu T, Yan J (2013) Porous CNTs/chitosan composite with lamellar structure prepared by icetemplating. In: Proceedings of SPIE-the international society for optical engineering, 8923, art no 89233A
Wu J, Lin J, Zhou M, Wei C (2000) Synthesis and properties of starch-graftpolyacrylamide/clay superabsorbent composite. Macromol Rapid Commun 21:1032–1034
Xiao W, Xu J, Liu X, Hu Q, Huang J (2013) Antibacterial hybrid materials fabricated by nanocoating of microfibril bundles of cellulose substance with titania/chitosan/ silver-nanoparticle composite films. J Mater Chem B 1:3477–3485
Xie Q, Hu X, Hu T, Xiao P, Xu Y, Leffew KW (2015) Polytrimethylene terephthalate: an example of an industrial polymer platform development in China. Macromol React Eng 9:401–408
Yu JG, Zhao XH, Yang H et al (2014) Aqueous adsorption and removal of organic contaminants by carbon nanotubes. Sci Total Environ 482–483:241–251
Yuceer M, Caner C (2014) Antimicrobial lysozyme-chitosan coatings affect functional properties and shelf life of chicken eggs during storage. J Sci Food Agric 94:153–162
Zabihzadeh SM (2010) Water uptake and flexural properties of natural filler/hdpe composites. BioResources 5:316–323
Zhang Y, Zhang M, Yang H (2015) Postharvest chitosan-g-salicylic acid application alleviates chilling injury and preserves cucumber fruit quality during cold storage. Food Chem 174: 558–563
Zhou Q, Malm E, Nilsson H et al (2009) Nanostructured biocomposites based on bacterial cellulosic nanofibers compartmentalized by a soft hydroxyethylcellulose matrix coating. Soft Matter 5:4124–4130
Zhuang X, Cheng B, Kang W, Xu X (2010) Electrospun chitosan/gelatin nanofibers containing silver nanoparticles. Carbohydr Polym 82:524–527
Zuraida A, Yusliza Y, Anuar H, Mohd Khairul Muhaimin R (2012) The effect of water and citric acid on sago starch bio-plastics. Int Food Res J 19:715–719

Recent Developments in Chitosan-Based Nanocomposites

Swaroopini Ramachandran, Malladi Rajinipriya, Jeremie Soulestin and Malladi Nagalakshmaiah

Abstract Chitosan is the second most abundant natural biopolymer available on the earth. Chitosan possesses its own identity among various other biopolymers due to its remarkable structural, chemical and mechanical properties. Recently, chitosan-based materials have emerged as ideal candidates for a wide range of applications such as biomedical, tissue engineering, filter and composite fields. This is mainly due to their accessibility for surface modification along with their non-toxic, biocompatible and biodegradable properties. Chitosan-incorporated composites appeared as most astonishing in their physical and mechanical properties. This chapter highlights the structural chemistry, extraction process and the different preparation methods of chitosan-based composites, along with most interesting, advanced studies developed from last decade.

Keywords Chitosan · Structural chemistry · Nanocomposites · Wound dressing and biomedical applications

1 Introduction

Chitosan is a naturally occurring polymer derived from chitin, prepared through thermochemical deacetylation in alkaline conditions. Chitin occurs as ordered crystalline microfibrils forming structural components in crustaceans, insects and fungi.

S. Ramachandran
Faculty of Agricultural and Environmental Sciences, McGill University, Macdonald Campus, 21, 111 Lakeshore St Anne de Bellevue, Montreal, QC, Canada

M. Rajinipriya
Center for Innovation in Technological Eco-design (CITE), University of Sherbrooke, 730 rue Bernard, Granby, QC, Canada

J. Soulestin · M. Nagalakshmaiah (✉)
Department of Polymer and Composite Technology & Mechanical Engineering (TPCIM), Institut Mines-Telecom, IMT Lille Douai, 941 rue Charles Bourseul, CS 10838, F-59508 Douai, France
e-mail: Nagalakshmaiah.malladi@imt-lille-douai.fr

M. L. Sanyang and M. Jawaid (eds.), *Bio-based Polymers and Nanocomposites*, https://doi.org/10.1007/978-3-030-05825-8_9

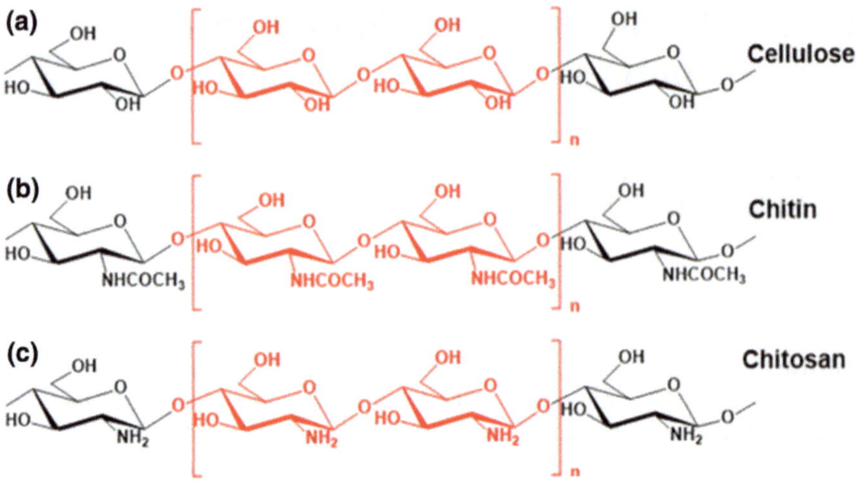

Fig. 1 Structures of **a** cellulose, **b** chitin and **c** chitosan. Adopted from Mohammed et al. (2017)

Commercially, chitin is sourced from the shell wastes of shrimp, lobster, krill and crabs. Chitin is the nature's second most abundant polymer, and several million tonnes of chitin are harvested annually, thus making it cheap and convenient (Gaf 1992). Chitin is a linear cationic heteropolymer with random residues of *N*-acetylglucosamine and glucosamine, which are linked with β-1-4 linkages. Chitobiose is represented by 4-*O*-(2-amino-2-deoxy-b-D-glucopyranosyl)-(1→4)-2-amino-2-deoxy-D glucose as the monomer (Yui et al. 1994). Chitin and cellulose share similar structures, with an acetamido group instead of the hydroxyl group at the C2 position as shown in Fig. 1. The different polymeric forms of chitin such as α, β and γ differ in their packaging and polarities of adjacent chains in successive sheets (Pillai et al. 2009).

In chitin, the degree of acetylation is usually 0.9, indicating that there are some amino groups, so the degree of *N*-acetylation or the ratio of 2-acetamido-2-deoxy-D-glucopyranose to 2-amino-2-deoxy-D-glucopyranose structural units has a major effects on its properties (Pillai et al. 2009). Chitin, when treated with an aqueous 40–45% w/v% of NaOH solution at 90–120 °C for 4–5 h, leads to deacetylation of chitin. The conditions that are used for deacetylation determines the molecular weight of the polymer and its degree of deacetylation. The major difference between chitin and chitosan (CS) is their solubility in different solvents (Pillai et al. 2009). This unsophisticated deacetylation reaction of chitin yields chitosan and is ideal for biofabrication of different nanocomposites in different applications such as biomedical, tissue engineering, polymer nanocomposites, membranes and wound dressing.

2 Chemistry and Properties

Chitosan has three major reactive sites: the amino group and two hydroxyl groups on the glucosidic residues as shown in Fig. 1. The amino group of chitosan is pH sensitive, making it responsible for the cationic nature and physiochemical properties, dictating the solubility and biological properties. Due to extensive hydrogen bonding, chitin is insoluble in water, low acid or organic solvents, and on the other hand, post-deacetylation, chitin, now in the form of chitosan, is readily soluble in dilute acidic solutions below a pH of 6.0, due to its basic nature, by virtue of presence of primary amino groups with a pKa of 6.3. At a low pH, protonation of these groups makes chitosan water soluble, and as pH increases, this deprotonation is reversed, and it is soluble (Cheaburu-Yilmaz et al. 2015). Protonation of the amine groups leads to the sorption of metal cations in acidic media because of competition between the metal cation and the amine group.

The porosity and material crystallinity of the material greatly affect the sorption properties. The sorption kinetics is thus dependent on the resistance to intra-particle diffusion and can be improved by cross-linking, allowing smaller sorbent particles and extra linkages between polymeric chains thus increasing solute transfer. Essentially, nanocomposites by virtue of their small size can enhance the sorption performance (Jaworska et al. 2003a, b; Piron et al. 1997). The solubility of chitin and chitosan has been shown in Table 1. Chitosan, a highly basic polysaccharide, displays superior viscosity, solubility in various solvent media, polyelectrolyte behaviour, film-forming capacity, metal chelation (Chelating agent), mucoadhesivity, polyoxysalt formation, optical and structural properties (Shukla et al. 2013). The antifungal or antimicrobial activities of the molecule are vital, and this phenomenon is possible because of the polycationic nature of chitosan (Shukla et al. 2013). The antimicrobial action of chitosan is hypothesized to be mediated by the electrostatic forces between the protonated amino group (NH2) in chitosan and the negative residues at cell surfaces (Tsai et al. 2002).

Table 1 Solubility of chitin and chitosan in the combination of organic solvents (adopted with permission from Mourya and Inamdar 2008)

Name of the biopolymer	Solvent combination
Chitin	5%LiCl in dimethylacetamide (DMA) and dimethylacetamide (DEA) LiCl in *N*-methyl-2-pyrrolidone (NMP) $CaCl_2$ in saturated methanol Hexafluoroisopropyl alcohol Hexafluoroacetone sesquihydrate Mixture of 1,2-dichloromethane and trichloroacetic acid (35:65) Saturated solution of lithium thiocyanate
Chitosan	Dilute aqueous organic or mineral acids below pH 6.5 Dimethyl sulphoxide (DMSO) p-Toluene sulphonic acid 10-camphoursulphonic acid

3 Extraction of Chitosan

Extraction of chitosan comprises three steps: (1) demineralization, (2) deproteinization and (3) deacetylation. The method could be in different or varied, involving demineralization subsequently, deproteinisation and deacetylation, or homogeneous where the sequence of events involves demineralization, deproteinisation and then deacetylation. A schematic representation of the chitosan extraction process is shown in Fig. 2.

Various factors such as chemical concentration, temperature used and reaction times play a role in the characteristics of the process (Samicho and Ramli 2011). Kurita (1998) found that the quality of chitosan, in terms of degree of deacetylation and molecular mass, was better when extracted heterogeneously (Kurita 1998). There have been a lot of studies that have investigated the parameters for optimal production of chitosan from various sources, and it has been summarized in the following table. Kurita and Abdou, Nagy et al. showed that the deacetylation process can be shortened by steeping in strong alkaline (sodium hydroxide) solution at room temperature prior to heating (Abdou et al. 2008; Kurita 1998).

3.1 *Demineralization*

Demineralization, as the name suggests, is the removal of minerals present in the raw materials (crab, shrimp, etc.). Raw materials such as crab, shrimp and lobster shells contain mainly calcium carbonate ($CaCO_3$) and some other minerals. Generally, in this stage, homogenized powder is treated with hydrochloric acid (HCl) solution for few hours in order to remove the minerals. Further, the remaining suspension is washed with water to remove the acid completely. The concentration of acid and duration of the reaction will differ depending on the source and biological background. Different conditions related to the demineralization step are listed in Table 2.

3.2 *Deproteinization*

Raw crab, shrimp and lobster shells have large amounts of protein (35–45%). In order to extract the chitin from the raw materials, protein must be removed. This

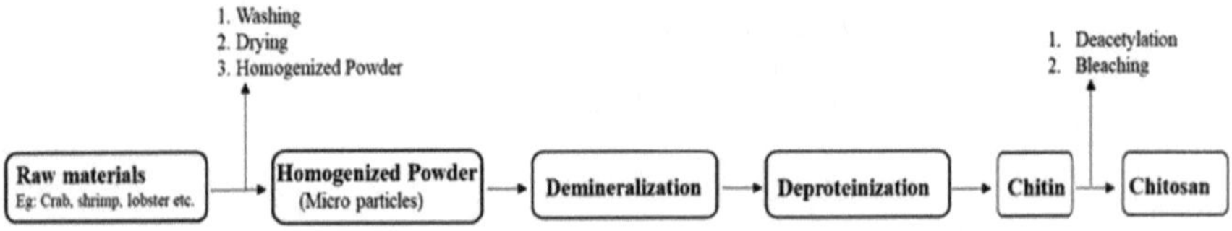

Fig. 2 Schematic representation of chitosan extraction process

Table 2 Conditions of demineralization process

Demineralization conditions	References
4% HCl, 25 °C, 24 h with a ratio of solid-to-solvent v/w equals to 15/1 (ml/g)	Seghir and Benhamza (2017)
1.0 N hydrochloric acid, HCl solution (1:15 solid-to-solvent ratio, w/v) with constant stirring for 1 h room temperature	Ahyat et al. (2017)
Room temperature using 1 M hydrochloric acid solution (40 millilitre per gram) for 3 h	Rasti et al. (2017)
0.73 mol/L hydrochloric acid for 132.61 min at room temperature	Tokatlı and Demirdöven (2017)
1 N HCl solution at room temperature for 6 h to remove	Yen et al. (2009)
1 N HCl solution for 2 h	Huang et al. (2018)
2-*N*-hydrochloric acid in the ratio 1:5 (w/v), allowed to stand for 1 h at 90 °C	Zahedi et al. (2018)

Table 3 Different conditions of deproteinization process

Deproteinization	References
3.5% sodium hydroxide, 1:5 HCl: sample (w/v), 1 h at 90 °C	Zahedi et al. (2018)
1 N NaOH at 80 °C for 1 h	Huang et al. (2018)
10% sodium hydroxide (NaOH) solution during 2 h at 70 °C and at a ratio of solvent to solid (v/w) of 15/1 ml/g	Seghir and Benhamza (2017)
1 M sodium hydroxide solution (20 millilitre per gram) at 70 °C	Rasti et al. (2017)
0.95 mol/L sodium hydroxide for 75.65 min at 60.49 °C	Tokatlı and Demirdöven (2017)
1.0 N sodium hydroxide, NaOH solution (1:15 solid-to-solvent ratio, w/v) at 80 °C for 6 h	Ahyat et al. (2017)

process is called deproteinization, which is crucial for the extraction of chitin. This step is also called as alkali treatment as raw materials are cooked with NaOH solution at different concentration and temperatures depending on the source of the raw materials. The major conditions reported in the literature were reported in Table 3.

3.3 Deacetylation

By treating the chitin with strong alkali solution, acetyl groups will be removed completely; hence this step is called deacetylation process. This reaction facilitates the formation of chitosan by forming the amine from amide group. This step was developed in different conditions, which are listed in Table 4.

Table 4 Different conditions reported on deacetylation process

Deacetylation	References
50% (w/v) of sodium hydroxide, NaOH at 1:20 solid-to-solvent ratio and stirred for 6 h at 110 °C	Ahyat et al. (2017)
50% sodium hydroxide at 100 °C for 720 min	Tokatlı and Demirdöven (2017)
45% sodium hydroxide solution (15 millilitre per gram) at 110 °C for 5, 15 and 24 h	Rasti et al. (2017)
100 °C, at a ratio of solvent to solid (v/w) of 10/1 (ml/g) and the duration reaction time is of 12 h, 40–50% NaOH	Seghir and Benhamza (2017)
50% sodium hydroxide, then heated using a hot plate at 90 °C for 80 min	Zahedi et al. (2018)

4 Preparation of Chitosan-Based Nanocomposites

The interest in the chitosan-based composites is increasing due to their remarkable biocompatible and antimicrobial properties. Chitosan-based composites are playing vital role in different applications such as tissue engineering, drug delivery, wound dressing, packaging, pharmaceutical drugs, electronic sensors and membranes. Figure 3 represents the different applications based on the chitin and chitosan in the modern world. Essentially, polymer nanocomposites based on chitosan are prepared by different methods such as microwave synthesis, spray drying, casting, heat curing, blending, melt processing, electrospinning, precipitation and phase inversion. The main objective of chitosan-based composites is to improve the barrier and antimicrobial, mechanical and biocompatible properties. In this context, polymer nanocomposites are prepared by different processing methods and the composition of the matrix-to-filler content will vary depending on the end-use applications.

Authors reported the recent developments (past five years) of the chitosan-based nanocomposites listed below. Table 5 consists of the composition of filler/matrix, processing methods and their end-use applications.

5 Applications

5.1 Biomedical Applications

Chitosan-based nanocomposites play a vital role in biomedical applications such as bone, cartilage, nerve, liver tissue engineering, drug delivery and wound dressing. This section highlights the importance of the chitosan in biomedical field. Tissue engineering is an interdisciplinary field that attractively combines life sciences and engineering to recuperate the tissues of the body that are damaged. It involves the use of in vitro seeding and proliferation of relevant cells in a biodegradable, non-toxic scaffold in various forms such as powders, gels and films (Dash et al. 2011).

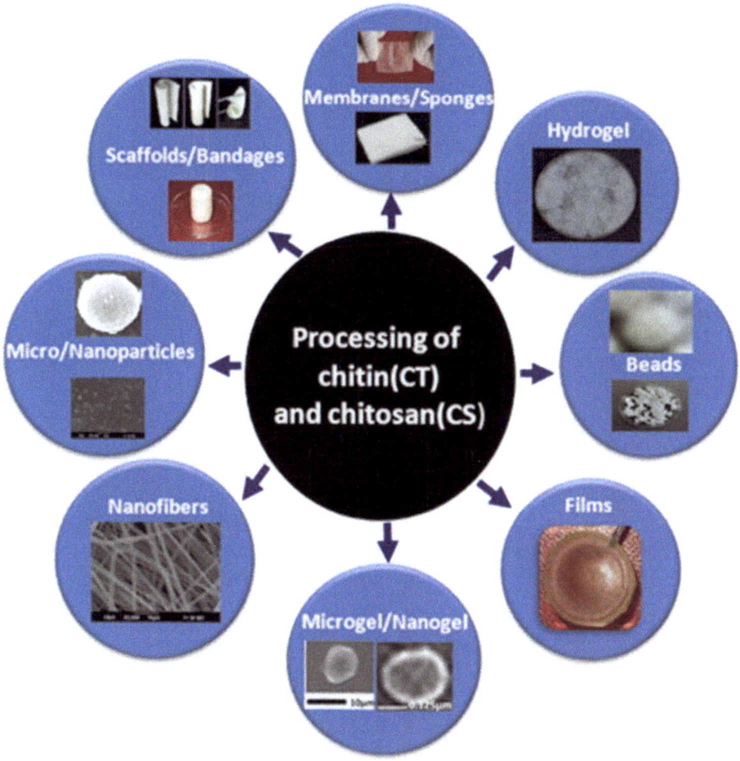

Fig. 3 Different applications based on chitin and chitosan (reprinted with permission from Anitha et al. 2014)

5.1.1 Bone Tissue Engineering

Chitosan has been extensively used in bone tissue engineering as it promotes cell growth and provides a mineral-rich matrix deposition by osteoblasts in culture (Seol et al. 2004). Chitosan is highly biocompatible (not harmful) and has been studied subcutaneously on rabbits, indicating that these chitosan-β tricalcium phosphate (TCP) composite scaffolds can be applied to non-loading bone regeneration (Yin et al. 2003); this allows the use of chitosan for different applications. However, chitosan and chitin are mechanically weak; thus, it is necessary to enhance the strength of the scaffold matrix by polymeric blends (Anitha et al. 2014). The mechanical strength of chitosan was enhanced by the addition of poly (lactic-co-glycolic)acid (PLAGA), and porous PLAGA/CS microsphere scaffolds were developed by sintering, which, unlike freeze drying, yielded a higher strength. Heparin-modified CS-PLAGA sintered scaffold was invented, in which the heparin was bound with recombinant human bone morphogenetic protein-2 (rhBMP-2), which enhances the osteoblastic proliferation and differentiation, thus facilitating bone formation (Abdel-Fattah et al. 2007; Jiang et al. 2006, 2010). Hydroxyapatite

Table 5 Composition of filler/matrix, processing methods and their end-use applications of the chitosan-based composites

S. no	System	Method	Size	Application	References
1.	Yolk shell Fe_3O_4 @PA Ni@Pd/ Chitosan nanocomposite			Sensitive determination of fluconazole in pharmaceutical preparations	Zad et al. (2018)
2.	ZnO nanoparticles (NPs), carboxymethyl chitosan (CMCS)	Spray drying	1–6 μm composites 10 nm size of ZnO	Antibacterial, improved biocompatible properties	Zhong et al. (2018)
3.	Chitosan and cassava starch nanostructured by montmorillonite (MMTNP) or bamboo nanofilters (BNF)	Casting	MMTNP: 315 ± 14 nm and BNF: 60 ± 3 nm		Llanos and Tadini (2018)
4.	Bentonite/chitosan		Surface area and pore volume of Bt 37.8 m^2/g, 0.1005 cm^3/g Bt/CS: 0.6 m^2/g, 0.0026 cm^3/g	Improvised thermal stability	Moussout et al. (2018)
5.	Chitosan/amino multi-walled carbon nanotubes (CS/AMWCNT) composite beads	Phase inversion	Specific surface area of 32.02 m^2/g for (CS/AMWCNT)	Bilirubin adsorption in hemoperfusion	Zong et al. (2018)
6.	Chitosan (CH)/polyvinyl alcohol (PVA) hydrogel and CH/PVA/ZnO nanocomposite		Size of ZnO nanoparticles is 30–35 nm	Antimicrobial properties	Abdeen et al. (2018)
7.	Chitosan (CH), PVA, ZnS, sulphosuccinic acid and plasticizers	Nanocomposite films were cross-linked by the heat curing process			Yun et al. (2017)
8.	Chitosan–bentonite nanocomposite (CBN) films			Wound dressing	Devi and Dutta (2017)

(continued)

Table 5 (continued)

S. no	System	Method	Size	Application	References
9.	Modified glassy carbon electrode with a chitosan nanocomposite film containing immobilized Mn(II)-IIP particles	Thermal polymerization		Electrochemical sensor for the determination of manganese (II) ion Mn(II) ions could be detected with a broad linear range from 2.0 μM to 9.0 μM	Roushani et al. (2017)
10.	Magnetic nanocomposite of clinoptilolite/maghemite/chitosan/urea		14 nm	Manganese removal from aqueous solution	Sareban and Javanbakht (2017)
11.	Carboxymethyl chitosan/CuO nanocomposite hydrogels		Size ranging from 20 to 50 nm	Antibacterial applications	Wahid et al. (2017)
12.	Chitosan/Soy protein isolate nanocomposite film reinforced by Cu nanoclusters		The average diameter of Cu NCs was approximately 50 nm	Packaging material	Li et al. (2017)
13.	Chitosan-poly(vinyl alcohol) nanocomposite films embedded with functionalized multi-walled carbon nanotube	Casting from aqueous solution		Adsorbents for the removal of Cd (II) from the aqueous solution	Mallakpour and Nezamzadeh Ezhieh (2017)
14.	Amine functional nanohydroxyapatite/chitosan bionanocomposite		58 ± 2 nm	Bone tissue engineering	Atak et al. (2017)
15.	Silver/g-carbon nitride/chitosan nanocomposite		20 nm	Possible applications in catalytic fields	An et al. (2017a)

(continued)

Table 5 (continued)

S. no	System	Method	Size	Application	References
16.	Nanocrystalline cellulose (NCC) containing chitosan-based nanocomposite films Dialdehyde cellulose (DAC) Cationized dialdehyde cellulose (CDAC)		NCC, DAC and CDAC was 283, 167 and 130 nm	Potential applications of the composite films in water-related fields	Tian et al. (2017)
17.	Preparation of chitosan/organoclay nanocomposite			silver(I) ion adsorbent	Nitayaphat and Jintakosol (2017)
18.	N-doped graphene natively grown on hierarchical porous carbon (N-PC-G/CS) nanocomposite	Ultrasonic method	BET surface area (541.627 m^2/g), pore size 4.1 nm	Novel sensor platform for determination of tartrazine (TT)	An et al. (2017b)
19.	Chitosan with PCL and clay nanocomposite films	Blending		Possibly food packaging	Rajesh and Harish Prashanth (2017)
20.	Chitosan/cloisite 10A nanocomposite		80 nm	Drug delivery systems	Roul et al. (2016)
21.	Haemoglobin immobilized on graphene quantum dots-chitosan (GQDs) nanocomposites	Hydrothermal method was used for the synthesis of GQDs		Sensitive and stable hydrogen peroxide biosensor	Mohammad-Rezaei and Razmi (2016)
22.	Chitosan-*g*-poly(acrylamide)/Zn (CPA-Zn) nanocomposite hydrogel	Synthesis in a microwave	10 nm	Controlled drug delivery of ofloxacin	Pathania et al. (2016)
23.	Chitosan–silver nanocomposite (CAgNC) v/s low molecular weight chitosan	Green synthesis through reduction method	Average sizes of LMW-chitosan and CAgNC were 1776 ± 23 nm and 240.1 ± 23.6 nm, respectively	Anticandidial activity	Dananjaya et al. (2016)

(continued)

Table 5 (continued)

S. no	System	Method	Size	Application	References
24.	Graphene oxide/chitosan/FeOOH nanocomposite	Modified Staudenmaier method		Removal of Pb(II) from aqueous solution	Sheshmani et al. (2015)
25.	Catechol-linked chitosan/carbon nanocomposite-modified electrode				Jirimali et al. (2015)
26.	Chitosan/polyrhodanine nanocomposite	Polymerization of rhodanine in chitosan solution	30 nm	Ni(II) ions removal from aqueous solution	Amiri et al. (2015)
27.	Na-montmorillonite (Na-MMT) intercalated medium molecular weight (MMW) chitosan		15.43 Å MMT basal spacing between layers	Drug delivery vehicle for in vitro release of theophylline sustained release tablet	Ainurofiq et al. (2014)
28.	TiO_2/Fe_2O_3/chitosan nanocomposite	Solution casting technology	20–30 nm	Photocatalytic degradation of Rhodamine B	Jia and Du (2015)
29.	Chitosan-g-poly acrylonitrile/silver nanocomposite	Graft copolymerization	15–20 nm		Hebeish et al. (2014)
30.	Nanocrystals of magnetite (Fe_3O_4), bare and coated with chitosan	alkaline precipitation	Both bare and coated MNPs (Fe_3O_4 and CS-Fe_3O_4) show particle size 21.8 ± 5.3 and 15.1 ± 5.0 nm respectively	In vivo hyperthermia therapy application	Shete et al. (2014)
31.	Chitosan–graphene oxide nanocomposite	Modified Hummer's method		Antibacterial activity	Sundar et al. (2014)
32.	Agarose–chitosan hybrid materials and nanocomposite ionogels		10–20 nm	Biotechnology and biomedical applications	Trivedi et al. (2014)
33.	Nanosized biosilica with chitosan as nanocomposites		20 nm	Absorption of azo dyes	Soltani et al. (2013)

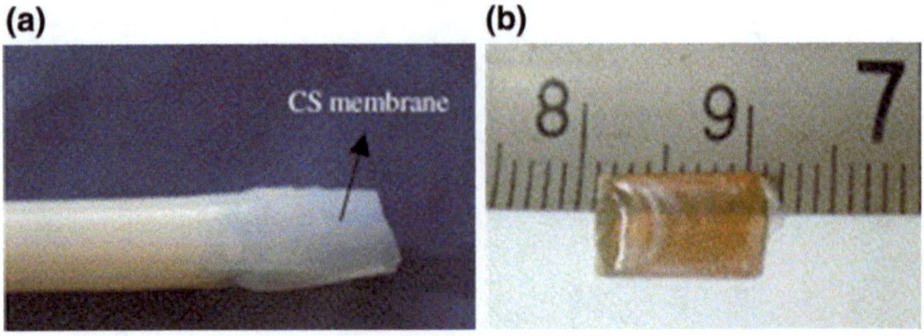

Fig. 4 Chitosan/hydroxy apatite composite wet gel rod (**a**) and dried rod (**b**) (reprinted with permission from Hu et al. 2004)

(HAp) is the major inorganic constituent in bone and teeth, and HAp or nano-HAp (nHAp) is being used to coat metallic implants or as ceramic composite matrices for bone tissue engineering applications (Glimcher 2006); further, these nHAp were incorporated into chitosan hydrogels to improve their mechanical strength (Chang et al. 2013). Hu et al. reported a chitosan/hydroxyapatite multi-layered nanocomposite (shown in Fig. 4) possessing excellent strength and bending modulus, thus making it suitable for internal fixation of long bone fractures (Hu et al. 2004). Several modifications have been made to enhance the properties of chitin/chitosan (Anitha et al. 2014).

By substituting chitosan with synthetic/natural polymers by different synthetic routes such as high temperature sintering, melt stretching and multilayer deposition for improved mechanical properties. Further, physical properties such as porosity and osteoconductivity growth factors were enhanced. Importantly, in vitro degradation significantly reduced.

5.1.2 Cartilage

Cartilage degeneration is often a result of genetic abnormalities, trauma or disease and may need intervention or replacement (Tuli et al. 2003). Cartilage engineering is aimed at complete healing, functional and scarless tissue and cell scaffold composites such as glycosaminoglycans (GAGs) which provide a stimulating environment for cartilage regeneration. Ideally, the substance that carries the cells should be similar to the natural environment, and cartilage-specific extracellular matrix components such as type II collagen and glycosaminoglycans are important in regulating chondrogenic development (Athanasiou et al. 2001; Kosher et al. 1973). Chitosan possesses structural similarity with GAG and hyaluronic acid, and this is being utilized in the form of microporous scaffolds by freeze drying and seeding porcine chondrocytes onto these constructs for chondrogenesis. Macroporous scaffolds could exploit the interconnecting porousness of chitosan scaffold for long-term cultures, as proven by Griffon et al., in some other study chitosan based sponges were used for developing a homogenous cell matrix deposit after dynamic culture for a month (Griffon et al. 2006; Nettles et al. 2002). An alginate/chitosan polymer fibre increased cell attachment and proliferation in vitro, when compared to alginate, exhibiting a high tensile strength (Iwasaki et al. 2004).

Chitosan-based PLLA composites showed an increased cell adhesion, biosynthetic activity and proliferative potential (Cui et al. 2003). Chitosan-based scaffolds provide controlled release of growth factors, which encourage the growth and biosynthetic ability. Transforming growth factor beta 1 (TGF β1) growth factor was incorporated into the porous collagen/chitosan/glycosaminoglycan (GAG) scaffold that leads to a controlled release of the growth factor and enhanced the regeneration and mechanical properties. A similar scaffold containing porous freeze-dried chitosan loaded with TGF β1 growth factor was used to treat cartilage defects (Lee et al. 2004). Hao et al. reported chitosan hydrogel which was formed for reconstructing cartilage and repair articular cartilage defects in sheep models shown in Fig. 5 (Hao et al. 2010).

Injectable hydrogel of chitosan and chitosan-based composites has shown promising effects on the regeneration of cartilage as they are structurally similar to the extracellular matrix. Direct injection of chitosan into injured cartilage of rat mimics the extracellular matrix (ECM) like other ECM mimickers such as chondroitin sulphate and hyaluronic acid which have showed infiltration of chondrocytes and thus enhance would healing (Jin et al. 2009; Lu et al. 1999; Park et al. 2013; Tan et al. 2009).

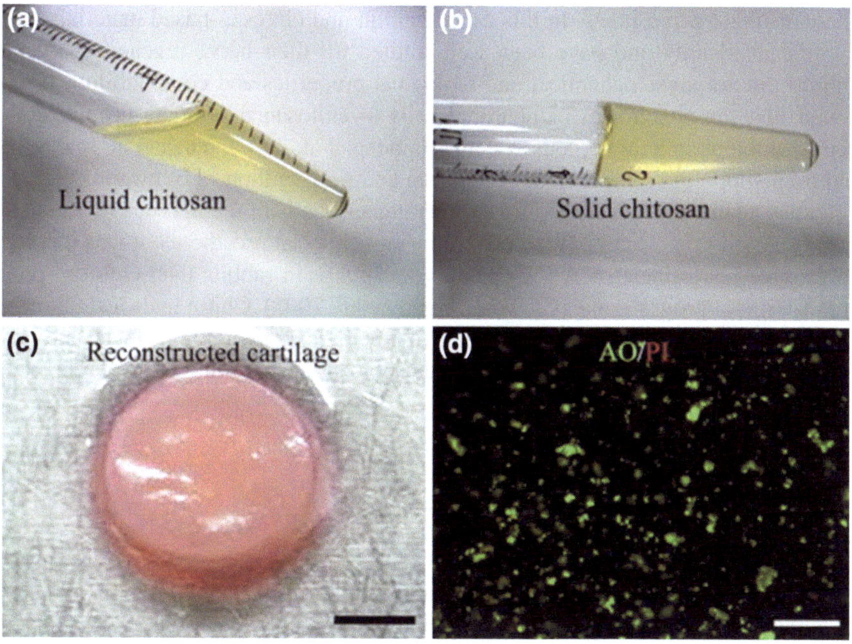

Fig. 5 Chitosan hydrogel for the reconstruction of the cartilage. **a** Chitosan hydrogels in liquid state at room temperature, **b** solidified chitosan hydrogels formed at 37 °C in 10–15 min, **c** the reconstructed cartilage in the 4-well plate, **d** the AO/PI staining showed that the chondrocytes remained >90% viable in the solid chitosan matrix after cultured 1 day in vitro, AO: green; PI: red. Bar = 0.5 cm (**c**), 100 μm (**d**). Reprinted with permission from (Hao et al. 2010)

5.1.3 Nerve, Liver Tissue Engineering

Lack of organ donors has prompted researchers and industries to find alternate therapies and treatment strategies. Bioartificial liver is a promising area of interest that can be used to treat fulminant hepatic failure (FHF) (Kim et al. 2008). Structural similarity of chitosan with the previously discussed ECM molecule GAG makes it a viable option for use in liver tissue engineering. A novel chitosan/collagen matrix was formed by using 1-ethyl-3-[3-dimethylaminopropyl] carbodiimide hydrochloride (EDC) as a cross-linking agent in *N*-hydroxy succinimide (NHS) buffer system. The cross-linked matrix provides good mechanical strength and compatibility with blood and hepatocytes. These scaffolds suffer from the risk of complexities such as thrombus formation, leading to occlusion and decreased membrane efficiency. Therefore, warranting more studies and exploring complex architectures will help to overcome these issues (Li et al. 2003a, b; Lindahl and Hook 1978; Wang et al. 2003). A similar chitosan/collagen/heparin matrix was developed for implantable bioartificial liver which exhibited good biocompatibility (Wang et al. 2005).

The goal of nerve repair strategies is to regenerate nerve fibres into endoneurial tubes, and there are two main ways: (a) bridging using grafting tubulization techniques and (b) end-to-end suturing of nerve stumps. However, grafting tubulization is essentially better as there is no tension at the repair sire (Ciardelli and Chiono 2006). Many biodegradable materials have been employed for the production of these artificial nerve tubes. In this aspect, chitin and chitosan-based materials have significant identity and have been well studied for their nerve regeneration capabilities due to physicochemical and biological properties and compatibility. It was found that the attachment of neuronal cells to chitosan membrane prompted the repair of the peripheral nervous system (Haipeng et al. 2000). Yuan et al. found that chitosan supported the adhesion, migration and proliferation of Schwann cells and provides a similar regeneration guide to Büngner bands in the nervous (Yuan et al. 2004). When progesterone was delivered using chitosan prostheses, it was reported to provide better facial nerve regenerative response in rabbits than chitosan prosthesis without progesterone (Chávez-Delgado et al. 2003). Chitin hydrogel tube was developed by Frier et al., without the use of any toxic cross-linking agents, and both chitin and chitosan indicated nerve cell adhesion and neurite outgrowth (Freier et al. 2005).

5.2 *Chitosan in Drug Delivery Applications*

Electrospun chitosan nanofibres have been used for drug delivery applications. It was found that a high-specific surface area and a short diffusion passage length provide a high release rate compared to the bulk material. The release profile can be controlled by variating the morphology, composition and porosity. Nanofibres of chitosan and other polymers such as polyvinyl alcohol (PVA), polyethylene glycol (PEG), poly(lactic-co-glycolic acid) (PLGA,) polyethylene glycol (PEO), polycaprolactone (PCL) and polylactic acid (PLA) have been used for drug release. Water-soluble carboxymethyl chitin (CMC) was used for drug delivery applications. CMC nanoparticles were prepared by cross-linking by using calcium chloride and ferric chloride, and these were generally non-toxic and found to have antibacterial behaviour. Further, due to their ferromagnetic behaviour it allows to use in drug trafficking systems. Hence, these kinds of scaffolds demonstrate promising future as a cancer drug delivery system (Jayakumar et al. 2010).

Another interesting study was reported by Dev et al. (2010) and developed poly (lactic acid) PLA/CS nanoparticle by emulsion method for anti-HIV drug delivery application (Dev et al. 2010). Recently, there have been attempts for target-specific drug delivery systems. In this framework, chitosan coated with doxorubicin has been used for treating liver cancer, and this system inhibits cell growth of liver cancer by promoting apoptosis and arresting cell cycle at G2/M phase through p53/PRC1 [Protein **Regulator of cytokinesis 1** (**PRC1**)] pathway (Ye et al. 2018).

5.3 *Chitosan-Based Wound Dressings*

Antibacterial resistance (AMR) of micro-organisms is one of the major problems that results in delayed wound healing and infections. Hence, new systems need to develop that are able to treat wounds and infections that are caused by AMR bacteria. Recently, number of studies reported the use of chitosan scaffolds and membranes for treatment of deep burns, wounds and cuts. Cellulose nanocrystal-loaded chitosan films with curcumin/Ag nanoparticles have been used as strong antimicrobial wound dressings which reduced wounds by 97.2% (Bajpai et al. 2017). Wound dressings often have similar disadvantages such as low flexibility, poor mechanical strength, lack of porosity and a tendency to adhere to the wound surface. Hydrogel-based wound dressings containing some antimicrobial agents provide a moist environment while preventing microbial infections are a promising alternative that have been explored. Hydrogel membranes were developed based on PVA, starch and chitosan with nano-ZnO. The results show that the dressing was effective in early stages of healing and it could be used for topical applications (Baghaie et al. 2017). A chitosan hydrogel/nano-ZnO bandage was developed by incorporation of zinc oxide nanoparticles into chitosan hydrogels. The nanocomposite bandage shows enhanced swelling, clotting and antimicrobial activity. In vivo evaluation in Sprague Dawley rats showed that they helped in faster recovery, re-epithelialization and collagen deposition as shown in Fig. 6. Thus, these could be used as bandages for burn wounds, chronic wounds and diabetic foot ulcers (Sudheesh Kumar et al. 2012).

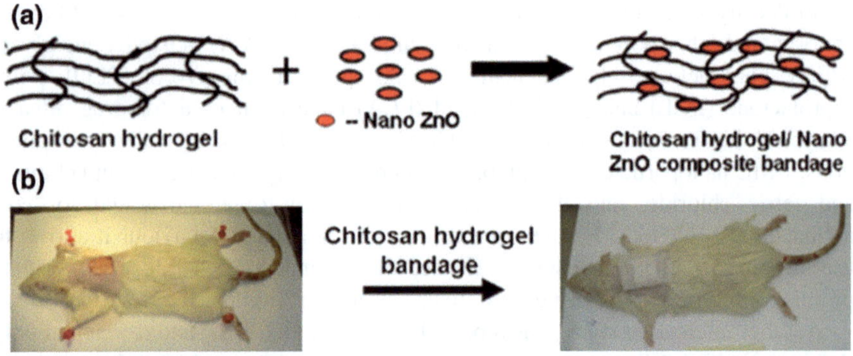

Fig. 6 Preparation of **a** chitosan/ZnO bandage and **b** in vivo wound healing study on rat (reprinted with permission from Sudheesh Kumar et al. 2012)

5.4 Antimicrobial Packaging Materials

5.4.1 Chitosan-Based Edible Films and Coatings

Chitin and chitosan are believed to be antimicrobial and antifungal because of their polycationic nature (Kim et al. 2003). Due to the electrostatic forces between the protonated amine group in chitosan and the electronegative charge at the cell surface was the key for the antimicrobial nature (Tsai et al. 2002). As the degree of deacetylation increases, the number of protonated amino groups present in chitosan increases. Hence, this deacetylation will help to improve the antimicrobial activity (Liu et al. 2004). Liu et al. also suggested that the bactericidal activity of chitosan is caused by the electrostatic interaction between amine groups and the phosphoryl group of the cell membrane phospholipids (Liu et al. 2004). Surprisingly, however, it was found that water-soluble chitosan supported and promoted the growth of *Candida albicans* even in acidic conditions, as against the water-insoluble counterpart, which exhibits inhibitory effects (Qin et al. 2006).

Chitosan layers formed are not compact when the chitosan molecules are very large, and the chitosan with lower molecular weight is difficult to precipitate and form a layer. The water-soluble chitosan is not able to form a layer which means that it has no antimicrobial activity. Even though the organism C. albicans has a pH optimum of growth at a low acid condition of 5.5, there is growth promotion in case of all water-soluble chitosan and chitooligomers at alkaline conditions! A possible reason for this is due to the structure of the cell wall, which contains chitin, hence having enhanced affinity with GlcN and GlcNAc residues, hence lowering its toxicity as well.

Chitosan films can be obtained by different methods such as casting, single or twin-screw extrusion. Generally, acidified solution of chitosan with or without plasticizer will help to form the films by casting evaporation (Cazón et al. 2017). Extrusion involves one or two rotating screws fitted in a barrel which progressively increase the pressure and mix the ingredients during the manufacture of films; upon completion, the mixture is expanded by passing through a dye (Espitia et al. 2014). Van den Broek et al. (2015) reported that extrusion is the most cost-efficient and solvent-free method. However, processing neat chitosan films by melt extrusion is not possible, as chitosan films are not flexible, thermally sealed like thermoplastics. Moreover, chitosan degrades before the melting point, and hence, processing neat chitosan films for commercial applications is not possible (Pelissari et al. 2011; van den Broek et al. 2015). In this context, researchers developed chitosan-based composites by blending with thermoplastics such as poly(butylene succinate), poly (butylene succinate adipate) and poly(butylene terephthalate adipate) (van den Broek et al. 2015). These blends help to improve the film-forming properties for commercial applications with enhanced mechanical and barrier properties.

Nevertheless, chitosan-based films are water permeable and thus limit usage in food applications (Aider 2010). In order to overcome these issues, scientist reported several strategies to improve the functionality and mechanical strength by changing the degree of deacetylation pH, solvent and plasticizers. Elsabee and Abdou et al. reported by mixing chitosan with other components such as proteins or polysaccharides (Elsabee and Abdou 2013), and cross-linking of chitosan with several reagents such as genipin, glutaraldehyde or formaldehyde has been used to prevent the dissolution or swelling of the chitosan-based films (Dutta et al. 2009).

Chitosan is inherently antimicrobial and can be used in tandem with antimicrobials or bioactive compounds, and has been studied broadly for the enhancement of food quality and safety (Ma et al. 2016). Encapsulation of essential oils in nanogels and nanoemulsions is a method that has been adopted in recent years to overcome issues with volatility, taste and odour and also help to improve the bioactivity of the essential oils through activation of passive cell absorption mechanisms (Rajaei et al. 2017). Such nanoemulsions and coatings have been used in the form of spray coatings (Rajaei et al. 2017; Severino et al. 2015) or have been brushed or sprayed (Ma et al. 2016) on the surface of the food to elicit antibacterial activity against foodborne pathogens. In a report by Sun, Sui and co-workers, chitosan films with different concentrations of beta-cyclodextrin (CD)/ trans-cinnamaldehyde, eugenol and carvacrol were used against *E. coli, S. typhimurium, Staphylococcus aureus* and *L. monocytogenes*. It was found that the films reduced bacterial growth, and the log reductions for β-CD/CIN CS and β-CD/CAR CS films were higher than β-CD/EUG CS films (Sun et al. 2014). Some interesting studies reported in the literature based on the Cinnamaldehyde (CIN) and Carvacrol (CAR) (Table 6).

5.5 *Chitosan-Based Membrane Technology*

By virtue of interesting properties such as described above, chitosan also finds applications in the field of membrane technology and is being used for many applications such as the preparation of adsorbent membranes for the removal of dyes, reactive metal ions and also for the creation of high-flux, antifouling filtration

Table 6 Chitosan-based systems and their properties

S. no	Formulation	Food matrix	Bacterial species	Important findings	References
1.	Clove essential oils encapsulated in chitosan–myristic acid nanogel.	Beef cutlet	*Salmonella enterica Enteritidis*	• Synergistic increase in the antioxidant activity of clove essential oil and chitosan–myristic acid nanogel was observed • MIC and MBC were 5 and 10 mg/l against the tested strain • Nanogel-encapsulated CEO more efficacious in in vivo studies (2 mg/L) versus just the essential oil (100 mg/L)	Rajaei et al. (2017)
2.	Chitosan (CS)–benzoic acid (BA) nanogel *Rosmarinus officinalis* essential oils	Beef cutlet	*Salmonella typhimurium*	• Mean diameter of nanoparticles was 100 nm • MIC and MBC were 100 and 200 mg/ml • There was an immediate reduction in the bacterial load post inoculation when compared to the control, where there was an increase during the 12-day storage period at 4 °C • Significant increase in the paleness of the beef cutlets observed	Hadian et al. (2017)
3.	Chitosan and gelatin films incorporating *Ziziphora clinopodioides* essential oil and ethanolic grape seed extract		• *Listeria monocytogenes* • *Salmonella typhimurium* • *Escherichia coli O157:H7*	• Reduced swelling index, tensile strength, puncture force and puncture deformation • Antimicrobial activities seen • Antibacterial effects against all tested bacteria were seen but there was no inhibition of growth	Shahbazi (2017)
4.	Lemongrass oil/chitosan emulsion (1% chitosan)	Grape berries	*Salmonella typhimurium*	• High shear mixing (HSM) and dynamic high pressure processing (DHP) yielded in particles sizes of 461.9–632.6 nm and 204.2–378.8 nm, respectively, and the DHP emulsions were more stable, further exhibiting a higher initial inhibition of Salmonella and better inhibition of total mesophilic aerobes, yeasts and moulds, retention of colour, total soluble solid content and antioxidant activity during storage at 25 °C • DHP minimized any sensory interferences when compared with uncoated berries and HSM • Nanoemulsions with and smaller droplet size were more efficacious when compared to the larger ones	Oh et al. (2017)

(continued)

Table 6 (continued)

S. no	Formulation	Food matrix	Bacterial species	Important findings	References
5.	Coating cantaloupes with 0.1% lauric arginate, 0.1% EDTA and 1% Cinnamon oil Chitosan-based coating	Cantaloupes	• *L. monocytogenes* • *S. enterica and* • *E. coli O157:H7*	• *S. enterica* counts reduced below detection limit after day 1, and no recovery was observed during storage • The redness and yellowness of uncoated cantaloupes were significantly higher than coated ones from day 6 • The firmness of uncoated cantaloupes and those coated with chitosan only was significantly lower than other treatments from day 10	Ma et al. (2016)
6.	Cumin and eucalyptus oil in chitosan	Chicken meat	• *Shigella dysenteriae* • *Listeria monocytogenes* • *Salmonella typhi* *Streptococcus pyogenes*	• MIC and MBC against *Salmonella* was found to be 1 and 2 mg/mL, which is significantly lower than that observed for the cumin and eucalyptus essential oils at 5.625 and 11.25 mg/mL • Lowest overall acceptability was for 2% cumin essential oil, while others did have an overall low acceptability against the control	Sharafati Chaleshtori et al. (2016)
7.	Fish gelatin containing 6% w/w chitosan nanoparticles with oregano essential oil		• *Staphylococcus aureus,* • *Listeria monocytogenes,* • *Salmonella enteritidis and* • *Escherichia coli*	1.2% OEO effectively inhibited the growth of tested microbes by forming zones between 27 and 33 mm	Hosseini et al. (2016)
8.	Chitosan incorporated with eucalyptus and cuminum essential oils	*Oncorhynchus mykiss* (rainbow trout)	• *Shigella dysenteriae,* • *Listeria monocytogenes,* • *Streptococcus pyogenes, and* • *Salmonella typhi*	• Minimum inhibitory concentration (MIC) and minimum bactericidal concentration (MBC) for the essential oils ranged between 0.351 and 11.25, and 1.406 and 11.25 mg/ml • Synergistic action between chitosan and the essential oils	Chaleshtori et al. (2016)
9.	Biocomposite films based on fish gelatin and chitosan containing Origanum vulgare L. essential oil (OEO)		• *Staphylococcus aureus* • *Listeria monocytogenes,* • *Salmonella enteritidis* • *Escherichia Coli*	• Highest test concentration inhibited all kinds of microbial growth producing inhibition zones between 32.0 and 34.1 mm • Decrease in the tensile strength and elastic modulus, increase in water solubility of the film	Hosseini et al. (2015)

(continued)

Table 6 (continued)

S. no	Formulation	Food matrix	Bacterial species	Important findings	References
10.	Chitosan-based coatings containing nanoemulsion of essential oils (EOs), gamma irradiation, modified atmosphere packaging (MAP) Carvacrol, lemon, bergamot and mandarin nanoemulsions 3% modified chitosan coatings	Green Beans	• *E. coli* • *S. typhimurium*	• MIC was 0.5 g/l for all four EOs for both the bacteria tested. • Use of this coating under MAP caused a synergistic effect with an increased radiosensitivity and reduction of microbial populations to level below detectable during the storage period for *E. coli* and from day 7 through end of storage for *Salmonella typhimurium*	Severino et al. (2015)
11.	Edible chitosan–cassava starch (CH-CS) coatings containing a mixture of *Lippia gracilis Schauer* genotypes	Guavas	• *Pseudomonas aeruginosa* • *Staphylococcus aureus* • *Bacillus cereus* • *Bacillus subtilis* • *Serratia marcescens* • *Escherichia coli* • *Listeria monocytogenes* • *Salmonella enteritidis*	• Formulations containing 2.0% cassava starch, 2.0% chitosan and 1.0, 2.0 or 3.0% EOM found to be the best inhibitors • Increased shelf life, lower titratable acidity and colour differences noticed	De Aquino et al. (2015)
12.	B-CD films incorporated with trans-cinnamaldehyde, eugenol and carvacrol		• *E. coli* • *S. typhimurium* • *Staphylococcus aureus* • *L. monocytogenes*	• Log reductions for β-CD/CIN CS and β-CD/CAR CS films were higher than β-CD/EUG CS films • The elongation at break (EB) decreased with the increasing concentrations of essential oils. Inclusion of the complex of β-cyclodextrin and 0.25% essential oils also significantly decreased water vapour permeability (WVP) of chitosan films	Sun et al. (2014)

(continued)

Table 6 (continued)

S. no	Formulation	Food matrix	Bacterial species	Important findings	References
13.	Cassava starch, chitosan and *Lippia gracilis* Schauer essential oil	Strawberries	• *Pseudomonas aeruginosa* • *Staphylococcus aureus* • *Bacillus cereus* • *Bacillus subtilis* • *Serratia marcescens* • *Escherichia coli* • *Enterococcus faecalis and* • *Salmonella Enteritidis*	• The formulation that showed the largest inhibition zone was: • (1.6% cassava starch, 0.6% chitosan and 2.4% oil) and 14 (1.0% cassava starch, 1.5% chitosan and 3.0% oil) enriched with LGRA106 or LGRA107 • Edible coating enriched with LGRA106 genotype (1.6% cassava starch, 0.6% chitosan and 2.4% essential oil) proved the potential to maintain the microbiological quality of strawberries throughout the storage period of a week at storage at 4 °C	Azevedo et al. (2014)
14.	Chitosan films incorporated with 1,8-cineole		*Salmonella Anatum*	No growth was observed in the film containing 4% w/w 1,8 cineole after the first day of incubation	Thangvaravut et al. (2012)
15.	Starch–chitosan films incorporated with orégano essential oil (OEO)		• *Bacillus cereus,* • *Escherichia coli,* • *Salmonella enteritidis, and* • *Staphylococcus aureus*	Effective inhibition against all four tested microbes, improved flexibility	Pelissari et al. (2009)

membranes with multifarious applications. This section gives an account of the latest developments in the field and the applicability of chitosan in various settings in the past 4 years.

5.5.1 Dye Removal from Wastewater

In modern world, water scarcity is growing and on the other hand, purification of the wastewater is becoming more challenging to researchers and industries. In this context, bio-based membranes generated from chitosan-based materials can be significant solution to resolve the wastewater purification issues. Wastewater polluted with dyes is a huge challenge because it causes serious damage to biological organisms. For the past few decades, this problem was treated by methods such as oxidation (Osugi et al. 2009), coagulation (He et al. 2015), adsorption (Habiba et al. 2017a), reverse osmosis (Su et al. 2016) and biodegradation (Guieysse and Norvill 2014). Absorption is one of the most widely used methods because of its high efficiency and economic feasibility. Frequently, Many adsorbent materials used in the different studies such as zeolites (Wang and Peng 2010), coal (Babel and Kurniawan 2004; Hui et al. 2005), powdered activated charcoal (Sarkar et al. 2006), waste ashes and post consumers (Zhang and Itoh 2003). Recently, focus towards chitosan has been increased because of its natural availability, non-toxic and biodegradable nature. Chitosan also possesses abundant amino and hydroxyl groups that contribute to adsorb dyes and heavy metal ions (Salehi et al. 2016). Chitosan can be moulded into different shapes including particles, fibres, films and beads, all of which are suited for adsorption (Salehi et al. 2016). Chitosan-based fibres show a lot of promise to adsorb dyes as it has a high-specific surface area and macroscale porosity, and currently, there are several methods such as electrospinning, phase separation, template synthesis and melt down that are being used to construct nanofibrous chitosan (Santhosh et al. 2016; Tan et al. 2015).

Major issues with decontamination of textile effluents are the removal of different colours. These colours arise from the incompletely fixed components onto textiles, which may cause disturbance in the ecology and undesirable water. Acid blue 113, acid yellow 49, reaction red 241, reactive blue 5, basic red 14 and basic blue 9 are some examples of dyes used in the textile industry (Santos et al. 2009). Li et al. (2018) used electrospun pure chitosan nanofibrous membranes with different diameters for the absorption of acid blue 113. The adsorption capacity of 377 mg/g was achieved by chitosan nanofibrous membrane with average fibre diameter of 86 nm. The absorption followed pseudo-second-order kinetics and followed Langmuir's isotherm model (Li et al. 2018). In another study, a hybrid membrane using oxidized dialdehyde starch and 3-aminopropyltriethoxysilane (APTES) as a cross-linking agent and chitosan/oxidized starch and chitosan/oxidized starch/silica, was studied. The adsorption properties were studied using to direct dyes: blue 71 and red 31. It was seen that adsorption capacity of chitosan/oxidized starch/silica hybrid membrane was optimal at pH of 9.82 and temperature 60 °C for blue 71 and red 31. The adsorption kinetics followed pseudo-second-order kinetics, and adsorption behaviour fitted well with Freundlich model (He et al. 2016).

Polyethersulphone nanofiltration membrane was synthesized by blending *O*-carboxymethyl chitosan/Fe_3O_4 nanoparticles via phase inversion. The carboxymethyl chitosan moiety was bound to the surface of magnetic Fe_3O_4 nanoparticles by co-precipitation, and this modified membrane was used for the removal of direct red 16. It was seen that the modification of the surface leads to a higher pure water flux and permeation when compared with the unmodified one. The surface hydrophilicity of the modified membrane was improved by virtue of an improvement in the surface water affinity. The rejection capability of the membrane was increased by the amendment of low amounts of chitosan-functionalized nanoparticles in the membrane casting solution. Nanofiltration performance of the membranes tested in terms of direct red 16 dye showed that the addition of *O*-carboxymethyl chitosan-coated Fe_3O_4 nanoparticles induced a negative charge on the membrane, thus leading to the increased rejection (Zahedi et al. 2018).

In some other study conducted by Habiba et al. (2017a, b), chitosan/polyvinyl alcohol nanofibrous membrane was fabricated using electrospinning process for the adsorption of Cr (VI), Fe (III) and methyl orange. Interestingly, it was found that the adsorption capacity of the nanofibre containing higher degree of deacetylation DD chitosan was higher for Ce (VI), but on the other hand, membrane containing lower DD showed higher adsorption for Fe (III) and methyl orange (Habiba et al. 2017b).

A synergistic approach to dye removal was proposed by Wen et al. (2015) using *Myrothecium verrucaria*, immobilized onto chitosan-Fe membrane. The laccase producing fungus was immobilized using a sodium alginate, oats mixture with chitosan-Fe on the membrane and oats providing nutritional supplements. Different dyes that were used to test the efficacy of these systems were first adsorbed onto the chitosan-Fe membrane, reducing the direct toxicity to the fungus, followed by laccase-mediated biodegradation (Wen et al. 2015).

Other membranes such as polyamide-6/chitosan nanofibrous membrane (Ghani et al. 2014), cellulose nanocrystals (Karim et al. 2014), organoclay/chitosan nanocomposite coated on the commercial polyvinylidene fluoride (PVDF) microfiltration membrane (Daraei et al. 2013), chitosan/hydroxyapatite composites (Shi et al. 2017) and chitosan/montmorillonite membranes (Nesic et al. 2012) have been used widely for the removal of dyes from water.

5.5.2 Membrane Technology for the Removal of Heavy Metals and Other Pollutants

Various operations have been used for the elimination of metals from aqueous solutions such as ion exchange, floatation, electro-coagulation, solvent extraction and adsorption etc. (Akhtar et al. 2007; Devi and Mishra 2010; Wang et al. 2006). However, most of these processes are not suitable for elimination of trace amounts of metal ions due to the slight driving force, inadequate reactivity and huge energy and material consumption, thus making their economic feasibility questionable. Membrane adsorption is a viable alternate that could be used for the removal of heavy metal ions from aqueous systems (He et al. 2008; Madaeni et al. 2011; Vatanpour et al. 2011). Membrane adsorption method is favourable due to severable advantages such as excellent removal efficacy, attractive hydrodynamics, high flux, low pressure drop, faster kinetics, reusability and no need for post-treatment and most importantly, the ease of scale up (Ghaee et al. 2012; Vieira et al. 2007). Choice of membrane is very essential to maintain high selectivity, low/no toxicity, high permeability, low-cost materials and fabrication, and overall stability of the system. Chitosan-based sorbents have been used now for a long time for their efficient heavy metal removal (Franco et al. 2004; Hadi 2013; Mohanasrinivasan et al. 2014; Ngah and Fatinathan 2008). Chitosan has also been used for removal of polyaromatic hydrocarbons (Crisafully et al. 2008). Carbon nanotubes were mixed with blend of chitosan/poly(vinyl alcohol) and cross-linked with silane, and this composite was used with naphthalene as a model polyaromatic hydrocarbons (PAH). The membrane showed strong chemical interaction and removal of up to 87% within 30 min. The green nature of this candidate makes it suitable for applications in future for PAH removal (Bibi et al. 2015).

Chitosan raw flakes (Gerente et al. 2010; Kwok et al. 2009), cross-linked chitosan beads, magnetic nanoparticles with chitosan beads (Wang et al. 2014) and titanium dioxide impregnated chitosan beads (Miller et al. 2011), molybdate impregnated chitosan beads (Dambies et al. 2000), and chitosan-immobilized sodium silicate have been used for adsorptive removal of arsenic from aqueous environments through adsorption. The performance of electrospun chitosan/hydroxyapatite composite nanofibre membrane was studied for the removal of lead, cobalt and nickel from aqueous solutions by Aliabadi and Irani (2014) (Aliabadi et al. 2014). It was found that the maximum adsorption capacity of the composite nanofibre membrane for the tested metal ions was in the order of lead (296.7 mg/g) > cobalt (213.8 mg/g) > nickel (180.2 mg/g) at equilibrium time of 30 min and temperature of 45 °C. The kinetics followed a pseudo-second-order kinetics and followed the Langmuir isotherm model. Copper and mercury ion adsorption on chitosan membranes was investigated in a batch mode by Vieira et al. (2007), and cross-linked chitosan was synthesized by immersion of chitosan in the epichlorohydrin solution, or in contact with glutaraldehyde solution followed by immersion with water; the membrane showed higher affinity for mercury than for copper ions. In a recent study, a mixed matrix of chitosan and PVDF was used to fabricate a novel membrane (CTS(Cd^{2+})/PVDF), via a one-step employing the Cd^{2+} through imprinting technology for the selective removal of cadmium ions (Tang et al. 2017).

6 Conclusions

Chitosan is a naturally occurring polysaccharide with promising properties eventually safe use in health care, environmental remediation and food packaging by virtue of its biocompatibility. However, use of neat chitosan films in certain biomedical and packaging applications is limited due to lack of mechanical properties. However, mechanical properties of the chitosan-based materials can be improved significantly through different blends and composites catering to the specific needs of each and every industry. Chitosan derivatives in the nanoscale level have interesting properties and are being investigated for their applicability as antimicrobials, drug release, enzyme immobilization and higher cytotoxicity. Chitin and chitosan-based nanofibres have the potential to be developed in the future as they have excellent biocompatibility, biodegradability and antibacterial activity; these nanofibres are thus excellent candidates for enhancement of absorption and sustained release of drugs and wound dressings. However, most of these applications are now still in the laboratory level, and additional studies are needed prior to clinical trials and commercialization of such nanofibres.

7 Future Perspectives

Natural biopolymer-based biodegradable packaging materials are emerging as new generation of packaging for the heavily increasing demand of fresh minimally processed fruits and vegetables. The biodegradable packaging material has a plethora of applications and can potentially replace plastics in the near future for disposable plastic cups, bags, plates and containers; however, due to inherent lack of high hydrophobicity and mechanical strength when compared to the petroleum-based counterparts, they have not been accepted in the food packaging industry. The application of nanotechnology might provide solutions to such issues.

Bio-nanocomposites of chitosan have a very strong future, and though they are present in low levels of production with allied high costs, research that explores the nanoscale applications shows high promise. Montmorillonite/chitosan bio-nanocomposites that are now used for immobilization and removal of anionic pesticides from soil and water can be used in pharmaceutical and food applications and enzymes such as firefly luciferase which can be used in an encapsulated form for the detection of ATP, making them smart packages that prevent growth of bacteria and fungi by virtue of their inherent antimicrobial/antifungal nature, but also have smart indications on their packaging. Thus, chitosan nanocomposites are the future with remarkable mechanical, thermal, barrier properties, crystallization and degradation rate, and melt strength with a myriad of applications.

References

Abdeen ZI, El Farargy AF, Negm NA (2018) Nanocomposite framework of chitosan/polyvinyl alcohol/ZnO: preparation, characterization, swelling and antimicrobial evaluation. J Mo Liq 250:335–343. https://doi.org/10.1016/j.molliq.2017.12.032

Abdel-Fattah WI, Jiang T, El-Bassyouni GE-T, Laurencin CT (2007) Synthesis, characterization of chitosans and fabrication of sintered chitosan microsphere matrices for bone tissue engineering. Acta Biomaterialia 3:503–514

Abdou ES, Nagy KS, Elsabee MZ (2008) Extraction and characterization of chitin and chitosan from local sources. Biores Technol 99:1359–1367

Ahyat NM, Mohamad F, Ahmad A, Azmi AA (2017) Chitin and chitosan extraction from portunus pelagicus. Malaysian J Anal Sci 21:770–777

Aider M (2010) Chitosan application for active bio-based films production and potential in the food industry LWT-food. Sci Technol 43:837–842

Ainurofiq A, Nurcahyo I, Yulianto R (2014) Preparation, characterization and formulation of nanocomposite matrix na-montmorillonite intercalated medium molecular weight chitosan for theophylline sustained release tablet. Int J Pharm Pharm Sci 6:131–137

Akhtar K, Akhtar MW, Khalid AM (2007) Removal and recovery of uranium from aqueous solutions by *Trichoderma harzianum*. Water Res 41:1366–1378

Aliabadi M, Irani M, Ismaeili J, Najafzadeh S (2014) Design and evaluation of chitosan/hydroxyapatite composite nanofiber membrane for the removal of heavy metal ions from aqueous solution. J Taiwan Inst Chem Eng 45:518–526

Amiri A, Ghorbani M, Jahangiri M (2015) A novel chitosan/polyrhodanine nanocomposite: preparation, characterisation and application for Ni(II) ions removal from aqueous solution. J Exp Nanosci 10:1374–1386. https://doi.org/10.1080/17458080.2015.1014871

An J, Yang Q, Luo Q, Li X, Yin R, Liu F, Wang D (2017a) Preparation and characterization of silver/g-carbon nitride/chitosan nanocomposite with photocatalytic activity. Integr Ferroelectr 180:52–60. https://doi.org/10.1080/10584587.2017.1337463

An ZZ et al (2017b) Preparation of chitosan/N-doped graphene natively grown on hierarchical porous carbon nanocomposite as a sensor platform for determination of tartrazine. Chin Chem Lett 28:1492–1498. https://doi.org/10.1016/j.cclet.2017.02.014

Anitha A et al (2014) Chitin and chitosan in selected biomedical applications. Prog Polym Sci 39:1644–1667

Atak BH, Buyuk B, Huysal M, Isik S, Senel M, Metzger W, Cetin G (2017) Preparation and characterization of amine functional nano-hydroxyapatite/chitosan bionanocomposite for bone tissue engineering applications. Carbohydr Polym 164:200–213. https://doi.org/10.1016/j.carbpol.2017.01.100

Athanasiou KA, Shah AR, Hernandez RJ, LeBaron RG (2001) Basic science of articular cartilage repair. Clin Sports Med 20:223–247

Azevedo AN, Buarque PR, Cruz EMO, Blank AF, Alves PB, Nunes ML, de Aquino Santana LCL (2014) Response surface methodology for optimisation of edible chitosan coating formulations incorporating essential oil against several foodborne pathogenic bacteria. Food Control 43:1–9

Babel S, Kurniawan TA (2004) Cr (VI) removal from synthetic wastewater using coconut shell charcoal and commercial activated carbon modified with oxidizing agents and/or chitosan. Chemosphere 54:951–967

Baghaie S, Khorasani MT, Zarrabi A, Moshtaghian J (2017) Wound healing properties of PVA/starch/chitosan hydrogel membranes with nano Zinc oxide as antibacterial wound dressing material. J Biomater Sci Polym Ed 28:2220–2241. https://doi.org/10.1080/09205063.2017.1390383

Bajpai SK, Ahuja S, Chand N, Bajpai M (2017) Nano cellulose dispersed chitosan film with Ag NPs/Curcumin: an in vivo study on Albino Rats for wound dressing. Int J Biol Macromol 104:1012–1019. https://doi.org/10.1016/j.ijbiomac.2017.06.096

Bibi S, Yasin T, Hassan S, Riaz M, Nawaz M (2015) Chitosan/CNTs green nanocomposite membrane: synthesis, swelling and polyaromatic hydrocarbons removal. Mater Sci Eng C 46:359–365

Cazón P, Velazquez G, Ramírez JA, Vázquez M (2017) Polysaccharide-based films and coatings for food packaging: a review. Food Hydrocolloids 68:136–148

Chaleshtori RS, Taghizadeh M, Khanalizadeh A, Hesami S, Heidaryan Z, Sahebjami P, Khatami M (2016) The effects of chitosan incorporated with eucalyptus and cuminum essential oils on storage time of oncorhynchus mykiss. J Mazandaran Univ Med Sci 25:150–161

Chang C, Peng N, He M, Teramoto Y, Nishio Y, Zhang L (2013) Fabrication and properties of chitin/hydroxyapatite hybrid hydrogels as scaffold nano-materials. Carbohydr Polym 91:7–13

Chávez-Delgado M et al (2003) Facial nerve regeneration through progesterone-loaded chitosan prosthesis. A preliminary report. J Biomed Mater Res Part B Appl Biomater 67:702–711

Cheaburu-Yilmaz CN, Yilmaz O, Vasile C (2015) Eco-friendly chitosan-based nanocomposites: chemistry and applications. In: Eco-friendly polymer nanocomposites. Springer, Berlin, pp 341–386

Ciardelli G, Chiono V (2006) Materials for peripheral nerve regeneration. Macromol Biosci 6:13–26

Crisafully R, Milhome MAL, Cavalcante RM, Silveira ER, De Keukeleire D, Nascimento RF (2008) Removal of some polycyclic aromatic hydrocarbons from petrochemical wastewater using low-cost adsorbents of natural origin. Biores Technol 99:4515–4519

Cui YL, Di Qi A, Liu WG, Wang XH, Wang H, Ma DM, De Yao K (2003) Biomimetic surface modification of poly(L-lactic acid) with chitosan and its effects on articular chondrocytes in vitro. Biomaterials 24:3859–3868

Dambies L, Guibal E, Roze A (2000) Arsenic (V) sorption on molybdate-impregnated chitosan beads. Colloids Surf A Physicochem Eng Aspects 170:19–31

Dananjaya SHS, Kulatunga DCM, Godahewa GI, Lee J, De Zoysa M (2016) Comparative study of preparation, characterization and anticandidal activities of a chitosan silver nanocomposite (CAgNC) compared with low molecular weight chitosan (LMW-chitosan). RSC Adv 6:33455–33461. https://doi.org/10.1039/c6ra03917k

Daraei P, Madaeni SS, Salehi E, Ghaemi N, Ghari HS, Khadivi MA, Rostami E (2013) Novel thin film composite membrane fabricated by mixed matrix nanoclay/chitosan on PVDF microfiltration support: preparation, characterization and performance in dye removal. J Membr Sci 436:97–108. https://doi.org/10.1016/j.memsci.2013.02.031

Dash M, Chiellini F, Ottenbrite R, Chiellini E (2011) Chitosan—a versatile semi-synthetic polymer in biomedical applications. Prog Polym Sci 36:981–1014

De Aquino AB, Blank AF, De Aquino Santana LCL (2015) Impact of edible chitosan-cassava starch coatings enriched with Lippia gracilis Schauer genotype mixtures on the shelf life of guavas (*Psidium guajava* L.) during storage at room temperature. Food Chem 171:108–116. https://doi.org/10.1016/j.foodchem.2014.08.077

Dev A, Binulal N, Anitha A, Nair S, Furuike T, Tamura H, Jayakumar R (2010) Preparation of poly(lactic acid)/chitosan nanoparticles for anti-HIV drug delivery applications. Carbohydr Polym 80:833–838

Devi N, Dutta J (2017) Preparation and characterization of chitosan-bentonite nanocomposite films for wound healing application. Int J Biol Macromol 104:1897–1904. https://doi.org/10.1016/j.ijbiomac.2017.02.080

Devi NB, Mishra S (2010) Solvent extraction equilibrium study of manganese (II) with Cyanex 302 in kerosene. Hydrometallurgy 103:118–123

Dutta P, Tripathi S, Mehrotra G, Dutta J (2009) Perspectives for chitosan based antimicrobial films in food applications. Food Chem 114:1173–1182

Elsabee MZ, Abdou ES (2013) Chitosan based edible films and coatings: a review. Mater Sci Eng C 33:1819–1841

Espitia PJP, Du WX, de Jesús Avena-Bustillos R, Soares NDFF, McHugh TH (2014) Edible films from pectin: physical-mechanical and antimicrobial properties—a review. Food Hydrocolloids 35:287–296

Franco LDO, Maia RDCC, Porto ALF, Messias AS, Fukushima K, Campos-Takaki GMD (2004) Heavy metal biosorption by chitin and chitosan isolated from Cunninghamella elegans (IFM 46109). Braz J Microbiol 35:243–247

Freier T, Montenegro R, Koh HS, Shoichet MS (2005) Chitin-based tubes for tissue engineering in the nervous system. Biomaterials 26:4624–4632

Gaf R (1992) Chitin chemistry. Macmillan, London
Gerente C, Andres Y, Mckay G, Le Cloirec P (2010) Removal of arsenic (V) onto chitosan: from sorption mechanism explanation to dynamic water treatment process. Chem Eng J 158:593–598
Ghaee A, Shariaty-Niassar M, Barzin J, Zarghan A (2012) Adsorption of copper and nickel ions on macroporous chitosan membrane: equilibrium study. Appl Surf Sci 258:7732–7743
Ghani M, Gharehaghaji AA, Arami M, Takhtkuse N, Rezaei B (2014) Fabrication of electrospun polyamide-6/chitosan nanofibrous membrane toward anionic dyes removal. J Nanotechnol. https://doi.org/10.1155/2014/278418
Glimcher MJ (2006) Bone: nature of the calcium phosphate crystals and cellular, structural, and physical chemical mechanisms in their formation. Rev Mineral Geochem 64:223–282
Griffon DJ, Sedighi MR, Schaeffer DV, Eurell JA, Johnson AL (2006) Chitosan scaffolds: interconnective pore size and cartilage engineering. Acta Biomater 2:313–320
Guieysse B, Norvill ZN (2014) Sequential chemical–biological processes for the treatment of industrial wastewaters: review of recent progresses and critical assessment. J Hazard Mater 267:142–152
Habiba U, Afifi AM, Salleh A, Ang BC (2017a) Chitosan/(polyvinyl alcohol)/zeolite electrospun composite nanofibrous membrane for adsorption of Cr6+, Fe3+ and Ni^{2+}. J Hazard Mater 322:182–194
Habiba U, Siddique TA, Talebian S, Lee JJL, Salleh A, Ang BC, Afifi AM (2017b) Effect of deacetylation on property of electrospun chitosan/PVA nanofibrous membrane and removal of methyl orange, Fe(III) and Cr(VI) ions. Carbohydr Polym 177:32–39. https://doi.org/10.1016/j.carbpol.2017.08.115
Hadi AG (2013) Synthesis of chitosan and its use in metal removal. Chem Mater Res 3(3)
Hadian M, Rajaei A, Mohsenifar A, Tabatabaei M (2017) Encapsulation of *Rosmarinus officinalis* essential oils in chitosan-benzoic acid nanogel with enhanced antibacterial activity in beef cutlet against Salmonella typhimurium during refrigerated storage. LWT-Food Sci Technol 84:394–401
Haipeng G, Yinghui Z, Jianchun L, Yandao G, Nanming Z, Xiufang Z (2000) Studies on nerve cell affinity of chitosan-derived materials. J Biomed Mater Res 52:285–295
Hao T et al (2010) The support of matrix accumulation and the promotion of sheep articular cartilage defects repair in vivo by chitosan hydrogels. Osteoarthritis and Cartilage 18:257–265
He Z-Y, Nie H-L, Branford-White C, Zhu L-M, Zhou Y-T, Zheng Y (2008) Removal of Cu^{2+} from aqueous solution by adsorption onto a novel activated nylon-based membrane. Bioresour Technol 99:7954–7958
He K, Lou T, Wang X, Zhao W (2015) Preparation of lignosulfonate–acrylamide–chitosan ternary graft copolymer and its flocculation performance. Int J Biol Macromol 81:1053–1058
He X, Du M, Li H, Zhou T (2016) Removal of direct dyes from aqueous solution by oxidized starch cross-linked chitosan/silica hybrid membrane. Int J Biol Macromol 82:174–181. https://doi.org/10.1016/j.ijbiomac.2015.11.005
Hebeish AA, Ramadan MA, Montaser AS, Farag AM (2014) Preparation, characterization and antibacterial activity of chitosan-g-poly acrylonitrile/silver nanocomposite. Int J Biol Macromol 68:178–184. https://doi.org/10.1016/j.ijbiomac.2014.04.028
Hosseini SF, Rezaei M, Zandi M, Farahmandghavi F (2015) Bio-based composite edible films containing *Origanumvulgare* L. essential oil. Ind Crops Prod 67:403–413
Hosseini SF, Rezaei M, Zandi M, Farahmandghavi F (2016) Development of bioactive fish gelatin/chitosan nanoparticles composite films with antimicrobial properties. Food Chem 194:1266–1274
Hu Q, Li B, Wang M, Shen J (2004) Preparation and characterization of biodegradable chitosan/hydroxyapatite nanocomposite rods via in situ hybridization: a potential material as internal fixation of bone fracture. Biomaterials 25:779–785
Huang C-Y, Kuo C-H, Wu C-H, Ku M-W, Chen P-W (2018) Extraction of crude chitosans from squid (*Illex argentinus*) pen by a compressional puffing-pretreatment process and evaluation of their antibacterial activity. Food Chem 254:217–223

Hui K, Chao CYH, Kot S (2005) Removal of mixed heavy metal ions in wastewater by zeolite 4A and residual products from recycled coal fly ash. J Hazard Mater 127:89–101

Iwasaki N et al (2004) Feasibility of polysaccharide hybrid materials for scaffolds in cartilage tissue engineering: evaluation of chondrocyte adhesion to polyion complex fibers prepared from alginate and chitosan. Biomacromolecules 5:828–833

Jaworska M, Kula K, Chassary P, Guibal E (2003a) Influence of chitosan characteristics on polymer properties: II. Platinum sorption properties. Polym Int 52:206–212

Jaworska M, Sakurai K, Gaudon P, Guibal E (2003b) Influence of chitosan characteristics on polymer properties. I: Crystallographic properties. Polym Int 52:198–205

Jayakumar R, Prabaharan M, Nair S, Tamura H (2010) Novel chitin and chitosan nanofibers in biomedical applications. Biotechnol Adv 28:142–150

Jia J, Du P (2015) Preparation of $TiO_2Fe_2O_3$ chitosan nanocomposite films and its photocatalytic degradation of Rhodamine B. Asian J Chem 27:1889–1893. https://doi.org/10.14233/ajchem.2015.18382

Jiang T, Abdel-Fattah WI, Laurencin CT (2006) In vitro evaluation of chitosan/poly(lactic acid-glycolic acid) sintered microsphere scaffolds for bone tissue engineering. Biomaterials 27:4894–4903

Jiang T et al (2010) Chitosan–poly(lactide-co-glycolide) microsphere-based scaffolds for bone tissue engineering: in vitro degradation and in vivo bone regeneration studies. Acta Biomater 6:3457–3470

Jin R, Teixeira LM, Dijkstra PJ, Karperien M, Van Blitterswijk C, Zhong Z, Feijen J (2009) Injectable chitosan-based hydrogels for cartilage tissue engineering. Biomaterials 30:2544–2551

Jirimali HD, Saravanakumar D, Shin W (2015) Preparation of catechol-linked chitosan/carbon nanocomposite-modified electrode and its applications. Bull Korean Chem Soc 36:1289–1291. https://doi.org/10.1002/bkcs.10232

Karim Z, Mathew AP, Grahn M, Mouzon J, Oksman K (2014) Nanoporous membranes with cellulose nanocrystals as functional entity in chitosan: removal of dyes from water. Carbohydr Polym 112:668–676. https://doi.org/10.1016/j.carbpol.2014.06.048

Kim KW, Thomas R, Lee C, Park HJ (2003) Antimicrobial activity of native chitosan, degraded chitosan, and *O*-carboxymethylated chitosan. J Food Prot 66:1495–1498

Kim I-Y, Seo S-J, Moon H-S, Yoo M-K, Park I-Y, Kim B-C, Cho C-S (2008) Chitosan and its derivatives for tissue engineering applications. Biotechnol Adv 26:1–21

Kosher RA, Lash JW, Minor RR (1973) Environmental enhancement of in vitro chondrogenesis: IV. Stimulation of somite chondrogenesis by exogenous chondromucoprotein. Dev Biol 35:210–220

Kurita K (1998) Chemistry and application of chitin and chitosan. Polym Degradation Stab 59:117–120

Kwok KC, Lee VK, Gerente C, McKay G (2009) Novel model development for sorption of arsenate on chitosan. Chem Eng J 151:122–133

Lee JE et al (2004) Effects of the controlled-released TGF-β1 from chitosan microspheres on chondrocytes cultured in a collagen/chitosan/glycosaminoglycan scaffold. Biomaterials 25:4163–4173

Li J, Pan J, Zhang L, Guo X, Yu Y (2003a) Culture of primary rat hepatocytes within porous chitosan scaffolds. J Biomed Mater Res Part A 67:938–943

Li J, Pan J, Zhang L, Yu Y (2003b) Culture of hepatocytes on fructose-modified chitosan scaffolds. Biomaterials 24:2317–2322

Li K, Jin S, Liu X, Chen H, He J, Li J (2017) Preparation and characterization of chitosan/soy protein isolate nanocomposite film reinforced by Cu nanoclusters. Polymers 9. https://doi.org/10.3390/polym9070247

Li C, Lou T, Yan X, Long Y-Z, Cui G, Wang X (2018) Fabrication of pure chitosan nanofibrous membranes as effective absorbent for dye removal. Int J Biol Macromol 106:768–774. https://doi.org/10.1016/j.ijbiomac.2017.08.072

Lindahl U, Hook M (1978) Glycosaminoglycans and their binding to biological macromolecules. Annu Rev Biochem 47:385–417

Liu H, Du Y, Wang X, Sun L (2004) Chitosan kills bacteria through cell membrane damage. Int J Food Microbiol 95:147–155

Llanos JHR, Tadini CC (2018) Preparation and characterization of bio-nanocomposite films based on cassava starch or chitosan, reinforced with montmorillonite or bamboo nanofibers. Int J Biol Macromolecules 107:371–382. https://doi.org/10.1016/j.ijbiomac.2017.09.001

Lu JX, Prudhommeaux F, Meunier A, Sedel L, Guillemin G (1999) Effects of chitosan on rat knee cartilages. Biomaterials 20:1937–1944

Ma Q, Zhang Y, Critzer F, Davidson PM, Zhong Q (2016) Quality attributes and microbial survival on whole cantaloupes with antimicrobial coatings containing chitosan, lauric arginate, cinnamon oil and ethylenediaminetetraacetic acid. Int J Food Microbiol 235:103–108

Madaeni S, Zinadini S, Vatanpour V (2011) Convective flow adsorption of nickel ions in PVDF membrane embedded with multi-walled carbon nanotubes and PAA coating. Sep Purif Technol 80:155–162

Mallakpour S, Nezamzadeh Ezhieh A (2017) Preparation and characterization of chitosan-poly (vinyl alcohol) nanocomposite films embedded with functionalized multi-walled carbon nanotube. Carbohydr Polym 166:377–386. https://doi.org/10.1016/j.carbpol.2017.02.086

Miller SM, Spaulding ML, Zimmerman JB (2011) Optimization of capacity and kinetics for a novel bio-based arsenic sorbent, TiO_2-impregnated chitosan bead. Water Res 45:5745–5754

Mohammad-Rezaei R, Razmi H (2016) Preparation and characterization of hemoglobin immobilized on graphene quantum dots-chitosan nanocomposite as a sensitive and stable hydrogen peroxide biosensor. Sens Lett 14:685–691. https://doi.org/10.1166/sl.2016.3691

Mohammed MO, Hussain KS, Haj NQ (2017) Preparation and bioactivity assessment of chitosan-1-acetic acid-5-flurouracil conjugates as cancer prodrugs. Molecules 22:1629

Mohanasrinivasan V, Mishra M, Paliwal JS, Singh SK, Selvarajan E, Suganthi V, Devi CS (2014) Studies on heavy metal removal efficiency and antibacterial activity of chitosan prepared from shrimp shell waste. 3 Biotech 4:167–175

Mourya V, Inamdar NN (2008) Chitosan-modifications and applications: opportunities galore. React Funct Polym 68:1013–1051

Moussout H, Ahlafi H, Aazza M, Amechrouq A (2018) Bentonite/chitosan nanocomposite: preparation, characterization and kinetic study of its thermal degradation. Thermochim Acta 659:191–202. https://doi.org/10.1016/j.tca.2017.11.015

Nesic AR, Velickovic SJ, Antonovic DG (2012) Characterization of chitosan/montmorillonite membranes as adsorbents for Bezactiv Orange V-3R dye. J Hazard Mater 209–210:256–263. https://doi.org/10.1016/j.jhazmat.2012.01.020

Nettles DL, Elder SH, Gilbert JA (2002) Potential use of chitosan as a cell scaffold material for cartilage tissue engineering. Tissue Eng 8:1009–1016

Ngah WW, Fatinathan S (2008) Adsorption of Cu (II) ions in aqueous solution using chitosan beads, chitosan–GLA beads and chitosan–alginate beads. Chem Eng J 143:62–72

Nitayaphat W, Jintakosol T (2017) Preparation of chitosan/organoclay nanocomposite as silver(I) ion adsorbent. Asian J Chem 29:683–690. https://doi.org/10.14233/ajchem.2017.20313

Oh YA, Oh YJ, Song AY, Won JS, Song KB, Min SC (2017) Comparison of effectiveness of edible coatings using emulsions containing lemongrass oil of different size droplets on grape berry safety and preservation. LWT Food Sci Technol 75:742–750. https://doi.org/10.1016/j.lwt.2016.10.033

Osugi ME, Rajeshwar K, Ferraz ER, de Oliveira DP, Araújo ÂR, Zanoni MVB (2009) Comparison of oxidation efficiency of disperse dyes by chemical and photoelectrocatalytic chlorination and removal of mutagenic activity. Electrochim Acta 54:2086–2093

Park H, Choi B, Hu J, Lee M (2013) Injectable chitosan hyaluronic acid hydrogels for cartilage tissue engineering. Acta Biomater 9:4779–4786

Pathania D, Gupta D, Kothiyal NC, Sharma G, Eldesoky GE, Naushad M (2016) Preparation of a novel chitosan-g-poly(acrylamide)/Zn nanocomposite hydrogel and its applications for

controlled drug delivery of ofloxacin. Int J Biol Macromol 84:340–348. https://doi.org/10.1016/j.ijbiomac.2015.12.041
Pelissari FM, Grossmann MVE, Yamashita F, Pined EAG (2009) Antimicrobial, mechanical, and barrier properties of cassava starch-chitosan films incorporated with oregano essential oil. J Agric Food Chem 57:7499–7504. https://doi.org/10.1021/jf9002363
Pelissari FM, Yamashita F, Grossmann MVE (2011) Extrusion parameters related to starch/chitosan active films properties. Int J Food Sci Technol 46:702–710
Pillai C, Paul W, Sharma CP (2009) Chitin and chitosan polymers: chemistry, solubility and fiber formation. Prog Polym Sci 34:641–678
Piron E, Accominotti M, Domard A (1997) Interaction between chitosan and uranyl ions. Role of physical and physicochemical parameters on the kinetics of sorption. Langmuir 13:1653–1658
Qin C, Li H, Xiao Q, Liu Y, Zhu J, Du Y (2006) Water-solubility of chitosan and its antimicrobial activity. Carbohydr Polym 63:367–374
Rajaei A, Hadian M, Mohsenifar A, Rahmani-Cherati T, Tabatabaei M (2017) A coating based on clove essential oils encapsulated by chitosan-myristic acid nanogel efficiently enhanced the shelf-life of beef cutlets. Food Packag Shelf Life 14:137–145
Rajesh M, Harish Prashanth KV (2017) Preparation & characterization of chitosan-poly-ε-caprolactone based nanocomposite blend films. J Polym Mater 34:235–248
Rasti H, Parivar K, Baharara J, Iranshahi M, Namvar F (2017) Chitin from the mollusc chiton: extraction, characterization and chitosan preparation. Iran J Pharm Res IJPR 16:366
Roul J, Mohapatra R, Sahoo SK (2016) Antimicrobial activity of novel chitosan/cloisite 10A nanocomposite: preparation, optimization, characterization and drug delivery behavior. Pak J Pharm Sci 29:1145–1150
Roushani M, Saedi Z, Hamdi F, Dizajdizi BZ (2017) Preparation an electrochemical sensor for detection of manganese (II) ions using glassy carbon electrode modified with multi walled carbon nanotube-chitosan-ionic liquid nanocomposite decorated with ion imprinted polymer. J Electroanal Chem 804:1–6. https://doi.org/10.1016/j.jelechem.2017.09.038
Salehi E, Daraei P, Shamsabadi AA (2016) A review on chitosan-based adsorptive membranes. Carbohydr Polym 152:419–432
Samicho Z, Ramli A (2011) Extraction of chitosan & its film-forming properties: a review. In: 2011 IEEE symposium on business, engineering and industrial applications (ISBEIA), pp 576–580
Santhosh C, Velmurugan V, Jacob G, Jeong SK, Grace AN, Bhatnagar A (2016) Role of nanomaterials in water treatment applications: a review. Chem Eng J 306:1116–1137
Santos VP, Pereira MF, Faria P, Órfão JJ (2009) Decolourisation of dye solutions by oxidation with H_2O_2 in the presence of modified activated carbons. J Hazard Mater 162:736–742
Sareban Z, Javanbakht V (2017) Preparation and characterization of a novel nanocomposite of clinoptilolite/maghemite/chitosan/urea for manganese removal from aqueous solution. Korean J Chem Eng 34:2886–2900. https://doi.org/10.1007/s11814-017-0216-9
Sarkar B, Chakrabarti P, Vijaykumar A, Kale V (2006) Wastewater treatment in dairy industries—possibility of reuse. Desalination 195:141–152
Seghir BB, Benhamza M (2017) Preparation, optimization and characterization of chitosan polymer from shrimp shells. J Food Meas Charact 11:1137–1147
Seol Y-J et al (2004) Chitosan sponges as tissue engineering scaffolds for bone formation. Biotech Lett 26:1037–1041
Severino R, Ferrari G, Vu KD, Donsì F, Salmieri S, Lacroix M (2015) Antimicrobial effects of modified chitosan based coating containing nanoemulsion of essential oils, modified atmosphere packaging and gamma irradiation against *Escherichia coli* O157:H7 and Salmonella Typhimurium on green beans. Food Control 50:215–222. https://doi.org/10.1016/j.foodcont.2014.08.029
Shahbazi Y (2017) The properties of chitosan and gelatin films incorporated with ethanolic red grape seed extract and Ziziphora clinopodioides essential oil as biodegradable materials for active food packaging. Int J Biol Macromol 99:746–753. https://doi.org/10.1016/j.ijbiomac.2017.03.065

Sharafati Chaleshtori F, Taghizadeh M, Rafieian-kopaei M, Sharafati-chaleshtori R (2016) Effect of chitosan incorporated with cumin and eucalyptus essential oils as antimicrobial agents on fresh chicken meat. J Food Process Preserv 40:396–404

Sheshmani S, Akhundi Nematzadeh M, Shokrollahzadeh S, Ashori A (2015) Preparation of graphene oxide/chitosan/FeOOH nanocomposite for the removal of Pb(II) from aqueous solution. Int J Biol Macromol 80:475–480. https://doi.org/10.1016/j.ijbiomac.2015.07.009

Shete PB, Patil RM, Thorat ND, Prasad A, Ningthoujam RS, Ghosh SJ, Pawar SH (2014) Magnetic chitosan nanocomposite for hyperthermia therapy application: preparation, characterization and in vitro experiments. Appl Surf Sci 288:149–157. https://doi.org/10.1016/j.apsusc.2013.09.169

Shi C, Lv C, Wu L, Hou X (2017) Porous chitosan/hydroxyapatite composite membrane for dyes static and dynamic removal from aqueous solution. J Hazard Mater 338:241–249. https://doi.org/10.1016/j.jhazmat.2017.05.022

Shukla SK, Mishra AK, Arotiba OA, Mamba BB (2013) Chitosan-based nanomaterials: a state-of-the-art review. Int J Biol Macromol 59:46–58

Soltani RDC, Khataee AR, Safari M, Joo SW (2013) Preparation of bio-silica/chitosan nanocomposite for adsorption of a textile dye in aqueous solutions. Int Biodeterior Biodegradation 85:383–391. https://doi.org/10.1016/j.ibiod.2013.09.004

Su CX-H, Low LW, Teng TT, Wong YS (2016) Combination and hybridisation of treatments in dye wastewater treatment: a review. J Environ Chem Eng 4:3618–3631

Sudheesh Kumar PT et al (2012) Flexible and microporous chitosan hydrogel/nano ZnO composite bandages for wound dressing: in vitro and in vivo evaluation. ACS Appl Mater Interfaces 4:2618–2629. https://doi.org/10.1021/am300292v

Sun X et al (2014) Antimicrobial and mechanical properties of β-cyclodextrin inclusion with essential oils in chitosan films. J Agric Food Chem 62:8914–8918

Sundar K, Harikarthick V, Swarna Karthika V, Ravindran A (2014) Preparation of chitosan-graphene oxide nanocomposite and evaluation of its antimicrobial activity. J Bionanosci 8:207–212. https://doi.org/10.1166/jbns.2014.1223

Tan H, Chu CR, Payne KA, Marra KG (2009) Injectable in situ forming biodegradable chitosan–hyaluronic acid based hydrogels for cartilage tissue engineering. Biomaterials 30:2499–2506

Tan KB, Vakili M, Horri BA, Poh PE, Abdullah AZ, Salamatinia B (2015) Adsorption of dyes by nanomaterials: recent developments and adsorption mechanisms. Sep Purif Technol 150:229–242

Tang X, Gan L, Duan Y, Sun Y, Zhang Y, Zhang Z (2017) A novel Cd^{2+}-imprinted chitosan-based composite membrane for Cd^{2+} removal from aqueous solution. Mater Lett 198:121–123

Thangvaravut H, Chiewchan N, Devahastin S (2012) Inhibitory effect of chitosan films incorporated with 1,8-cineole on SALMONELLA attached on model food surface, vol 506. https://doi.org/10.4028/www.scientific.net/AMR.506.599

Tian X, Yan D, Lu Q, Jiang X (2017) Cationic surface modification of nanocrystalline cellulose as reinforcements for preparation of the chitosan-based nanocomposite films. Cellulose 24:163–174. https://doi.org/10.1007/s10570-016-1119-3

Tokatlı K, Demirdöven A (2017) Optimization of chitin and chitosan production from shrimp wastes and characterization. J Food Process Preserv 42(2):e13494

Trivedi TJ, Rao KS, Kumar A (2014) Facile preparation of agarose-chitosan hybrid materials and nanocomposite ionogels using an ionic liquid via dissolution, regeneration and sol-gel transition. Green Chem 16:320–330. https://doi.org/10.1039/c3gc41317a

Tsai GJ, Su WH, Chen HC, Pan CL (2002) Antimicrobial activity of shrimp chitin and chitosan from different treatments and applications of fish preservation. Fish Sci 68:170–177

Tuli R, Li W-J, Tuan RS (2003) Current state of cartilage tissue engineering. Arthritis Res Ther 5:235

van den Broek LA, Knoop RJ, Kappen FH, Boeriu CG (2015) Chitosan films and blends for packaging material. Carbohydr Polym 116:237–242

Vatanpour V, Madaeni SS, Zinadini S, Rajabi HR (2011) Development of ion imprinted technique for designing nickel ion selective membrane. J Membr Sci 373:36–42

Vieira R, Guibal E, Silva E, Beppu M (2007) Adsorption and desorption of binary mixtures of copper and mercury ions on natural and crosslinked chitosan membranes. Adsorption 13:603–611

Wahid F, Wang HS, Lu YS, Zhong C, Chu LQ (2017) Preparation, characterization and antibacterial applications of carboxymethyl chitosan/CuO nanocomposite hydrogels. Int J Biol Macromol 101:690–695. https://doi.org/10.1016/j.ijbiomac.2017.03.132

Wang S, Peng Y (2010) Natural zeolites as effective adsorbents in water and wastewater treatment. Chem Eng J 156:11–24

Wang X et al (2003) Crosslinked collagen/chitosan matrix for artificial livers. Biomaterials 24:3213–3220

Wang X, Yan Y, Lin F, Xiong Z, Wu R, Zhang R, Lu Q (2005) Preparation and characterization of a collagen/chitosan/heparin matrix for an implantable bioartificial liver. J Biomater Sci Polym Ed 16:1063–1080

Wang X-J, Xia S-Q, Ling C, Zhao J-F, Chovelon J-M, Nicole J-R (2006) Biosorption of cadmium (II) and lead (II) ions from aqueous solutions onto dried activated sludge. J Environ Sci 18:840–844

Wang J, Xu W, Chen L, Huang X, Liu J (2014) Preparation and evaluation of magnetic nanoparticles impregnated chitosan beads for arsenic removal from water. Chem Eng J 251:25–34

Wen Y, Liang Y, Shen C, Wang H, Fu D, Wang H (2015) Synergistic removal of dyes by *Myrothecium verrucaria* immobilization on a chitosan-Fe membrane. RSC Adv 5:68200–68208. https://doi.org/10.1039/c5ra11320b

Ye BL, Zheng R, Ruan XJ, Zheng ZH, Cai HJ (2018) Chitosan-coated doxorubicin nano-particles drug delivery system inhibits cell growth of liver cancer via p 53/PRC1 pathway. Biochem Biophys Res Commun 495:414–420. https://doi.org/10.1016/j.bbrc.2017.10.156

Yen M-T, Yang J-H, Mau J-L (2009) Physicochemical characterization of chitin and chitosan from crab shells. Carbohydr Polym 75:15–21

Yin Y, Ye F, Cui J, Zhang F, Li X, Yao K (2003) Preparation and characterization of macroporous chitosan–gelatin/β-tricalcium phosphate composite scaffolds for bone tissue engineering. J Biomed Mater Res Part A 67:844–855

Yuan Y, Zhang P, Yang Y, Wang X, Gu X (2004) The interaction of Schwann cells with chitosan membranes and fibers in vitro. Biomaterials 25:4273–4278

Yui T, Kobayashi H, Kitamura S, Imada K (1994) Conformational analysis of chitobiose and chitosan. Biopolymers 34:203–208

Yun YH, Youn HG, Shin JY, Yoon SD (2017) Preparation of functional chitosan-based nanocomposite films containing ZnS nanoparticles. Int J Biol Macromol 104:1150–1157. https://doi.org/10.1016/j.ijbiomac.2017.07.016

Zad ZR, Davarani SSH, Taheri A, Bide Y (2018) A yolk shell Fe_3O_4@PA-Ni@Pd/Chitosan nanocomposite-modified carbon ionic liquid electrode as a new sensor for the sensitive determination of fluconazole in pharmaceutical preparations and biological fluids. J Mol Liq 253:233–240. https://doi.org/10.1016/j.molliq.2018.01.019

Zahedi S, Ghomi JS, Shahbazi-Alavi H (2018) Preparation of chitosan nanoparticles from shrimp shells and investigation of its catalytic effect in diastereoselective synthesis of dihydropyrroles. Ultrason Sonochem 40:260–264

Zhang F-S, Itoh H (2003) Adsorbents made from waste ashes and post-consumer PET and their potential utilization in wastewater treatment. J Hazard Mater 101:323–337

Zhong Q, Tian J, Liu T, Guo Z, Ding S, Li H (2018) Preparation and antibacterial properties of carboxymethyl chitosan/ZnO nanocomposite microspheres with enhanced biocompatibility. Mater Lett 212:58–61. https://doi.org/10.1016/j.matlet.2017.10.062

Zong W et al (2018) Preparation of chitosan/amino multiwalled carbon nanotubes nanocomposite beads for bilirubin adsorption in hemoperfusion. J Biomed Mater Res Part B Appl Biomater 106:96–103. https://doi.org/10.1002/jbm.b.33806

Structural Properties of Protein and Their Role in Polymer Nanocomposites

Malladi Rajinipriya, Florent Gauvin, Mathieu Robert, Saïd Elkoun and Malladi Nagalakshmaiah

Abstract The growing concern over the environment raises the question of biodegradability and renewability in numerous fields, including materials and energy. Therefore, industries and researchers focus more on the development of biomaterials generated from natural sources. In this context, proteins and their unique properties made them ideal candidates to be used in different novel applications and are not limited to films, hydrogels, biological tissue engineering, and polymer composites. In this chapter, authors highlight the different sources of protein and their structural properties. Then, the extraction methods are discussed in detail. Further, the different processing methods to prepare the protein-based composites were explained. In overall, this chapter aims to highlight the recent developments of protein-based materials in different fields, based on the literature.

Keywords Proteins · Structural properties · Tissue engineering Nanocomposites · Hydrogels

1 Introduction

Due to the growing environmental concerns, the interest toward bio-based materials has considerably increased in recent years. Predominantly, the advantages of using these natural materials are mainly due to their renewability, biodegradability, and abundance (Belgacem and Gandini 2008). Consequently, both industries and

M. Rajinipriya · M. Robert · S. Elkoun · M. Nagalakshmaiah (✉)
Center for Innovation in Technological Eco design (CITE), University of Sherbrooke, Granby, QC, Canada
e-mail: Nagalakshmaiah.Malladi@usherbrooke.ca

F. Gauvin
Department of the Built Environment, Eindhoven University of Technology, Eindhoven, The Netherlands

M. L. Sanyang and M. Jawaid (eds.), *Bio-based Polymers and Nanocomposites*,
https://doi.org/10.1007/978-3-030-05825-8_10

researchers are paying attention toward the bio-based feedstocks to generate value-added products in different applications such as energy and composite materials. The most frequently used biopolymers are cellulose, lignin, starch, chitosan, bacterial cellulose (BC), and proteins accessible through the plant, animal, and vegetal biomass (Belgacem and Gandini 2008; Silva et al. 2014). These biopolymers are regularly used as alternatives for petroleum-based polymers as a matrix and blend to create novel materials.

In order to reduce the dependency on the petroleum-based polymers, researchers often reinforce polymers with bio-based materials in order to enhance the mechanical properties and biodegradability of the polymer composites. The most frequent method to process the composites is the blending of virgin fibers or crystals with matrices. Among the numerous biopolymers, proteins have their own credentials in order to create the novel functional and green materials (Zhang and Zeng 2008). Proteins are macromolecules with versatile functional groups in living organisms, and it can play a vital role fundamentally in all the biological processes like the transportation of cells, tissue generation, control drug release, mechanical stiffness, enzymatic catalysis, and immune protection (Ashok 2006).

Proteins are biological polymers with the backbone of different amino acids with long peptide bonds, and they naturally exist in different sources such as plants, animals, and bacteria. The primarily used proteins in the creation of the biomaterials are silks, collagen, elastin, resilins, and keratins (Hu et al. 2012b).

The recent development of protein-based materials allows applications such as biosensor, films, polymer composites, hydrogels, adhesives, foams, and tissue generation (Kumar et al. 2002; Belgacem and Gandini 2008; Hu et al. 2012b). This is mainly due to their biocompatibility, high toughness, and flexibility in functionality. However, the compatibility is still a major issue in the case of the polymer composites. It is limiting the usage of the protein in composites.

2 Structural Properties of Proteins

Proteins are polymers, which occur naturally and often called as polypeptides. They are made up of either similar or different amino acids. Indeed, proteins are different from one to another mainly in the sequential arrangement of the amino acids. Proteins are vital in the biological systems and play an important role in the process of storing the genetic information in the deoxyribonucleic acid (DNA). Principally, these polypeptide macromolecules help to function the transportation of the molecules, organisms, and the replication of DNA as well as catalyzing the metabolic reactions. The synthesis of the protein is naturally occurred inside the cell through various steps. Initially, the synthesis of the messenger ribonucleic acid (mRNA) is initiated by the transcription process through the definite sequence of the DNA templates (Lake 1983; Kozak 1983). Secondly, the translation of the mRNA to amino acids ensues through polypeptide according to the transcripted DNA sequence.

As described, proteins are present in different natural sources (like plants, animals, and bacteria). However, each protein has a different sequence of the amino acids with long polypeptide chains. For instance, the hierarchical configuration of the macrostructure of a silkworm to its atomic structure (amide bond) is shown in Fig. 1. Any source of protein is present in the form of long chains, and each chain of the protein is arranged with different amino acids and is attached to each other through amide bonds. The size of the silkworm to amide bond will vary from few centimeters to few angstroms (Å). The sequence of the amino acids arranged in the macromolecular chains varies depending on the source. Mainly, the chains of the protein are composed with 20 different amino acids and they are divided into aliphatic, aromatic, cationic, and anionic amino acids. These amino acids are shown in Fig. 2a. The amino acids adopt zwitter ionic form, which is due to the weak base and acid present in the amino acids. The structural arrangement of the protein chains is in four different stages like primary, secondary, tertiary, and quaternary structures (Belgacem and Gandini 2008; Food Chemistry 2009) as shown in Fig. 2b.

As shown in Fig. 2b, the structure of a protein is highly linked to its amino acid sequence, especially in the primary and important level of the sequence. This stage of the sequence is arranged with different amino acids by acid–amine coupling. The interactions like hydrogen, ionic, and Van der Waals forces between the macromolecular chains lead to the formation of the secondary structure. Importantly, the secondary structure of the protein is made of α-helix, β-sheets, and β-turns structures. Protein domains have different structural assumptions; among them, α-helix appears to be the most frequently shaped through hydrogen bonds formed between the hydrogen in the amide bond and oxygen of the fourth amino acid (Silva et al. 2014). The two other forms of structures are β-sheets and β-turns formed due to the hydrogen bonding between two different linear polypeptide chains. These structures can be parallel, if two strands are in similar direction; if two strands are in opposite path, then it will be anti-parallel. Particularly, β-turns are formed when the first and fourth amino acids formed by hydrogen bonding with second and third amino acids in the polypeptide chains (Silva et al. 2014).

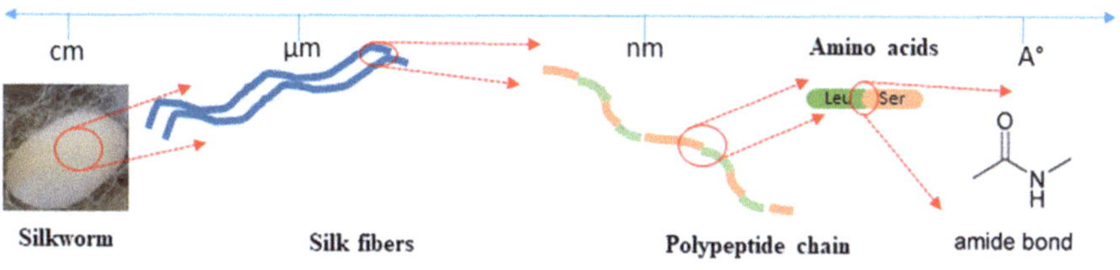

Fig. 1 Hierarchical representation of protein chains in silkworm

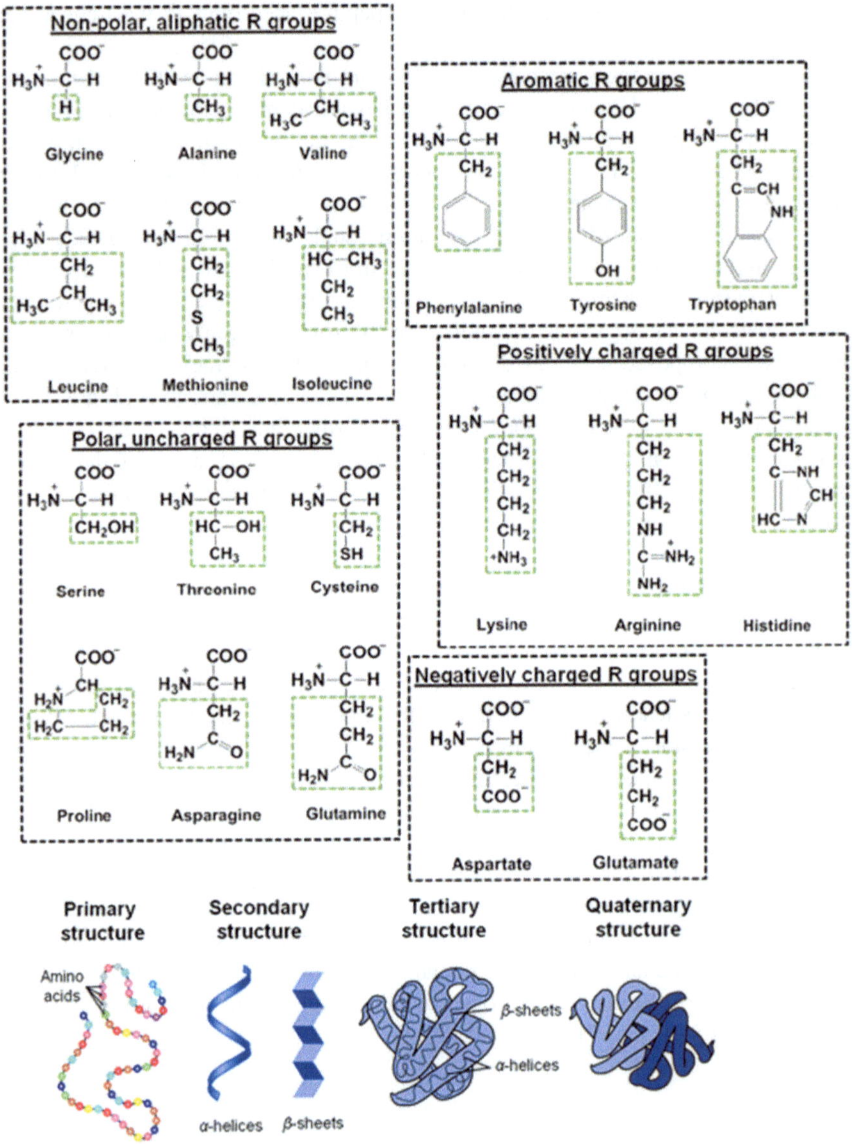

Fig. 2 **a** Different amino acids present in proteins, and **b** structural organization of the protein chains. Reprinted with permission from Silva et al. (2014)

3 Source and Isolation of Protein

Figure 3 shows the different sources of protein that are produced by bacteria, animals, and plants. However, the extraction process of the protein is very intense and time-taking process. Mainly, the extraction method completely depends on the source and end-use application. There are two important steps which aim to isolate the protein, that is, (i) cell lysis and (ii) purification (Biorad). Cell lysis is a process which aims to destruct the cell by different methods like chemical, mechanical, enzymatic, and multiple cycles of freeze-thawing. The resulted material is separated by centrifugation process. The successive recovered materials are purified by precipitation and characterized by different chromatographic techniques depend on the requirements of the end-use properties (Toldra 2013).

The progression and interest of using protein-based materials forced researchers to explore different extraction methods. The recently emerged technique is solvent extraction process, for instance, use of ionic liquids, as this provides the best results in terms of unfolding, thermal stability, crystallization, and shelf life. Particularly, the folding can stimulate the irreversible agglomeration through highly ordered amyloid states (Byrne and Angell 2009). The role of protein in innovative biomaterials is increasing in the composites field. For instance, soy protein-based adhesives are used for particleboard (Khosravi et al. 2010) or thermoplastic elastomers (Wang et al. 2010). Blends of soy protein with natural rubber (Tian et al. 2010) and some interesting studies are also reported on the benefit of using proteins for tissue engineering applications (Hu et al. 2012b; Gomes et al. 2012; Werkmeister and Ramshaw 2012; Jonker et al. 2012; DiMarco and Heilshorn

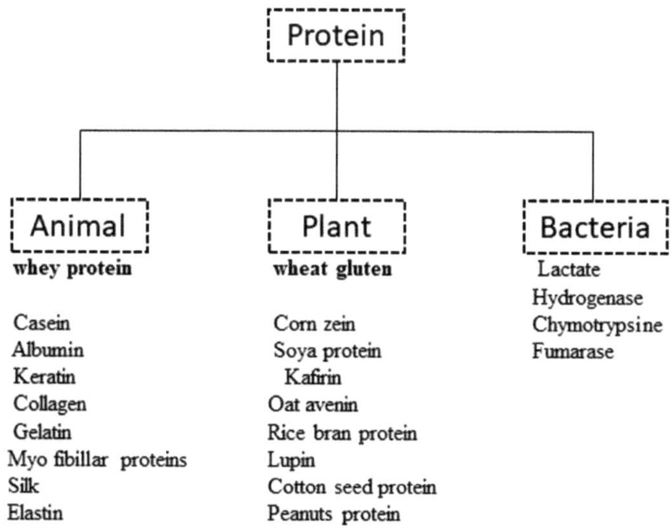

Fig. 3 Different proteins with respect to source

2012). The isolation process of the protein is not necessarily the same for all the sources and recent advancements in this field led to novel methods. Hence, specific proteins and their detailed extraction process are explained individually in the next sections.

3.1 Silk

Silk fibers have diverse structures, and their amino acid composition varies according to the used source. Silk fibroin fiber's diameter is about 10–25 nm long, and these thin fibers stick together by hydrophilic proteins named sericins in multilayers and are called Bombyx mori cocoon (Altman et al. 2003; Vepari and Kaplan 2007). In order to isolate silk fibers from sericins, the cocoons need to be boiled in aqueous alkaline solution. This process can also be done by the more advanced methods like plasma and proteolytic enzyme treatments (Li et al. 2012a). The silk fibers are composed with three amino acids, mainly serine (Ser), alanine (Ala), and glycine (Gly), accounting for 12, 30, and 43%, respectively (Altman et al. 2003).

3.2 Soy Protein Isolate

Soy protein isolate (SPI) is mainly extracted from plant-based materials like oil-seeds and grains. The major source of the soy is soybean plant (*Glycine max L. Merr.*). The composition of the soybeans is protein, oil, and carbohydrates in 40–45, 18–20, and 25–30%, respectively (Swain et al. 2004). The isolation of protein involves multiple steps like the separation of defatted soy flour. Further, continuous washing steps with soluble constituents will yield the SPI-rich solid with 90% protein content (Singh et al. 2008).

3.3 Elastin

Elastin is a highly cross-linked polypeptide with very distinctive chemical composition as compared to other polypeptides. It is present mainly in aorta (57%), vascular vessels (50%), elastic ligaments (32%), and also skin (Vrhovski and Weiss 1998). Due to their complex structural arrangement of polypeptide chains, isolation of elastin is very challenging (Daamen et al. 1997; Mecham 2008). However, different enzymatic and chemical treatments with cyanogen bromide and 2-mercaptoethanol are permitting to isolate the pure elastin. Indeed, recently, elastin was extracted from horse ligamentum nuchae and human aorta (Silva et al. 2014) for the first time.

4 Protein-Based Composite Materials

Recent year's interest toward bio-based materials increased owing to their remarkable properties. Different natural polymers such as cellulose, lignin, chitosan, and protein have drawn a lot of attention from researchers in order to deteriorate the petro-sourced polymers. For the development of more sustainable materials, protein-based materials hauled the topmost interest for various biomedical applications such as tissue engineering, polypeptide biomedical hydrogels, foams, gels, films, and polymer composites (Skopinska-Wisniewska et al. 2009). The current interests and some future emerging applications are shown in Fig. 4.

The most distinctive proteins used in recent years are collagen, silk, keratin, milk protein, and soy protein due to their abundance and interesting mechanical and biodegradable properties. Due to their remarkable strength and toughness, silk fibers are widely used in textile applications. However, the colossal potentiality of silk fibers is essential to use in the contemporary tissue engineering application and it is the most investigated field in recent years. Collagen, elastin, and keratin are most extensively used for biomaterial applications and more precisely, tissue repairing, scaffolding, skin substitutes, sheets, and sponge-like materials (Rnjak et al. 2009; Rnjak-Kovacina et al. 2012; Abou Neel et al. 2013). The processing of these biomaterials has mainly involved in two different processes like wet and dry methods.

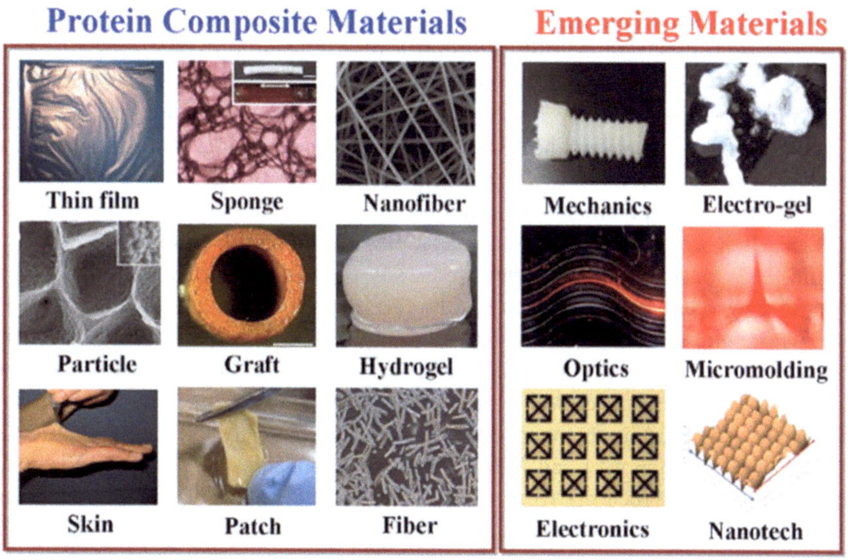

Fig. 4 The diverse applications of protein-based materials. Reprinted with permission from Hu et al. (2012b)

4.1 Processing and Performance of Protein-Based Composite Materials

4.1.1 Processing

As described above, proteins exhibit multiple applications like biocompatible hydrogels, tissue engineering, and polymer composites for advanced biomedical applications. Fig. 4 shows the diverse and newly emerged applications in the field of protein-based materials.

Fundamentally, these biomaterials are prepared by solvent casting, electrospinning, melt processing, etc. As any polymer nanocomposites (PNC), reinforced protein composites can be manufactured by various ways, affecting significantly the properties of the material. Among the parameters, which can be set during the manufacture, the nature and composition of the protein matrix, the type of processing, and the type of reinforcements as well as the admixture added to the blend are highlighted as potential factors influencing the final properties of the PNC. In this section, different applications based on protein and their performances are described.

4.1.2 Challenges

As compared to traditional thermoplastic polymers, using protein as a matrix brings problems, notably the brittleness of the resulting composite. Thus, the processing of protein-based films requires the addition of plasticizer in order to reduce its brittleness. Plasticizers weaken the intermolecular forces between polymer chains, resulting in higher ductility but also lower stiffness and resistance of the films (Banker 1966). Moreover, the temperature of the processing also influences the properties of the composite. Low temperature increases the resistance of the film, whereas higher temperatures result in a film with a rougher and less homogeneous surface (da Rocha et al. 2013). Various studies show that the optimum formulation would be a concentration of protein [e.g., soy protein (SPI)] of 8.65% and plasticizer (e.g., glycerol) concentration of 60% (Nandane and Jain 2014). Moreover, SPI and wheat gluten (WG) have a low solubility in a conventional solvent such as water or ethanol that is problematic in the casting process. Researchers show that the solubility of SPI and WG varies with the pH, where the processing of the composite is limited by the poor dispersion of the protein around the isoelectric pH region of SPI (pH = 4.5) and WG (pH = 7.6) (Gennadios et al. 1993; Nishinari et al. 2014). At higher pH ($6 > \text{pH} > 11$ for the SPI and >11 for the WG), the manufactured composites have higher tensile strength thanks to a better dispersion of the protein molecular chains during the solvent casting of films. The limitations for the other protein are not clear due to limited work in polymer nanocomposites.

5 Applications of Protein-Based Composite Materials

5.1 Polypeptide Hydrogels

Protein-based hydrogels consist of exceptional properties in terms of biocompatibility and biodegradability which eventually allow them ideal material for biomedical applications. Hydrogels have a tendency to hold bulk quantities of water or biological fluids, thanks to their three-dimensional (3D) cross-linking network. The self-assembly or chemical cross-linking of the small molecules into a 3D polymeric network made them to be insoluble in hydrophilic medium (Peppas et al. 2000; Hoffman 2012). This section highlights the synthesis of the hydrogels from proteins like silk fibroin, collagen, and elastin.

In the literature, some interesting studies were reported for tissue engineering applications. Foss et al. have the detailed preparation of 3D porous sponge-like hydrogels with silk and hyaluronic acid (HA). In this study, the influence of the HA addition on the physical properties of sponges, with and without cross-linking with genipin, was investigated (Foss et al. 2013). In some other study, silk/HA hydrogels are prepared and used for the drug delivery carrier. In this study, poly(ethylene glycol)-diacrylate was cross linked with thiol-modified HA and this composite has shown favorable mechanical properties and easily controllable degradation properties (Elia et al. 2013). Then, Hu et al. have reported the replacement of the nucleus pulposus with silk/polyurethane. In this study, the preparation and characterization of the mechanical, rheological, and biological behavior of novel-injectable composite hydrogels were investigated (Hu et al. 2012a). Similar hydrogels were prepared by Huang et al. and were used for controlled drug delivery applications (Huang et al. 2013). In this study, the swelling and de-swelling performance at different conditions like temperature, ionic strength, and pH were characterized (Huang et al. 2013).

Elastin is another protein can also be used for drug carrier applications, bioactuators, and implantable sensors (Silva et al. 2014). Moreover, Wang et al. have reported an interesting approach on light-controlled graphene-elastin composite hydrogel actuators. Hydrogels prepared with graphene and elastin have shown that the solvent or temperature conditions have a strong influence on the shaping and shrinking properties. These nanocomposites (NC) expressed tunable and rapid motions controlled by the light position and intensity (Wang et al. 2013a), and the results can be seen in Fig. 5b. These protein-based nanocomposites proved that sensibly designed proteins lead to novel and functional materials for extensive applications. In some other study, 3D structure of poly ε-caprolactone (PCL)/elastin cross-linked hydrogels (Fig. 5a) was prepared and characterized, showing the influence of the physical parameters like processing time, pressure, temperature, and depressurization rate (Annabi et al. 2011).

Xian Li et al. demonstrated the chitosan/collagen-based hydrogels for the tissue reconstruction. In this study, the preparation and morphology of the hydrogels were characterized before and after degradation, meaning that these hydrogels were used

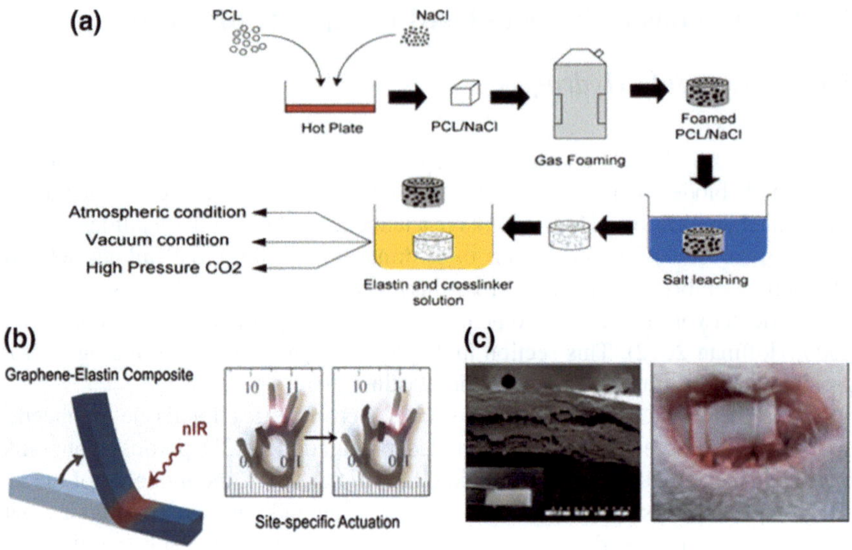

Fig. 5 Different protein-based composite materials. **a** PCL-elastin hydrogels, reprinted with permission from Annabi et al. (2011), **b** light-controlled graphene-elastin composite hydrogel, reprinted with permission from Wang et al. (2013a), **c** silk fibroin/BC-based hydrogels for the tissue reconstruction in vivo compatibility, reprinted with slight modification from Silva et al. (2014)

for the histocompatibility in vivo was evaluated (Li et al. 2012b). Juncosa-Melvin et al. have reported that collagen sponges can be used to repair the tendon defects in rabbits and results shown the 70% average repair in the rabbit tissue (Juncosa-Melvin et al. 2006).

5.2 *Protein-Based Materials for Tissue Engineering Applications*

One of the main explored fields in protein-based composites is tissue engineering, which is possible regardless of the protein source. Mainly, silk fibroin, collagen, and elastin are certainly the utmost investigated proteins. In the past decade, these composites were mainly used for tissue repair, artificial blood vessels, and composite plates for cortical bone repair (Silva et al. 2014). The most common method for the preparation of artificial blood vessels is realized by using collagen/elastin nanofiber meshes synthetized by electrospinning and stabilized with cross-linking agents in order to get better mechanical properties (Buttafoco et al. 2006a). In some other study, Buttafoco et al. have reported the homogeneous distribution of the porous network that is suitable for tissue engineering applications. Initially, collagen/elastin was cross linked in the presence of several reagents in polar-solvent (water/ethanol) medium and further freeze-dried (Buttafoco et al. 2006b).

Some interesting studies were reported by using the silk/gelatin composites for liver tissue engineering. Yang et al. have reported the in vitro and in vivo studies of these composites, by increasing the gelatin content, which results in higher compatibility with cell adhesion and faster degradation rate (Yang et al. 2012). Lee et al. have reported that the composites prepared with silk fibroin/bacterial cellulose (BC) for zygomatic arch factures regeneration, and these biomaterials were also tested in vitro and in vivo studies (Lee et al. 2013), and results can be seen in Fig. 5c. For cortical bone and cartilage repair applications, mechanical properties are significant. Choi et al. have reported that the composite plates prepared with silk/BC for cortical bone applications, where the influence of the BC on mechanical properties was studied (Choi et al. 2013).

In some other study, composites were prepared with cellulose/silk in ionic liquids. Further biological results of these composites shown that these blends can be used for in vitro tissue engineering applications (Singh et al. 2013). Furthermore, the regenerated blends in hydrophilic ionic liquids with cellulose/silk and their physical properties were investigated (Shang et al. 2011). It is well known that nanocellulose is ultimately good in mechanical properties (Nagalakshmaiah et al. 2016). In order to obtain adequate mechanical properties for tissue engineering applications, researchers have prepared composites with cellulose and collagen. Aji P Mathew et al. reported the cross-linking between the nanocellulosic fibers and collagen by using glutaraldehyde. Upon cross-linking the implantable composite films shown significant improvement in mechanical properties (Mathew et al. 2012). In some other study, cellulose was assimilated with collagen for better mechanical behavior and cell adhesion (Steele et al. 2013).

5.3 *Protein as a Reinforcing Phase with Polymers*

Another most explored application from protein is polymer nanocomposites by using protein as a filler or as a matrix due their good mechanical properties, biodegradability, and biocompatibility. This section deals with polymer composites prepared with different proteins such as silk, soy, elastin, or collagen for different applications like conductive composites and biomedical applications. Predominantly, silk fibers are essentially ideal candidates for conductive materials. For example, electrodes have been developed with poly(3,4-ethylenedioxythiophene)-Poly(styrene-sulfonate)-conductive copolymer and silk fiber bundles, and these composites are biocompatible with living cells (Tsukada et al. 2012). In some other studies, silk fibroin was either coated by chemical polymerization with polypyrrole for conductive materials (Aznar-Cervantes et al. 2012). Researchers have also shown that the highest young's modulus and tensile strength achieved were with poly (lactide-co-glycolide) (PLGA)/gelatin/elastin blends. Moreover, these blends have shown the supportive characteristics of metabolization of human endothelial cells (Han et al. 2011).

Soy and collagen are the major proteins used to be incorporated with another biopolymer like chitosan, BC, cellulose, and starch. For example, soy protein has been used to prepare the blends and composites with chitosan (Wang et al. 2012), wheat-bran cellulose (Wang et al. 2013b), Corn flour (Yu et al. 2012), and cellulose (Luo et al. 2010). Highly stiff nanofibrous composites with Collagen/chitosan have been produced by electrospinning (Chen et al. 2012). In some other study, chitosan 10–30 wt% was used as a filler with collagen, while increasing the filler content mechanical properties was improved (Kumar et al. 2012). Uriarte et al. have also reported the preparation of chitosan and collagen biofilms which have a higher elongation at break and lower tensile strength as compared to the neat chitosan films (Uriarte-Montoya et al. 2010).

6 Conclusion and Future Perspective

Proteins in composite materials have a promising and bright future. Indeed, as discussed in this chapter, numerous interesting studies have emerged recently, showing that the proteins properties (e.g., good mechanical properties, biocompatible, biodegradable) can match various high-end applications. Among them, the field of biomedical engineering seems to be the most promising. Protein hydrogels or polymer blends made of silk fibroin, collagen, and elastin have shown great potential for tissue engineering (e.g., tissue and bone repairing, and artificial blood vessel), biological scaffolding or also skin substitute. In overall, protein-based composites have a great versatility as well as the required properties to play a major role as a biomedical material in a very near future.

Acknowledgements The authors want to thank National Science and Engineering Research Council (NSERC) of Canada, the "Centre québecois des matériaux fonctionnels" (CQMF), and the "Consortium de Recherché et Innovations en Bioprocédés Industriels du Québec" (CRIBIQ) of Fonds de recherche du Québec − Nature et technologies (FRQNT).

References

Abou Neel EA, Bozec L, Knowles JC, Syed O, Mudera V, Day R, Hyun JK (2013) Collagen—emerging collagen based therapies hit the patient. Adv Drug Deliv Rev 65:429–456. https://doi.org/10.1016/j.addr.2012.08.010

Altman GH, Diaz F, Jakuba C, Calabro T, Horan RL, Chen J, Lu H, Richmond J, Kaplan DL (2003) Silk-based biomaterials. Biomaterials 24:401–416

Annabi N, Fathi A, Mithieux SM, Weiss AS, Dehghani F (2011) Fabrication of porous PCL/ elastin composite scaffolds for tissue engineering applications. J Supercrit Fluids 59:157–167
Ashok K (2006) Protein biotechnology. Discovery Publishing House, New Delhi
Aznar-Cervantes S, Roca MI, Martinez JG, Meseguer-Olmo L, Cenis JL, Moraleda JM, Otero TF (2012) Fabrication of conductive electrospun silk fibroin scaffolds by coating with polypyrrole for biomedical applications. Bioelectrochemistry 85:36–43. https://doi.org/10.1016/j.bioelechem.2011.11.008
Banker GS (1966) Film coating theory and practice. J Pharm Sci 55:81–89
Belgacem MN, Gandini A (eds) (2008) Monomers, polymers and composites from renewable resources, 1st edn. Elsevier, Amsterdam
Biorad Protein Expression and Purification Series. Introduction to protein purification, recombinant protein expression and purification
Buttafoco L, Engbers-Buijtenhuijs P, Poot AA, Dijkstra PJ, Daamen WF, van Kuppevelt TH, Vermes I, Feijen J (2006a) First steps towards tissue engineering of small-diameter blood vessels: preparation of flat scaffolds of collagen and elastin by means of freeze drying. J Biomed Mater Res B Appl Biomater 77B:357–368. https://doi.org/10.1002/jbm.b.30444
Buttafoco L, Kolkman NG, Engbers-Buijtenhuijs P, Poot AA, Dijkstra PJ, Vermes I, Feijen J (2006b) Electrospinning of collagen and elastin for tissue engineering applications. Biomaterials 27:724–734. https://doi.org/10.1016/j.biomaterials.2005.06.024
Byrne N, Angell CA (2009) Formation and dissolution of hen egg white lysozyme amyloid fibrils in protic ionic liquids. Chem Commun 1046. https://doi.org/10.1039/b817590j
Chen ZG, Wang PW, Wei B, Mo XM, Cui FZ (2012) Electrospun collagen–chitosan nanofiber: a biomimetic extracellular matrix for endothelial cell and smooth muscle cell. Acta Biomater 6:372–382. https://doi.org/10.1016/j.actbio.2009.07.024
Chen Q, Liang S, Thouas GA (2013) Elastomeric biomaterials for tissue engineering. Prog Polym Sci 38:584–671. https://doi.org/10.1016/j.progpolymsci.2012.05.003
Choi Y, Cho SY, Heo S, Jin H-J (2013) Enhanced mechanical properties of silk fibroin-based composite plates for fractured bone healing. Fibers Polym 14:266–270. https://doi.org/10.1007/s12221-013-0266-5
da Rocha M, Loiko MR, Gautério GV, Tondo EC, Prentice C (2013) Influence of heating, protein and glycerol concentrations of film-forming solution on the film properties of Argentine anchovy (*Engraulis anchoita*) protein isolate. J Food Eng 116:666–673. https://doi.org/10.1016/j.jfoodeng.2013.01.004
Daamen WF, Hafmans T, Veerkamp JH, van Kuppevelt TH (1997) Comparison of "ve procedures for the puri" cation of insoluble elastin. Biomaterials 22:2005
DiMarco RL, Heilshorn SC (2012) Multifunctional materials through modular protein engineering. Adv Mater 24:3923–3940. https://doi.org/10.1002/adma.201200051
Elia R et al (2013) Silk–hyaluronan-based composite hydrogels: a novel, securable vehicle for drug delivery. J Biomater Appl 27(6):749–762
Food Chemistry (2009) Springer, Berlin
Foss C, Merzari E, Migliaresi C, Motta A (2013) Silk fibroin/hyaluronic acid 3D matrices for cartilage tissue engineering. Biomacromolecules 14:38–47. https://doi.org/10.1021/bm301174x
Gennadios A, Brandenburg AH, Weller CL, Testin RF (1993) Effect of pH on properties of wheat gluten and soy protein isolate films. J Agric Food Chem 41:1835–1839
Gomes S, Leonor IB, Mano JF, Reis RL, Kaplan DL (2012) Natural and genetically engineered proteins for tissue engineering. Prog Polym Sci 37:1–17. https://doi.org/10.1016/j.progpolymsci.2011.07.003
Han J, Lazarovici P, Pomerantz C, Chen X, Wei Y, Lelkes PI (2011) Co-electrospun blends of PLGA, gelatin, and elastin as potential nonthrombogenic scaffolds for vascular tissue engineering. Biomacromolecules 12:399–408. https://doi.org/10.1021/bm101149r
Hoffman AS (2012) Hydrogels for biomedical applications. Adv Drug Deliv Rev 64:18–23. https://doi.org/10.1016/j.addr.2012.09.010

Hu J, Chen B, Guo F, Du J, Gu P, Lin X, Yang W, Zhang H, Lu M, Huang Y, Xu G (2012a) Injectable silk fibroin/polyurethane composite hydrogel for nucleus pulposus replacement. J Mater Sci Mater Med 23:711–722. https://doi.org/10.1007/s10856-011-4533-y

Hu X, Cebe P, Weiss AS, Omenetto F, Kaplan DL (2012b) Protein-based composite materials. Mater Today 15:208–215

Huang Y, Zhang B, Xu G, Hao W (2013) Swelling behaviours and mechanical properties of silk fibroin–polyurethane composite hydrogels. Compos Sci Technol 84:15–22. https://doi.org/10.1016/j.compscitech.2013.05.007

Jonker AM, Löwik DWPM, van Hest JCM (2012) Peptide- and protein-based hydrogels. Chem Mater 24:759–773. https://doi.org/10.1021/cm202640w

Juncosa-Melvin N, Shearn JT, Boivin GP, Gooch C, Galloway MT, West JR, Nirmalanandhan VS, Bradica G, Butler DL (2006) Effects of mechanical stimulation on the biomechanics and histology of stem cell–collagen sponge constructs for rabbit patellar tendon repair. Tissue Eng 12:2291–2300

Khosravi S, Khabbaz F, Nordqvist P, Johansson M (2010) Protein-based adhesives for particleboards. Ind Crops Prod 32:275–283. https://doi.org/10.1016/j.indcrop.2010.05.001

Kozak M (1983) Comparison of initiation of protein synthesis in prokaryotes, eucaryotes, and organelles

Kumar R, Choudhary V, Mishra S, Varma IK, Mattiason B (2002) Adhesives and plastics based on soy protein products. Ind Crops Prod 16:155–172

Kumar BS, Aigal S, Ramesh DV (2012) Air-dried 3D-collagen–chitosan biocomposite scaffold for tissue engineering application. Polym Compos 33(11):2029–2035

Lake JA (1983) Ribosome evolution: the structural bases of protein synthesis in archaebacteria, eubacteria, and eukaryotes

Lee JM, Kim JH, Lee OJ, Park CH (2013) The fixation effect of a silk fibroin–bacterial cellulose composite plate in segmental defects of the zygomatic arch: an experimental study. JAMA Otolaryngol Neck Surg 139:629–635

Li G, Liu H, Li T, Wang J (2012a) Surface modification and functionalization of silk fibroin fibers/fabric toward high performance applications. Mater Sci Eng C 32:627–636. https://doi.org/10.1016/j.msec.2011.12.013

Li X, Ma X, Fan D, Zhu C (2012b) New suitable for tissue reconstruction injectable chitosan/collagen-based hydrogels. Soft Matter 8:3781. https://doi.org/10.1039/c2sm06994f

Luo L-H, Zhang Y-F, Wang X-M, Wan Y, Chang PR, Anderson DP, Chen Y (2010) Preparation, characterization, and in vitro and in vivo evaluation of cellulose/soy protein isolate composite sponges. J Biomater Appl 24:503–526. https://doi.org/10.1177/0885328208099337

Mathew AP, Oksman K, Pierron D, Harmad M-F (2012) Crosslinked fibrous composites based on cellulose nanofibers and collagen with in situ pH induced fibrillation. Cellulose 19:139–150. https://doi.org/10.1007/s10570-011-9624-x

Mecham RP (2008) Methods in elastic tissue biology: elastin isolation and purification. Methods 45:32–41. https://doi.org/10.1016/j.ymeth.2008.01.007

Nagalakshmaiah M, El Kissi N, Dufresne A (2016) Ionic compatibilization of cellulose nanocrystals with quaternary ammonium salt and their melt extrusion with polypropylene. ACS Appl Mater Interfaces 8:8755–8764. https://doi.org/10.1021/acsami.6b01650

Nandane AS, Jain R (2014) Study of mechanical properties of soy protein based edible film as affected by its composition and process parameters by using RSM. J Food Sci Technol. https://doi.org/10.1007/s13197-014-1417-4

Nishinari K, Fang Y, Guo S, Phillips GO (2014) Soy proteins: a review on composition, aggregation and emulsification. Food Hydrocoll 39:301–318. https://doi.org/10.1016/j.foodhyd.2014.01.013

Peppas NA, Bures P, Leobandung W, Ichikawa H (2000) Hydrogels in pharmaceutical formulations. Eur J Pharm Biopharm 50:27–46

Rnjak J, Li Z, Maitz PKM, Wise SG, Weiss AS (2009) Primary human dermal fibroblast interactions with open weave three-dimensional scaffolds prepared from synthetic human elastin. Biomaterials 30:6469–6477. https://doi.org/10.1016/j.biomaterials.2009.08.017

Rnjak-Kovacina J, Wise SG, Li Z, Maitz PKM, Young CJ, Wang Y, Weiss AS (2012) Electrospun synthetic human elastin:collagen composite scaffolds for dermal tissue engineering. Acta Biomater 8:3714–3722. https://doi.org/10.1016/j.actbio.2012.06.032

Shang S, Zhu L, Fan J (2011) Physical properties of silk fibroin/cellulose blend films regenerated from the hydrophilic ionic liquid. Carbohydr Polym 86:462–468. https://doi.org/10.1016/j.carbpol.2011.04.064

Silva NHCS, Vilela C, Marrucho IM, Freire CSR, Pascoal Neto C, Silvestre AJD (2014) Protein-based materials: from sources to innovative sustainable materials for biomedical applications. J Mater Chem B 2:3715. https://doi.org/10.1039/c4tb00168k

Singh P, Kumar R, Sabapathy SN, Bawa AS (2008) Functional and edible uses of soy protein products. Compr Rev Food Sci Food Saf 7:14–28

Singh N, Rahatekar SS, Koziol KKK, Ng TS, Patil AJ, Mann S, Hollander AP, Kafienah W (2013) Directing chondrogenesis of stem cells with specific blends of cellulose and silk. Biomacromolecules 14:1287–1298. https://doi.org/10.1021/bm301762p

Skopinska-Wisniewska J, Sionkowska A, Kaminska A, Kaznica A, Jachimiak R, Drewa T (2009) Surface characterization of collagen/elastin based biomaterials for tissue regeneration. Appl Surf Sci 255:8286–8292. https://doi.org/10.1016/j.apsusc.2009.05.127

Steele TWJ, Huang CL, Nguyen E, Sarig U, Kumar S, Widjaja E, Loo JSC, Machluf M, Boey F, Vukadinovic Z, Hilfiker A, Venkatraman SS (2013) Collagen–cellulose composite thin films that mimic soft-tissue and allow stem-cell orientation. J Mater Sci Mater Med 24:2013–2027. https://doi.org/10.1007/s10856-013-4940-3

Swain SN, Biswal SM, Nanda PK, Nayak PL (2004) Biodegradable soy-based plastics: opportunities and challenges. J Polym Environ 12:35–42

Tian H, Zhang L, Wu Q, Wang X, Chen Y (2010) Creation of hydrophobic materials fabricated from soy protein and natural rubber: surface, interface, and properties. Macromol Mater Eng NA-NA. https://doi.org/10.1002/mame.200900366

Toldrá F, Nollet LM (eds) (2012) Proteomics in foods: principles and applications. Springer Science & Business Media. http://b-ok.cc/book/2084540/1b3de1

Tsukada S, Nakashima H, Torimitsu K (2012) Conductive polymer combined silk fiber bundle for bioelectrical signal recording. PLoS ONE 7:e33689. https://doi.org/10.1371/journal.pone.0033689

Uriarte-Montoya MH, Arias-Moscoso JL, Plascencia-Jatomea M, Santacruz-Ortega H, Rouzaud-Sández O, Cardenas-Lopez JL, Ezquerra-Brauer JM (2010) Jumbo squid (Dosidicus gigas) mantle collagen: extraction, characterization, and potential application in the preparation of chitosan–collagen biofilms. Biores Technol 101(11):4212–4219

Vepari C, Kaplan DL (2007) Silk as a biomaterial. Prog Polym Sci 32:991–1007. https://doi.org/10.1016/j.progpolymsci.2007.05.013

Vrhovski B, Weiss AS (1998) Biochemistry of tropoelastin. FEBS J 258:1–18

Wang W, Guo Y, Otaigbe JU (2010) Synthesis, characterization and degradation of biodegradable thermoplastic elastomers from poly(ester urethane)s and renewable soy protein isolate biopolymer. Polymer 51:5448–5455. https://doi.org/10.1016/j.polymer.2010.09.056

Wang X, Shi N, Chen Y, Li C, Du X, Jin W, Chen Y, Chang PR (2012) Improvement in hemocompatibility of chitosan/soy protein composite membranes by heparinization. Biomed Mater Eng 22:143–150

Wang E, Desai MS, Lee S-W (2013a) Light-controlled graphene-elastin composite hydrogel actuators. Nano Lett 13:2826–2830. https://doi.org/10.1021/nl401088b

Wang Z, Sun X, Lian Z, Wang X, Zhou J, Ma Z (2013b) The effects of ultrasonic/microwave assisted treatment on the properties of soy protein isolate/microcrystalline wheat-bran cellulose film. J Food Eng 114:183–191. https://doi.org/10.1016/j.jfoodeng.2012.08.004

Werkmeister JA, Ramshaw JAM (2012) Recombinant protein scaffolds for tissue engineering. Biomed Mater 7:012002. https://doi.org/10.1088/1748-6041/7/1/012002

Yang Z, Xu LS, Yin F, Shi YQ, Han Y, Zhang L, Jin HF, Nie YZ, Wang JB, Hao X, Fan DM, Zhou XM (2012) In vitro and in vivo characterization of silk fibroin/gelatin composite scaffolds for liver tissue engineering: SF/G as liver tissue engineering. J Dig Dis 13:168–178. https://doi.org/10.1111/j.1751-2980.2011.00566.x

Yu L, Ramaswamy HS, Boye J (2012) Twin-screw extrusion of corn flour and soy protein isolate (SPI) blends: a response surface analysis. Food Bioprocess Technol 5:485–497. https://doi.org/10.1007/s11947-009-0294-8

Zhang L, Zeng M (2008). Proteins as sources of materials. In: Monomers, polymers and composites from renewable resources. pp 479–493

Polylactic Acid (PLA)-Based Nanocomposites: Processing and Properties

Ahmed Sharif, Sudipta Mondal and Md Enamul Hoque

Abstract Polylactic acid (PLA) is the most comprehensively explored biodegradable and renewable thermoplastic polyester. It has the capacity to substitute polymers from fossil fuel-based resources. However, certain properties such as uncontrolled degradation, poor thermal properties, gas permeability, and profound brittleness characteristic may restrict the wide-scale utilization of PLA. Consequently, nano-based reinforcements have been widely exploited to upgrade some of the shortcomings. With the advancement in the field of nanotechnology, PLA nanocomposites have emerged as an excellent material. These materials have a big potential for applications in food packaging, medical applications, and tissue cultures. This chapter mainly assesses the different types of nanocomposites from PLA in terms of reinforcing materials. The resultant composite structures are correlated for different processing methods with the quality and factors of dispersion of reinforcements.

Keywords PLA · Nanocomposite · Biodegradable · Renewable Polyester · Food packaging · Medical applications

1 Introduction

Lately, the use of biopolymers has been increased due to more environmentally aware consumers and the concern about global warming. These polymers are recognized biomaterials that may also be used in our daily life as a replacement of conventional polymers and plastics (Albertsson and Karlsson 1994; Ali et al. 1994).

A. Sharif · S. Mondal
Department of Materials and Metallurgical Engineering, Bangladesh University of Engineering and Technology (BUET), Dhaka 1000, Bangladesh

M. E. Hoque (✉)
Department of Biomedical Engineering, Military Institute of Science and Technology (MIST), Mirpur Cantonment, Dhaka 1216, Bangladesh
e-mail: enamul1973@gmail.com

M. L. Sanyang and M. Jawaid (eds.), *Bio-based Polymers and Nanocomposites*,
https://doi.org/10.1007/978-3-030-05825-8_11

Due to the widespread reliance on polymers, scientific and engineering efforts are carried out for the innovation, advancements, and reformations of these materials. Polylactic acid (PLA), a compostable polymer derived from endless sources (mainly starch and sugar), has been considered as a promising material to decrease the solid waste disposal and suitable material for household goods and packaging applications (Luo et al. 2018). PLAs are aliphatic polyesters of lactic acid used mainly in the biomedical field such as implant devices, internal sutures, and tissue scaffolds, because of its low availability, high cost, and limited molecular weight (Rasal et al. 2010; Michael et al. 2016). It has some properties similar to conventional plastics (like polyethylene or polyethylene terephthalate). Moreover, the processing techniques are quite similar. Different forms of fibers, films, plastic containers, and other household products are manufactured from PLA granulates.

2 Overview of PLA

The structural formula of PLA is given in Fig. 1. Lactic acid (2-hydroxypropionic acid) is a simple chiral molecule that exists as two enantiomers, L- and D-lactic acid. In this way, the term "polylactic acid" refers to a family of polymers: poly-L-lactic acid (PLLA), poly-D-lactic acid (PDLA), and poly-D,L-lactic acid (PDLLA). Furthermore, DL-PLA is a syndiotactically alternating D,L-copolymer or a copolymer having L and D units. PLLA has achieved enormous consideration for its brilliant biocompatibility and physical properties. However, its lengthy degradation periods and high crystallinity of its fragments are the major concerns for widespread applications.

The chemistry of PLA involves the processing and polymerization of lactic acid monomer (Hoque et al. 2005). PLA with higher than 90% PLLA tends to be crystalline, while the lower optically pure is amorphous. The melting temperature and the glass transition temperature of PLA increase with the amount of PLLA. The density of amorphous and crystalline PLLA has been reported as 1.248 and 1.290 g/ml, respectively. Generally, PLA is soluble in acetonitrile, dioxane, chloroform, methylene chloride, 1,1,2-trichloroethane, and dichloroacetic acid. On the other hand, PLA is partially soluble at room temperature and completely soluble at boiling temperature in ethyl benzene, toluene, acetone, and tetrahydrofuran. The major beneficial properties of PLA are given below.

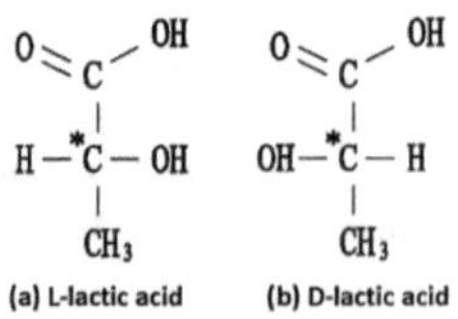

Fig. 1 Structural formulas of **a** L-lactic acid and **b** D-lactic acid

- Transparency: PLA has an excellent property of transparency. Due to this property, PLAs are used in the radio control toys, drones, aerospace, automotive, prototyping, and many more.
- Eco-friendly: Besides being derived from renewable resources (e.g., corn, wheat, or rice), PLA is biodegradable, compostable, and recyclable (Drumright et al. 2000; Sawyer 2003). Its synthesis also utilizes less CO_2 (Dorgan et al. 2001; Hutmacher et al. 2008). These eco-pleasant characteristics make PLA an attractive biopolymer.
- Biocompatibility: The most striking property of PLA is its biocompatibility. PLA does not produce poisonous or cancer-causing consequences in local tissues (Athanasiou et al. 1996). When PLAs are implanted in living organisms including the human body, they hydrolyze to hydroxy acid. It is then integrated into the tricarboxylic acid cycle and expelled (Kimura et al. 1988). Moreover, PLA degradation outcomes are harmless (up to a specific composition) which is the main reason for biomedical applications (Athanasiou et al. 1996; Eling et al. 1982; Kimura et al. 1988). The Food and Drug Administration (FDA) has also approved PLA for direct contacting with biological fluids (Gupta et al. 2007).
- Processability: PLAs can be manufactured by various techniques. PLAs can also be synthesized industrially with novel economical polymerization systems. Some common methods include extrusion, injection molding, injection stretch blow molding, casting, thermoforming, blown film, foaming, blending, fiber spinning, and compounding (Linnemann et al. 2003; Okada 2002; Schugens et al. 1995).
- Energy savings: PLA demands 25–55% less energy to produce than conventional polymers, and this requirement can be further cut down to less than 10% in the near future (Vink et al. 2003). The energy-saving characteristics of PLA synthesis are substantially strategic with respect to cost.

Although PLA has some prospective advantages, it has shortcomings as well, which restrain its exploitation in specific uses. These disadvantages are listed below.

- Brittleness—PLA is a very brittle material. In tension, it shows under 10% of elongation at fracture (Hiljanen-Vainio et al. 1996; Rasal and Hirt 2009). The tensile strength of PLA films ranges between 50 and 70 MPa with an elongation of 4% at break and an elastic modulus of ~ 3 GPa. It shows an impact strength of 2.5 kJ/m^2 (Anderson et al. 2008). These mechanical characteristics are more or less comparable to polyethylene.
- Slow degradation rate—The degradation rate of PLA is influenced by the PLA crystallinity, molecular mass and its arrangement, permeability, surface characteristics, and the stereoisomeric content. The slow degradation rate of PLA raises some concerns. In some cases, degradation could follow several years (Bergsma et al. 2006). There was report of a second surgery almost 3 years after implantation to remove a PLA-based implant (Incardona et al. 1996).

- Hydrolysis rate and thermal resistance—As the hydrolysis rate is fast and thermal resistance is poor, the PLA-based food package is inappropriate for hot liquid.
- Hydrophobicity—PLA is somewhat hydrophobic in nature. The contact angle between PLA and stagnant water is around 80°. Also, PLA is less coherent to biological cell. It is also reported that it could stimulate an incendiary reaction with anatomical liquids at straight association (Burg et al. 1999; Ratner 1995).
- Lack of reactive side chain groups—PLA is chemically passive without any lateral chain clusters for reaction. That is why exterior and volume reformations through chemical reaction are difficult activity for PLA.

Due to those limitations, blending PLAs with flexible polymers or reinforcing with inorganic or natural fillers are common approaches to enhance the mechanical characteristics of PLAs.

3 Processing of PLA

The monomeric building block of PLA is lactic acid. Lactic acid can be produced by bacterial fermentation as well as petrochemical route. Sugar or starch obtained from vegetable sources (e.g., corn, wheat, or rice) (Hoque et al. 2013a) is used for these processes. Commercial PLAs are copolymers of both poly-L-lactic acid (PLLA) and poly-D, L-lactic acid (PDLLA). They are manufactured from L-lactides and D, L-lactides (Auras et al. 2004; Nampoothiri et al. 2010; Saeidlou et al. 2012). The L-lactic acid spins the plane of polarized light clockwise; on the contrary, D-lactic acid turns it counterclockwise. Lactic acid produced by petrochemical routes is a 50/50 mixture of the D and L forms. That is why it becomes optically inactive. Since the fermentation method is eco-friendlier, it has been exploited significantly to synthesis PLA since the 1990 (Anderson et al. 2008). In the bacterial fermentation process, an augmented breed of Lactobacillus is employed to transform corn starch into lactic acid (Dorgan et al. 2000). The L-isomer is the main product of naturally occurring PLAs in the fermentation process (99.5% of the L-isomer and 0.5% of the D-isomer) (Anderson et al. 2008; Vink et al. 2007).

Polymers of lactic acid can be constructed by a number of methods, including azeotropic dehydrative condensation, direct condensation polymerization, and chain formation via lactide development (Fig. 2). Notably, economically viable high molecular weight PLA resins are produced via the lactide ring-opening polymerization route (Garlotta 2001; Ikada and Tsuji 2000; Martin and Averous 2001).

The optical integrity of PLA dictates the structural, barrier, thermal, and mechanical properties. Ninety percentage or more L-isomer-containing PLA polymers become crystalline, while depleted L-isomer comprising PLAs is amorphous. For PLA blends with less than 1% of D-isomer, injection molded techniques can be exploited. However, with less crystallinity PLA combinations with greater D-isomer composition (i.e., 4–8%) are convenient for extrusion, thermoforming, and blow

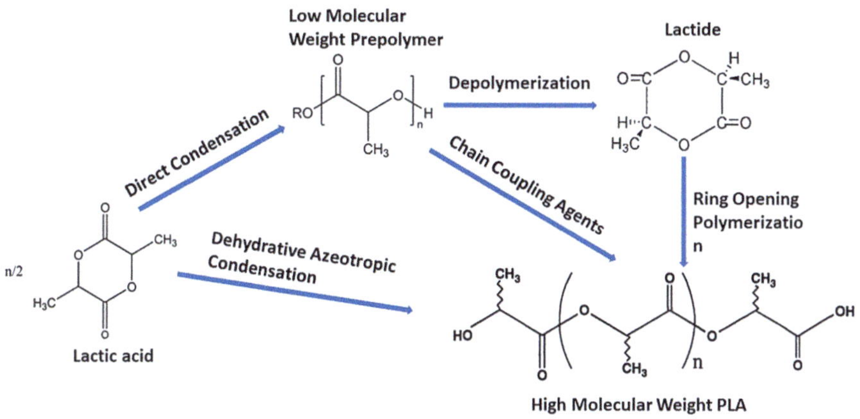

Fig. 2 Synthesis of PLA from L- and D-lactic acids. Adapted from Auras et al. (2004)

molding. Pure PLA typically is poor heat- and shock-resistant. Therefore, reinforcements, plasticizers, and other mixtures are added to augment their properties.

4 PLA-Based Nanocomposites

Mechanical and thermal properties of the PLA matrix are enhanced by incorporating inorganic and organic materials. Macro- , micro- , and nano-reinforcements are commonly employed to make PLA-based composites. The main advantage of nano-reinforcements is the quantity required to produce desired properties in PLA. Typically, PLA nanocomposites with only 0.5–8 wt% reinforcement are adequate for this purpose. Different types of nano-fillers that are used to make PLA-nanocomposites include carbon nanotubes (CNT), layered silicates, hydroxyapatite, layered titanate, and aluminum hydroxide. In PLA nanocomposites, the nanofiller is in nanoscale (i.e., <100 nm) at least in one dimension. Recently, Raquez et al. (2013) comprehensively reviewed PLA-based nanocomposites. Nanofillers can be classified based on their shape:

i. Layered nanofillers: They are nanolayered materials. Their thickness is approximately 1 nm. Customary one-dimensional nanofillers are sheets of silicates, different clays, and graphene.
ii. Whiskers or nanofibers: These shapes have an average diameter of below 100 nm. Carbon nanotubes (CNTs) and nanocellulose are the familiar instances.
iii. Nanoparticles: This three-dimensional shape should have an average diameter of below 100 nm as well. Silica particles, polyhedral oligomeric silsesquioxane (POSS), and metal oxides are the most common.

These nanofillers mainly improve mechanical and thermal properties. But, some other properties may also be improved. These include flame retardant (FR), anti-static, antibacterial effect, conductive, UV protection, and gas barrier. Nanofillers provide a way to extend and improve the existing properties of PLA (Raquez et al. 2013).

4.1 Layered Ceramic/PLA Nanocomposite

As PLA is chiefly hydrophobic in nature, the homogenous dispersion of the hydrophilic ceramic-layered structures is a major concern. They are often unable to be uniformly dispersed in the polymer matrix. Figure 3 shows different types of intercalation/dispersions behavior of layered ceramics in a PLA-based matrix. Commonly, adaption is needed to amplify the interaction between the layered ceramics and PLA by the addition of organic minerals during dispersion (Dash et al. 2008; Parida et al. 2006).

PLA/layered ceramics nanocomposites are mainly prepared by intercalation with solvent, melt, or in situ. Among them, solvent intercalation is one of the simplest procedures to synthesise nanocomposites. In this method, a solvent plays the major role by dissolving the PLA for easy dispersion of the ceramic nanofiller in the solution (Cumkur et al. 2015). Here, the polymer is first allowed to make a solution with suitable dissolvable. And then, the nanofillers are added and stirred for dispersion in the solution. High-speed cyclone and/or sonication techniques are used for adding the nanofiller suspension to the polymer solution. In melt intercalation, a mixture of PLA and layered ceramics is first heated above the glass transition or melting temperature of PLA. Then, the heated mixture is blended using shear forces in an extruder. Nevertheless, the interaction of the layered ceramics with hydrophobic PLA is improved usually through surface modification of the ceramic nanofillers.

In in situ polymerization technique, polylactides are mixed with layered ceramic fillers with a catalyst, and then, in situ polymerization by ring opening takes place to form PLA nanocomposites. Paul et al. (2005) produced PLLA/organo-modified

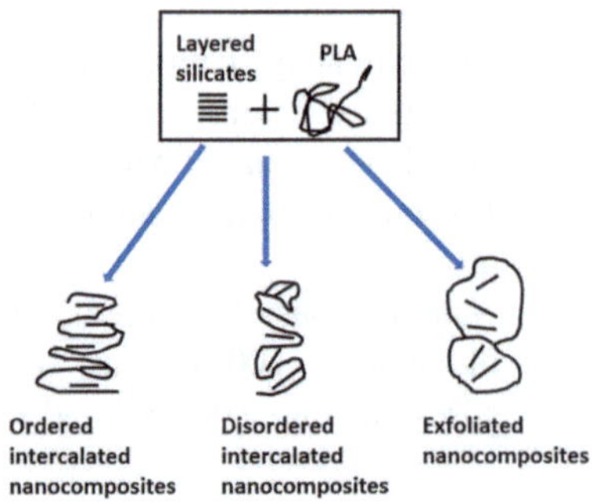

Fig. 3 Structures of layered silicate–polymer nanocomposites (the concept was redrawn from Raquez et al. 2013)

montmorillonite nanocomposites with both intercalated and exfoliated structures by employing the in situ ring-opening polymerization of poly-L-lactides. They realized that the modification of nanofiller acted a major role in dispersion phenomenon. When modified montmorillonite Na was exploited, exfoliation of PLA on layered structure took place (Paul et al. 2005).

Krikorian and Pochan (2004) reported that the degree of clay miscibility with the matrix and the clay distribution condition in the PLLA matrix both considerably affected the crystallinity and ultimate structure of the nanocomposites. They suggested that the nucleating efficiency of PLA crystals with organoclays during intercalation was greater than that during exfoliation. As a consequence, overall bulk crystallization was much higher in the intercalated system than that in the exfoliated system. In addition, they concluded that the spherulite growth was significantly greater in the fully exfoliated nanocomposite due to lower nucleating efficiency (Krikorian and Pochan 2004).

Nanosilica fillers' addition shows advancements in different properties of PLA. The improved properties include tensile, flexural, heat distortion temperature, gas permeability, crystallization rate, biodegradability, and so on. These properties were mainly dictated by the structure of the composites. In this case, the amount and the quality distribution of the reinforcement in the PLA matrix are the two major variables. Different polymeric compatibilizers are added to increase the interaction of silica/PLA for the uniform dispersion of the silica nanofillers (Sinha Ray et al. 2002). Mainly, epoxy and maleic anhydride-based polymeric compatibilizers are used.

In case of clay-based nanoparticles, the organic modifiers were used to increase the intimate contact between clay nanoparticles and the PLA matrix (Tehrani et al. 2014). Comparatively, clays are inexpensive and can provide considerable developments in mechanical and thermal properties of PLA (Lagarón and Cabedo 2014). That is why they are very common nowadays. Most often, a simple melt blending method was used for the fabrication of the composite structure. For the clay-based nanoparticles, Li et al. (2014) investigated PLLA/PDLA/clay nanocomposites. They suggest that dispersion can be improved by enhancing shear force during melt blending (Li et al. 2014). According to their study, clay fillers normally yield higher mechanical strength. So, a rigid structure can be prepared. In an interesting study, Lai et al. (2014) investigated the *N*,*N*-bis(2-hydroxyethyl)-*N*-methyl-*N*-tallow ammonium chloride modified nanoclay addition in PDLLA matrix in melt intercalation process. They reported a tremendous increase in ductility (i.e., with an elongation of 208% at break) of PLA/clay nanocomposites. The extensively exfoliated and uniformly dispersed clay platelets along with the highly plasticized interphase around these clay platelets were observed in their study. The high ductility may be evolved for high level of exfoliation and high dispersion which results in multiple shear banding in the amorphous PLA structure (Lai et al. 2014).

4.2 Layered Double Hydroxide/PLA Nanocomposites

In recent years, layered double hydroxide (LDH) as a layered nanostructure became very popular due to its various unique properties which are not found in traditional layered silicates. LDH has mainly two layers. One layer is the positively charged metal hydroxide sheets, and another is the intercalated anions. Water molecules are also attached to the structure (Mandal et al. 2009). The general formula is recorded as $[M^{2+}{}_{1-x}M^{3+}{}_{x}(OH)_2]A^{n-}{}_{x/n}yH_2O$, where M^{2+} and M^{3+} are divalent and trivalent metal cations like Mg^{2+} and Al^{3+}, respectively, and A^{n-} are interlayer anions like $CO_3{}^{2-}$, Cl^-, and $NO_3{}^-$ (Costa et al., 2007). The benefits of using LDHs in nanocomposites include the suitability of organo-LDHs for intercalation of hydrophobic polymers and absence of trace heavy metals. LDH platelets have a high aspect ratio, tunable layer charge density, and low production cost. As there are strong electrostatic interactions between highly charged hydroxide layers and intercalated anions, it is difficult to efficiently disperse and to some extent to exfoliate (Chiang and Wu 2010).

In order to simplify the intercalation of polymeric chains between LDH layers, sometimes a pre-treatment of the nanofiller is required. For these, organic or polymeric anions like alkyl carboxylates, alkyl sulfates, and acrylates can be readily incorporated into the LDH interlayers by ion exchange or in situ polymerization reactions (Manzi-Nshuti et al. 2009). Another improvement of dispersion of LDH was studied by Katiyar et al. (2010). They explored the synthesis of PLA/LDH nanocomposites by ring-opening polymerization (ROP) of LA as shown in Fig. 4 (Katiyar et al. 2010). In this study, two types of LDH carbonate (LDH-CO_3) and laurate-modified LDH (LDH-C12) were investigated as LDH nanofillers. LDH-C12 was used to obtain exfoliated nanocomposites. But there was a problem with molecular weight of PLA which was significantly reduced after in situ polymerization. The possible reason could be chain termination promoted via LDH surface hydroxyl groups or metal-catalyzed degradation.

Katiyar et al. (2011) also examined the barrier properties of resulting PLA/LDH nanocomposites manufactured by extrusion technology in the presence of laurate-intercalated LDH (LDH-C12). They also compared with MMT clay, i.e., Cloisite 30B which is an organo-modified montmorillonite (Katiyar et al. 2011). Two different approaches for nanocomposite synthesis were compared, i.e., pot addition and masterbatch way by an in situ intercalative ROP. It was found that PLA/Cloisite 30B nanocomposites exhibited significant enhancement in oxygen and water vapor barrier properties comparing to those of PLA/LDH-C12 nanocomposites.

4.3 Glass/PLA Nanocomposites

Glass particle-reinforced PLA nanocomposites were developed for better mechanical properties. It was reported that inclusion of a glass microsphere

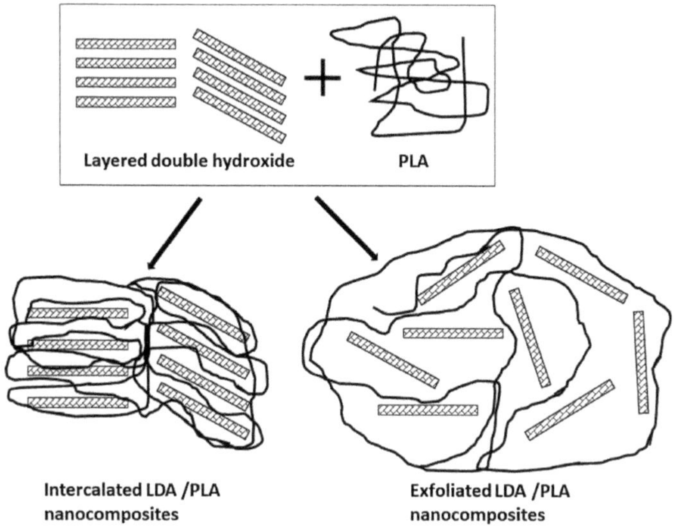

Fig. 4 Schematic illustration of in situ ring-opening polymerization of LA in the presence of LDHs. Reproduced from Katiyar et al. (2010)

uniformly throughout the matrix improves the Young's modulus, flexural modulus, impact strength, flexural strength, and thermal properties of PLA (Malinowski et al. 2015). As expected, the total elongation at fracture was decreased in tension test. Surface free energy was significantly influenced owing to the composite admixing (Malinowski et al. 2015). In this context, another fully biodegradable composite structure with PLA and calcium phosphate soluble glass particles were developed by Navarro et al. (2005). They suggested that the incorporation of phosphate glass particles into PLA significantly accelerated the degradation of the PLA and induced the formation of calcium phosphate precipitates at the composite surface (Navarro et al. 2005). Huda et al. (2008) found that both mechanical and thermophysical properties of glass fiber-reinforced PLA composite showed improved performance with the corresponding properties of neat PLA.

4.4 Silica/PLA Nanocomposites

Generally, hard fillers such as nanosilica are used to improve the strength, adhesion, durability, and abrasion resistance of conventional polymer composites. Zhu et al. (2010) demonstrated that the oleic acid (OA)-modified SiO_2 nanoparticles were effective fillers for improving the flexibility of PLA even though tensile strength decreased the nanocomposites drastically than that of the pristine PLA. In another study, Yan et al. (2007) reported the surface-modified silica nanoparticles with a L-lactic acid oligomer by direct grafting onto the surface of the silica nanoparticles.

The loading of grafted SiO_2 nanoparticles in PLLA matrix greatly improved the toughness and tensile strength of the composite structure because of the uniform dispersion of nano-SiO_2 in the PLA matrix. They also found that the incorporation of ungrafted SiO_2 nanoparticles into PLLA actually deteriorated the mechanical properties of the composites (Yan et al. 2007). Other studies also found that the addition of 2-methacryloyloxyethyl isocyanate (MOI) or trimethyl hexamethylene diisocyanate (TMDI) significantly imparted superior mechanical properties of the SiO_2/PLA nanocomposites with much higher elongation compared to neat PLA (Chen et al. 2010, 2012). Another study demonstrated that simple melt-compounded PLA/spherical nano-SiO_2 composites showed better thermal stability than that of neat PLA (Wen et al. 2011). In this case, they proposed grafting of PLA chains with SiO_2 without any surface modification additives (Wen et al. 2011). Here, they did not measure the mechanical properties of the nanocomposites.

4.5 Titanium Dioxide/PLA Nanocomposites

Due to large surface areas, titanium dioxide nanoparticles possess significant photocatalytic and magnetic properties. They also showed exceptional hydrophilicity, great toughness, and antibacterial properties. PLA/TiO_2 nanocomposites were studied with a view to photodegradability (Nakayama and Hayashi 2007), improvement of toughness (Meng et al. 2011), the crystallization kinetics (Liao et al. 2007), and cancer cells treatment (Chen et al. 2007). Li et al. (2011) investigated the TiO_2 nanowire/PLA nanocomposites which showed covalent bonding between TiO_2 surface and PLA through in situ melt polycondensation. They claimed strong interfacial interaction and homogenous distribution of inorganic nanowires in PLA matrix (Li et al. 2011). This strong interaction between TiO_2 and PLA displayed significantly higher glass transition temperature and thermal stability, compared with pure PLA. Luo et al. (2009) investigated that the effect of initial grafting and polymerization with lactic acid on the surface of TiO_2 nanoparticles. They showed that the TiO_2/PLA nanocomposites prepared by melt blending with PLA and grafted TiO_2 resulted in improved mechanical properties (Luo et al. 2009). Zhuang et al. (2009) have found that the in situ polymerization of L-lactides in the presence of organically modified TiO_2 nanoparticles showed markedly improved thermal and mechanical properties of PLA matrix.

TiO_2 acts as a wonderful photocatalyst in the degradation of organic pollutants due to its effective photostability, reusability, non-toxicity, and cheap as well. Thus, the relatively low photodegradability of a PLA matrix can be extremely enhanced by TiO_2 nanoparticles addition. Besides photodegradation, TiO_2-coated substrates can be used as antibacterial materials (Cai et al. 2003). The antibacterial effect is common for biological cells. These cells can be successfully killed with photocatalytic activity of TiO_2 particles that are incorporated into the cells. The biological cells can also be cultured onto the surface of a substrate with TiO_2 coating. The same effect can be found here too.

4.6 Zinc Oxide/PLA Nanocomposites

Like TiO_2, zinc oxide nanoparticles have excellent antibacterial action for which they are used as nanofillers for PLAs. Another property of ZnO is intensive UV absorption. But, the addition of ZnO nanoparticles and other Zn derivatives into PLA at melt-processing temperatures leads to severe degradation of the polyester matrix. A sharp reduction of thermomechanical characteristics was noticed for the PLA nanocomposite with untreated ZnO reinforcements (Murariu et al. 2011). These problems can be overcome by various surface treatments of nanofillers with selected additives (i.e., stearic acid, stearates, fatty amides). Surface-treated ZnO with triethoxy caprylylsilane was found to be very effective for the improvement of thermomechanical characteristics of the PLA nanocomposite (Murariu et al. 2011). In another study, PLA–ZnO nanocomposites with surface-treated ZnO rod-like nanoparticles showed excellent barrier properties and a good antibacterial activity against gram-positive and gram-negative bacteria (Pantani et al. 2013). Bussiere et al. (2012) also discovered that surface-treated ZnO played an anti-nucleating role in the crystallization of PLA nanocomposites. Anžlovar et al. (2018) studied the nanocomposites of ZnO/PLA (prepared by melt processing) in terms of degradation behavior of PLA. They found that nano-ZnO performed as an obstructer of PLA crystallization process and shifted the polymer glass transition temperature to lower temperatures. They also confirmed that nano-ZnO participated as a reactant and an accelerator in the degradation reaction of PLA (Anžlovar et al. 2018).

4.7 Alumina/PLA Nanocomposites

Alumina/PLA composites have some unique applications. They have shared interaction via polar coupling and hydrogen bonding, which creates good adhesion between these two different materials. Alumina is a ceramic material. It is widely used clinically because it exhibits highly bioinert characteristics, excellent corrosion resistance, and improved strength. It is also used in a large number of dental and orthopedic applications. Sometimes, alumina coatings are used that provide tissue growth and maxillofacial reconstruction (Vallet-Regí et al. 1998). This is of practical interest in PLA, which contain carboxylic ester groups in the structure and can provide strong interactions with alumina particles.

4.8 Fe_3O_4/PLA Nanocomposites

Fe_3O_4 nanoparticles have attracted much attention due to their wide prospective functions in many areas, such as pigments, sorbents, coatings, flocculents, catalysts, gas sensors, ion exchangers, and lubricants (Liang et al. 2012; Mu et al. 2017;

Shabanian et al. 2015). Fe_3O_4 nanocomposites have exhibited promising usages in diverse fields such as magnetic recording, magnetic data storage devices, toners and inks for xerography, magnetite/polymer composite microbubbles, magnetic resonance imaging, wastewater treatment, bioseparation, and medicine (Prasad et al. 2018; Sajjadi et al. 2017; Xia et al. 2017). Furthermore, Fe_3O_4 nanoparticles can be easily produced by co-precipitation methods in the presence of appropriate surfactants (Chen et al. 2005). Zheng et al. took advantage of the super-paramagnetic properties of Fe_3O_4 nanoparticles in order to elaborate the mechanism behind magnetic-induced shape memory PDLLA nanocomposites (Zheng et al. 2009). The mechanism depicted seemed to be quite simple. Under magnetic field, magnetic nanoparticles became heated due to inductive heating within the composites that was applied to thermally induce a shape memory transition (Razzaq et al. 2012). The formulation of PDLLA/Fe_3O_4 nanocomposites used in shape memory outcome was also very clear cut. It was carried out by solution/precipitation methods (Zheng et al. 2009). They confirmed the strong interactions between PLA matrix and the magnetic nanoparticles via hydrogen bonding with different types of characterization techniques. Recently, Taccola et al. (2011) have reported on the elaboration of free-standing PLA nanofilms containing super-paramagnetic Fe_3O_4 nanoparticles. According to their analysis, the resulting nanofilms can be used in the biomedical field (Taccola et al. 2011).

4.9 Calcium Carbonate/PLA Nanocomposites

Calcium carbonate ($CaCO_3$) is widely used in the area of paint, paper, rubber, plastics, biomedical devices, etc. (Hoque et al. 2013b). It is one of the cheapest reinforcing materials. Researches have been done to prepare nanosized $CaCO_3$ with special shapes and a limited size distribution. Due to the problems of high surface energy and particle agglomeration, surface modifiers such as stearic acid, titanates, silanes, and zirconates are employed (Gu et al. 2009). A group of scientists from China investigated the impact of nano-$CaCO_3$ addition on the crystallization characteristics of PLA nanocomposite (Liang et al. 2013). According to their study, crystallization onset temperature, crystallization temperature, and crystallization end temperature of nano-$CaCO_3$/PLA nanocomposite were found to be at higher temperatures as compared to that of the neat PLA. The crystallinity of the nano-$CaCO_3$/PLA nanocomposites elevated with higher amount of the nano-$CaCO_3$. They speculated that this improvement in degree of crystallinity was related to the heterogeneous nucleation of PLA matrix on the nano-$CaCO_3$. Nekhamanurak et al. (2012a) investigate the influence of two plasticizers, polyethylene glycol and tributyl citrate, on the thermomechanical properties and fracture behavior of nano-$CaCO_3$/PLA nanocomposite. They showed that the incorporation of plasticizers improved the thermal properties of PLA. They also confirmed that the fracture behavior of nano-$CaCO_3$/PLA nanocomposites changed from brittle to ductile with the addition of plasticizers.

In another study, Nekhamanurak et al. found that the addition of silica in nano-$CaCO_3$/PLA nanocomposites had a profound effect on the mechanical and fracture behavior (Nekhamanurak et al. 2012b). They confirmed that compared to PLA-$CaCO_3$ nanocomposite, incorporating $CaCO_3$@SiO_2 at the same content increased elastic modulus, % elongation at break, and notched impact strength of PLA nanocomposites. They also concluded that surface modification of $CaCO_3$ with a SiO_2 layer was a beneficial option for better compatibility between PLA matrix and nanofillers. Kumar et al. (2014) also suggested that surface treatment by stearic acid was needed to improve inorganic nanofillers/polymer. The $CaCO_3$ modification induced homogeneous and fine dispersion of nanoparticles into PLA polymer as well as strong interfacial bonding between the two phases. An increase in the glass transition temperature (T_g) and storage modulus of the resultant nanocomposites was observed with the increase of $CaCO_3$ ratio. Shi et al. 2013 investigated the effect of poly(butylene adipate-co-terephthalate) (PBAT) addition on the mechanical properties of PLA/$CaCO_3$ nanocomposite. They observed that with the addition of PBAT impact strength increased; however, the tensile of the nanocomposite showed a decreasing tendency (Shi et al. 2013).

5 Other Nanocomposites

The nanofillers can improve material properties with the consideration of some controlling factors. The size, shape and amount of the filler, the filler/polymer interaction, and dispersion characteristics with polymeric matrices are the most important (Chow 1978). There are several nanofillers that have been investigated in the design of PLA-based nanocomposites in contrast to layered silicates or inorganic salts. Besides these, there are some other nanofillers which have not been commonly used for the preparation of PLA-based nanocomposites. Some of the examples are CNTs, polyhedral oligomeric silsesquioxanes (POSS), cellulose nanofibers, lignin nanoparticles, etc.

5.1 Carbon/PLA Nanocomposites

Carbonaceous materials like CNT are the most common nanofillers for composite fabrication. Graphite and carbon black have been also extensively used nanofillers. They enhance mechanical resilience and thermal/electrical conductivity. Carbon nanotubes (CNTs) are allotropes of carbon with a cylindrical nanostructure. The main property of CNTs is surface functionalization. This property is used to increase solubility and obtain uniform dispersion in the polymer matrix.

Solvent evaporation, in situ polymerization, and melt blending are the main processes for manufacturing polylactic acid–carbon nanotube blends (Chiu et al. 2008). Oxidation is the common chemical modification strategy on CNT surfaces. During oxidation, carboxylic acid or hydroxyl groups are introduced on the surface of CNTs (Fig. 5).

Graphite–PLA-based nanocomposites are also popular. Fukushima et al. report a silica–graphite–PLA-based nanohybrid made from expanded graphite and organically modified montmorillonite through a melt blending technique. The composite has excellent thermal stability, fire retardancy, and mechanical strength. The crystallization process is accelerated by graphite nanolayers. Montmorillonite is used for the enhancement of thermal resistance (Fukushima et al. 2010). Recently, a nanocomposite has been prepared by melt blending from PLAs and graphite. This nanocomposite not only showed improved thermomechanical and fire retardant properties, but also has higher Young's modulus as compared with PLAs (Murariu et al. 2010).

In short, the carbonaceous nanofillers enhance the mechanical strength of the PLA matrix. There are mainly two modes: (i) the physical interaction and intercalation of the polymer–filler–polymer are high and (ii) due to nucleation effect and enhancement of crystallinity. But higher cost of production limits their industrial application. But, the versatility and the improvement they provide to PLA are promising and so these are becoming popular day by day.

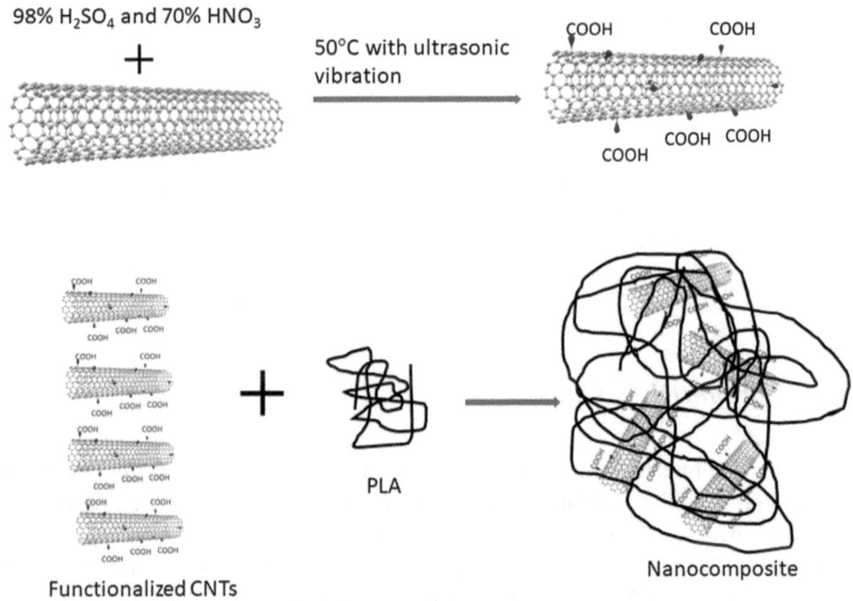

Fig. 5 Surface modification or functionalization of carbon nanotube-based materials. Redrawn from Chiu et al. (2008)

5.2 Cellulose/PLA Nanocomposites

Cellulose nanofibers (CNFs) originated from renewable biomass have drawn considerable attention as an option for reinforcing PLA. In this regard, a slurry of plasticizer and cellulose fibers was blended with the PLA melt in an extruder for composite making (Oksman et al. 2006). The results showed that the cellulose fibers improved the mechanical properties of the PLA. The scientists from Japan have also investigated cellulose microfibrils/PLA nanocomposite in terms of mechanical properties (Suryanegara et al. 2009; Iwatake et al. 2008). For the improvement of dispersion of microfibrils in PLA, an organic solvent was added in the batch. They reported that the resultant composites showed improved Young's modulus and tensile strength than neat PLA. Tingaut et al. (2009) reported well-dispersed acetylated MFC in a PLA matrix. The dispersion of nano- to submicron-wide web-like network fibers was possible because the grafted acetyl groups allowed reduced hydrogen bonding between cellulose microfibrils. They showed that acetylated MFC provided more translucent nanocomposites with reduced hygroscopicity and improved thermal stability versus unmodified MFC (Tingaut et al. 2009).

In another study, CNF/PLA nanocomposites were developed by twin screw extrusion without any surface modification treatment (Jonoobi et al. 2010). They showed that the tensile modulus and strength increased from 2.9 to 3.6 GPa and from 58 to 71 MPa, respectively, in 5 wt% CNF nanocomposites with respect to that in neat PLA. Nakagaito et al. (2009) reported a production process of microfibrillated cellulose (MFC)/PLA nanocomposites by stacking method without any organic additives. They concluded that the modulus, strength, and strain at fracture of the composite structure increased linearly with the MFC content. They confirmed that the enhancement in toughness in MFC/PLA composites was believed to be the major advantage.

5.3 Lignin/PLA Nanocomposites

Lignins are aromatic polymers which constitute a large component of the cell walls of plants. Several studies were completed on nanocomposites containing lignin and PLA. Gordobil et al. (2014) showed that the incorporation of organosolv process yielded lignin from almond shells and caused degradation of both failure strength and Young's modulus. However, acetylation of the lignin was found to improve the mechanical properties as compared to unmodified lignin (Gordobil et al. 2014). Spiridon et al. (2015) studied the mechanical and thermal properties of unmodified lignin/PLA composites. This study established a rise in mechanical properties and impact strength in PLA–lignin composites without modification treatment of the lignin–PLA system (Spiridon et al. 2015).

5.4 POSS/PLA Nanocomposites

Recently, polyhedral oligomeric silsesquioxanes (POSS) have attracted a lot of attention as reinforcing isotropic nanofiller. Their dimensions are in the range of 1–3 nm (Fina et al. 2010). These POSS nanoparticles consist of a cage-like siloxane structure surrounded by eight organic R groups. These R groups can be designed as either non-reactive or reactive. For polymer blending, non-reactive R groups are needed, and for copolymerization, reactive R groups are needed (Moniruzzaman and Winey 2006).

Comparing with commercially available and common nanofillers such as clays and CNTs, POSS nanoparticles have the smallest size, together with a low density. But the incorporation of POSS nanoparticles can be used in a wide range of polymers through simple blending or copolymerization. And POSS improved mechanical properties such as strength, modulus, rigidity, and reduced flammability (Kuo and Chang 2011). POSS nanostructures may also be used as scaffolds for drug delivery due to their nonvolatility (Ghanbari et al. 2011; Kawakami et al. 2010; Wu and Mather 2009).

6 Conclusions and Future Perspectives

The bio-based polymer industry is already competing with fossil fuel-based chemical industry, which has amplified in the last 20 years and so (Babu et al. 2013). The production of bio-based polymers and other chemicals from renewable resources has turned out to be a certainty. The current technological approach is driven toward cellulose-based feedstocks, stems, leaves, and solid municipal waste streams. More and more of these technologies are already in the development stage to introduce new bio-based polymers with higher performance and value (Carus et al. 2011).

Fantastically, encouraging progresses in the collaborative aspects are additionally expected by the achievement of hybrid PLA nanocomposites and attempts should be formulated in this direction. For instance, recently Pal et al. prepared cellulose nanocrystal (CNC) and reduced graphene oxide (rGO)-reinforced PLA hybrid composite and found good dispersion of CNC and rGO within the PLA matrix with higher tensile strength and thermal stability in addition to antibacterial response and better biocompatibility (Pal et al. 2017). In another case, Dhar et al. of Indian Institute of Technology Guwahati proposed a novel noninvasive approach of orienting non-toxic bioderived magnetic cellulose nanocrystal (CNC)/PLA composite in the presence of low magnetic fields, with potential applications for high-performance engineering applications (Fig. 6) (Dhar et al. 2016). For distinctive end-use applications in the biomedical to the engineering sectors, new PLA composites should be designed with the addition of combinations of fillers and reinforcing fibers of micro- and nanoranges, their combination with conventional

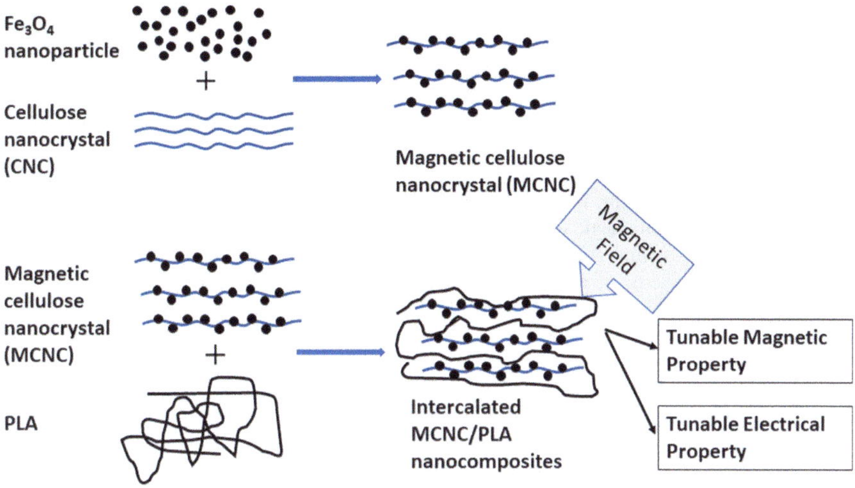

Fig. 6 Schematic illustration of magnetic cellulose nanocrystals-based PLA nanocomposite. The concept was redrawn from Dhar et al. (2016)

additives. The future development of novel PLA-based bio-nanocomposites with improved properties and multifunctionality can be visualized as a promising theme of research, as well as the advantage of their synergistic combination with inorganic nanosized solids. Their unique properties should be a solid base to develop new applications and opportunities for biocomposites in the twenty-first-century "green" materials world.

References

Albertsson AC, Karlsson S (1994) Chemistry and biochemistry of polymer biodegradation. In: Griffin GJL (ed) Chemistry and technology of biodegradable polymers. Blackie, Glasgow, p 7

Ali SAM, Doherty PJ, Williams DF (1994) The mechanisms of oxidative degradation of biomedical polymers by free radicals. J Appl Polym Sci 51(8):1389–1398

Anderson KS, Schreck KM, Hillmyer MA (2008) Toughening polylactide. Polym Rev 48(1):85–108

Anžlovar A, Kržan A, Žagar E (2018) Degradation of PLA/ZnO and PHBV/ZnO composites prepared by melt processing. Arab J Chem 11(3):343–352

Athanasiou KA, Niederauer GG, Agrawal CM (1996) Sterilization, toxicity, biocompatibility and clinical applications of polylactic acid/polyglycolic acid copolymers. Biomaterials 17(2):93–102

Auras R, Harte B, Selke S (2004) An overview of polylactides as packaging materials. Macromol Biosci 4(9):835–864

Babu RP, O'connor K, Seeram R (2013) Current progress on bio-based polymers and their future trends. Prog Biomater 2(1):8

Bergsma JE, Rozema FR, Bos RRM, Boering G, de Bruijn WC (2006) Late degradation tissue response to poly(L-lactide) bone plates and screws. In: The biomaterials: silver jubilee compendium, pp 101–107

Burg KJL, Holder WD, Culberson CR, Beiler RJ, Greene KG, Loebsack AB et al (1999) Parameters affecting cellular adhesion to polylactide films. J Biomater Sci Polym Ed 10 (2):147–161

Bussiere PO, Therias S, Gardette JL, Murariu M, Dubois P, Baba M (2012) Effect of ZnO nanofillers treated with triethoxy caprylylsilane on the isothermal and non-isothermal crystallization of poly(lactic acid). Phys Chem Chem Phys 14(35):12301–12308

Cai R, Kubota Y, Fujishima A (2003) Effect of copper ions on the formation of hydrogen peroxide from photocatalytic titanium dioxide particles. J Catal 219(1):214–218

Carus M, Carrez D, Kaeb H, Venus J (2011). Level playing field for bio-based chemistry and materials. Nova Institute, pp 04–18

Chen S, Feng J, Guo X, Hong J, Ding W (2005) One-step wet chemistry for preparation of magnetite nanorods. Mater Lett 59(8–9):985–988

Chen C, Lv G, Pan C, Song M, Wu C, Guo D et al (2007) Poly(lactic acid)(PLA) based nanocomposites—a novel way of drug-releasing. Biomed Mater 2(4):L1

Chen BK, Shen CH, Chen SC, Chen AF (2010) Ductile PLA modified with methacryloyloxyalkyl isocyanate improves mechanical properties. Polymer 51(21):4667–4672

Chen BK, Shen CH, Chen AF (2012) Preparation of ductile PLA materials by modification with trimethyl hexamethylene diisocyanate. Polym Bull 69(3):313–322

Chiang MF, Wu TM (2010) Synthesis and characterization of biodegradable poly(L-lactide)/ layered double hydroxide nanocomposites. Compos Sci Technol 70(1):110–115

Chiu WM, Chang YA, Kuo HY, Lin MH, Wen HC (2008) A study of carbon nanotubes/ biodegradable plastic polylactic acid composites. J Appl Polym Sci 108(5):3024–3030

Chow TS (1978) Effect of particle shape at finite concentration on the elastic moduli of filled polymers. J Polym Sci Polym Phys Ed 16(6):959–965

Costa FR, Saphiannikova M, Wagenknecht U, Heinrich G (2007) Layered double hydroxide based polymer nanocomposites. In: Wax crystal control nanocomposites stimuli-responsive polymers. Springer, Berlin, pp 101–168

Cumkur EA, Baouz T, Yilmazer U (2015) Poly(lactic acid)–layered silicate nanocomposites: the effects of modifier and compatibilizer on the morphology and mechanical properties. J Appl Polym Sci 132(38)

Dash S, Mishra S, Patel S, Mishra BK (2008) Organically modified silica: synthesis and applications due to its surface interaction with organic molecules. Adv Coll Interface Sci 140 (2):77–94

Dhar P, Kumar A, Katiyar V (2016) Magnetic cellulose nanocrystal based anisotropic polylactic acid nanocomposite films: influence on electrical, magnetic, thermal, and mechanical properties. ACS Appl Mater Interfaces 8(28):18393–18409

Dorgan JR, Lehermeier H, Mang M (2000) Thermal and rheological properties of commercial-grade poly(lactic acid)s. J Polym Environ 8(1):1–9

Dorgan JR, Lehermeier HJ, Palade LI, Cicero J (2001) Polylactides: properties and prospects of an environmentally benign plastic from renewable resources. In: Macromolecular symposia, vol 175, no 1. WILEY-VCH Verlag GmbH, Weinheim, pp 55–66

Drumright RE, Gruber PR, Henton DE (2000) Polylactic acid technology. Adv Mater 12 (23):1841–1846

Eling B, Gogolewski S, Pennings AJ (1982) Biodegradable materials of poly(l-lactic acid): 1. Melt-spun and solution-spun fibres. Polymer 23(11):1587–1593

Fina A, Monticelli O, Camino G (2010) POSS-based hybrids by melt/reactive blending. J Mater Chem 20(42):9297–9305

Fukushima K, Murariu M, Camino G, Dubois P (2010) Effect of expanded graphite/ layered-silicate clay on thermal, mechanical and fire retardant properties of poly(lactic acid). Polym Degrad Stab 95(6):1063–1076

Garlotta D (2001) A literature review of poly(lactic acid). J Polym Environ 9(2):63–84

Ghanbari H, Cousins BG, Seifalian AM (2011) A nanocage for nanomedicine: polyhedral oligomeric silsesquioxane (POSS). Macromol Rapid Commun 32(14):1032–1046

Gordobil O, Egüés I, Llano-Ponte R, Labidi J (2014) Physicochemical properties of PLA lignin blends. Polym Degrad Stab 108:330–338

Gu SY, Zou CY, Zhou K, Ren J (2009) Structure-rheology responses of polylactide/calcium carbonate composites. J Appl Polym Sci 114(3):1648–1655

Gupta B, Revagade N, Hilborn J (2007) Poly(lactic acid) fiber: an overview. Prog Polym Sci 32 (4):455–482

Hiljanen-Vainio M, Varpomaa P, Seppälä J, Törmälä P (1996) Modification of poly(L-lactides) by blending: mechanical and hydrolytic behavior. Macromol Chem Phys 197(4):1503–1523

Hoque ME, Hutmacher DW, Feng W, Li S, Huang MH, Vert M, Wong YS (2005) Fabrication using a rapid prototyping system and in vitro characterization of PEG-PCL-PLA scaffolds for tissue engineering. J Biomater Sci Polym Ed 16(12):1595–1610

Hoque ME, Ye TJ, Yong LC, Dahlan KM (2013a) Sago starch-mixed low-density polyethylene (LDPE) biodegradable polymer: synthesis and characterization. J Mater Article ID 365380, 7 pages. https://doi.org/10.1155/2013/365380

Hoque MdE, Shehryar M, Nurul Islam KMD (2013b) Processing and characterization of cockle shell calcium carbonate ($CaCO_3$) bioceramic for potential application in bone tissue engineering. J Mater Sci Eng 2(4):132

Huda MS, Drzal LT, Mohanty AK, Misra M (2008) Effect of fiber surface-treatments on the properties of laminated biocomposites from poly(lactic acid) (PLA) and kenaf fibers. Compos Sci Technol 68(2):424–432

Hutmacher DW, Hoque ME, Wong YS (2008) Design, fabrication and physical characterization of scaffolds made from biodegradable synthetic polymers in combination with RP systems based on melt extrusion. In: Bidanda B, Bartolo P (eds) virtual prototyping & bio manufacturing in medical applications. Springer, USA

Ikada Y, Tsuji H (2000) Biodegradable polyesters for medical and ecological applications. Macromol Rapid Commun 21(3):117–132

Incardona SD, Fambri L, Migliaresi C (1996) Poly-L-lactic acid braided fibres produced by melt spinning: characterization and in vitro degradation. J Mater Sci Mater Med 7(7):387–391

Iwatake A, Nogi M, Yano H (2008) Cellulose nanofiber-reinforced polylactic acid. Compos Sci Technol 68(9):2103–2106

Jonoobi M, Harun J, Mathew AP, Oksman K (2010) Mechanical properties of cellulose nanofiber (CNF) reinforced polylactic acid (PLA) prepared by twin screw extrusion. Compos Sci Technol 70(12):1742–1747

Katiyar V, Gerds N, Koch CB, Risbo J, Hansen HCB, Plackett D (2010) Poly L-lactide-layered double hydroxide nanocomposites via in situ polymerization of L-lactide. Polym Degrad Stab 95(12):2563–2573

Katiyar V, Gerds N, Koch CB, Risbo J, Hansen HCB, Plackett D (2011) Melt processing of poly (L-lactic acid) in the presence of organomodified anionic or cationic clays. J Appl Polym Sci 122(1):112–125

Kawakami Y, Kakihana Y, Miyazato A, Tateyama S, Hoque MA (2010) Polyhedral oligomeric silsesquioxanes with controlled structure: formation and application in new Si-based polymer systems. In: Silicon polymers. Springer, Berlin, pp 185–228

Kimura Y, Shirotani K, Yamane H, Kitao T (1988) Ring-opening polymerization of 3 (S)-[(benzyloxycarbonyl) methyl]-1, 4-dioxane-2, 5-dione: a new route to a poly(alpha-hydroxy acid) with pendant carboxyl groups. Macromolecules 21(11):3338–3340

Krikorian V, Pochan DJ (2004) Unusual crystallization behavior of organoclay reinforced poly (L-lactic acid) nanocomposites. Macromolecules 37(17):6480–6491

Kumar V, Dev A, Gupta AP (2014) Studies of poly(lactic acid) based calcium carbonate nanocomposites. Compos B Eng 56:184–188

Kuo SW, Chang FC (2011) POSS related polymer nanocomposites. Prog Polym Sci 36(12):1649–1696

Lagarón JM, Cabedo L (2014) Polylactide (PLA)/clay nano-biocomposites. Poly(lactic acid) science and technology: processing, properties, additives and applications, no 12, p 215

Lai SM, Wu SH, Lin GG, Don TM (2014) Unusual mechanical properties of melt-blended poly (lactic acid)(PLA)/clay nanocomposites. Eur Polym J 52:193–206

Li Y, Chen C, Li J, Sun XS (2011) Synthesis and characterization of bionanocomposites of poly (lactic acid) and TiO_2 nanowires by in situ polymerization. Polymer 52(11):2367–2375

Li Y, Han C, Zhang X, Xu K, Bian J, Dong L (2014) Poly(L-lactide)/Poly(D-lactide)/clay nanocomposites: enhanced dispersion, crystallization, mechanical properties, and hydrolytic degradation. Polym Eng Sci 54(4):914–924

Liang RP, Yao GH, Fan LX, Qiu JD (2012) Magnetic Fe_3O_4@ Au composite-enhanced surface plasmon resonance for ultrasensitive detection of magnetic nanoparticle-enriched α-fetoprotein. Anal Chim Acta 737:22–28

Liang JZ, Zhou L, Tang CY, Tsui CP (2013) Crystalline properties of poly(L-lactic acid) composites filled with nanometer calcium carbonate. Compos B Eng 45(1):1646–1650

Liao R, Yang B, Yu W, Zhou C (2007) Isothermal cold crystallization kinetics of polylactide/nucleating agents. J Appl Polym Sci 104(1):310–317

Linnemann B, Sri Harwoko M, Gries T (2003) Polylactide fibers (PLA). Chem Fibers Int 53 (6):426–433

Luo YB, Li WD, Wang XL, Xu DY, Wang YZ (2009) Preparation and properties of nanocomposites based on poly (lactic acid) and functionalized TiO_2. Acta Mater 57(11):3182–3191

Luo Y, Cao Y, Guo G (2018) Effects of TiO_2 nanoparticles on the photodegradation of poly (lactic acid). J Appl Polym Sci 135(30):46509

Malinowski R, Janczak K, Rytlewski P, Raszkowska-Kaczor A, Moraczewski K, Żuk T (2015) Influence of glass microspheres on selected properties of polylactide composites. Compos B Eng 76:13–19

Mandal S, Tichit D, Lerner DA, Marcotte N (2009) Azoic dye hosted in layered double hydroxide: physicochemical characterization of the intercalated materials. Langmuir 25(18):10980–10986

Manzi-Nshuti C, Songtipya P, Manias E, Jimenez-Gasco MM, Hossenlopp JM, Wilkie CA (2009) Polymer nanocomposites using zinc aluminum and magnesium aluminum oleate layered double hydroxides: effects of LDH divalent metals on dispersion, thermal, mechanical and fire performance in various polymers. Polymer 50(15):3564–3574

Martin O, Averous L (2001) Poly(lactic acid): plasticization and properties of biodegradable multiphase systems. Polymer 42(14):6209–6219

Meng B, Tao J, Deng J, Wu Z, Yang M (2011) Toughening of polylactide with higher loading of nano-titania particles coated by poly (ε-caprolactone). Mater Lett 65(4):729–732

Michael FM, Khalid M, Walvekar R, Ratnam CT, Ramarad S, Siddiqui H, Hoque ME (2016) Effect of nanofillers on the physico-mechanical properties of load bearing bone implants. Mater Sci Eng C 67:792–806

Moniruzzaman M, Winey KI (2006) Polymer nanocomposites containing carbon nanotubes. Macromolecules 39(16):5194–5205

Mu B, Tang J, Zhang L, Wang A (2017) Facile fabrication of superparamagnetic graphene/polyaniline/Fe_3O_4 nanocomposites for fast magnetic separation and efficient removal of dye. Sci Rep 7(1):5347

Murariu M, Dechief AL, Bonnaud L, Gallos A, Fontaine G, Bourbigot S, Dubois P (2010) The production and properties of polylactide composites filled with expanded graphite. Polym Degrad Stab 95(5):889–900

Murariu M, Doumbia A, Bonnaud L, Dechief AL, Paint Y, Ferreira M et al (2011) High-performance polylactide/ZnO nanocomposites designed for films and fibers with special end-use properties. Biomacromolecules 12(5):1762–1771

Nakagaito AN, Fujimura A, Sakai T, Hama Y, Yano H (2009) Production of microfibrillated cellulose (MFC)-reinforced polylactic acid (PLA) nanocomposites from sheets obtained by a papermaking-like process. Compos Sci Technol 69(7–8):1293–1297

Nakayama N, Hayashi T (2007) Preparation and characterization of poly(L-lactic acid)/TiO_2 nanoparticle nanocomposite films with high transparency and efficient photodegradability. Polym Degrad Stab 92(7):1255–1264
Nampoothiri KM, Nair NR, John RP (2010) An overview of the recent developments in polylactide (PLA) research. Biores Technol 101(22):8493–8501
Navarro MELBA, Ginebra MP, Planell JA, Barrias CC, Barbosa MA (2005) In vitro degradation behavior of a novel bioresorbable composite material based on PLA and a soluble CaP glass. Acta Biomater 1(4):411–419
Nekhamanurak B, Patanathabutr P, Hongsriphan N (2012a) Thermal–mechanical property and fracture behaviour of plasticised PLA–$CaCO_3$ nanocomposite. Plast Rubber Compos 41 (4–5):175–179
Nekhamanurak YB, Patanathabutr P, Hongsriphan N (2012b) Mechanical properties of hydrophilicity modified $CaCO_3$-poly(lactic acid) nanocomposite. Int J Appl Phys Math 2(2):98
Okada M (2002) Chemical syntheses of biodegradable polymers. Prog Polym Sci 27(1):87–133
Oksman K, Mathew AP, Bondeson D, Kvien I (2006) Manufacturing process of cellulose whiskers/polylactic acid nanocomposites. Compos Sci Technol 66(15):2776–2784
Pal N, Dubey P, Gopinath P, Pal K (2017) Combined effect of cellulose nanocrystal and reduced graphene oxide into poly-lactic acid matrix nanocomposite as a scaffold and its anti-bacterial activity. Int J Biol Macromol 95:94–105
Pantani R, Gorrasi G, Vigliotta G, Murariu M, Dubois P (2013) PLA-ZnO nanocomposite films: water vapor barrier properties and specific end-use characteristics. Eur Polymer J 49(11):3471–3482
Parida SK, Dash S, Patel S, Mishra BK (2006) Adsorption of organic molecules on silica surface. Adv Coll Interface Sci 121(1–3):77–110
Paul MA, Delcourt C, Alexandre M, Degée P, Monteverde F, Rulmont A, Dubois P (2005) (Plasticized) polylactide/(organo-) clay nanocomposites by in situ intercalative polymerization. Macromol Chem Phys 206(4):484–498
Prasad J, Singh AK, Shah J, Kotnala RK, Singh K (2018) Synthesis of MoS_2-reduced graphene oxide/Fe_3O_4 nanocomposite for enhanced electromagnetic interference shielding effectiveness. Mater Res Express 5(5):055028
Raquez JM, Habibi Y, Murariu M, Dubois P (2013) Polylactide (PLA)-based nanocomposites. Prog Polym Sci 38(10–11):1504–1542
Rasal, R. M., & Hirt, D. E. (2009). Toughness decrease of PLA-PHBHHx blend films upon surface-confined photopolymerization. J Biomed Mater Res Part A Official J Soc Biomater Jpn Soc Biomater Aust Soc Biomater Korean Soc Biomater 88(4):1079–1086
Rasal RM, Janorkar AV, Hirt DE (2010) Poly(lactic acid) modifications. Prog Polym Sci 35 (3):338–356
Ratner BD (1995) Surface modification of polymers: chemical, biological and surface analytical challenges. Biosens Bioelectron 10(9–10):797–804
Razzaq MY, Behl M, Lendlein A (2012) Magnetic memory effect of nanocomposites. Adv Func Mater 22(1):184–191
Saeidlou S, Huneault MA, Li H, Park CB (2012) Poly(lactic acid) crystallization. Prog Polym Sci 37(12):1657–1677
Sajjadi M, Nasrollahzadeh M, Sajadi SM (2017) Green synthesis of Ag/Fe_3O_4 nanocomposite using Euphorbia peplus Linn leaf extract and evaluation of its catalytic activity. J Colloid Interface Sci 497:1–13
Sawyer DJ (2003) Bioprocessing–no longer a field of dreams. In: Macromolecular symposia, vol 201, no 1. WILEY-VCH Verlag, Weinheim, pp 271–282
Schugens CH, Grandfils CH, Jérôme R, Teyssie PH, Delree P, Martin D et al (1995). Preparation of a macroporous biodegradable polylactide implant for neuronal transplantation. J Biomed Mater Res 29(11):1349–1362
Shabanian M, Khoobi M, Hemati F, Khonakdar HA, Wagenknecht U, Shafiee A (2015) New PLA/PEI-functionalized Fe_3O_4 nanocomposite: preparation and characterization. J Ind Eng Chem 24:211–218

Shi N, Cai J, Dou Q (2013) Crystallization, morphology and mechanical properties of PLA/PBAT/ $CaCO_3$ composites. In: Advanced materials research, vol 602. Trans Tech Publications, Switzerland, pp 768–771

Sinha Ray S, Yamada K, Okamoto M, Ueda K (2002) Polylactide-layered silicate nanocomposite: a novel biodegradable material. Nano Lett 2(10):1093–1096

Spiridon I, Leluk K, Resmerita AM, Darie RN (2015) Evaluation of PLA–lignin bioplastics properties before and after accelerated weathering. Compos B Eng 69:342–349

Suryanegara L, Nakagaito AN, Yano H (2009) The effect of crystallization of PLA on the thermal and mechanical properties of microfibrillated cellulose-reinforced PLA composites. Compos Sci Technol 69(7–8):1187–1192

Taccola S, Desii A, Pensabene V, Fujie T, Saito A, Takeoka S et al (2011) Free-standing poly(L-lactic acid) nanofilms loaded with superparamagnetic nanoparticles. Langmuir 27(9):5589–5595

Tehrani MA, Akbari A, Majumder M (2014) Polylactic acid (PLA) layered silicate nanocomposites. In: Handbook of polymernanocomposites. Processing, performance and application. Springer, Berlin, pp 53–67

Tingaut P, Zimmermann T, Lopez-Suevos F (2009) Synthesis and characterization of bionanocomposites with tunable properties from poly(lactic acid) and acetylated microfibrillated cellulose. Biomacromolecules 11(2):454–464

Vallet-Regí M, Granado S, Arcos D, Gordo M, Cabanas MV, Ragel CV et al (1998) Preparation, characterization, and in vitro release of ibuprofen from Al_2O_3/PLA/PMMA composites. J Biomed Mater Res Official J Soc Biomater Jpn Soc Biomater Aust Soc Biomater 39(3): 423–428

Vink ET, Rabago KR, Glassner DA, Gruber PR (2003) Applications of life cycle assessment to NatureWorks™ polylactide (PLA) production. Polym Degrad Stab 80(3):403–419

Vink ET, Glassner DA, Kolstad JJ, Wooley RJ, O'Connor RP (2007) The eco-profiles for current and near-future NatureWorks® polylactide (PLA) production. Ind Biotechnol 3(1):58–81

Wen X, Zhang K, Wang Y, Han L, Han C, Zhang H et al (2011) Study of the thermal stabilization mechanism of biodegradable poly(L-lactide)/silica nanocomposites. Polym Int 60(2):202–210

Wu J, Mather PT (2009) POSS polymers: physical properties and biomaterials applications. Polym Rev 49(1):25–63. https://doi.org/10.1080/15583720802656237

Xia Y, Fang J, Li P, Zhang B, Yao H, Chen J et al (2017) Solution-processed highly superparamagnetic and conductive PEDOT: PSS/Fe_3O_4 nanocomposite films with high transparency and high mechanical flexibility. ACS Appl Mater Interfaces 9(22):19001–19010

Yan S, Yin J, Yang Y, Dai Z, Ma J, Chen X (2007) Surface-grafted silica linked with L-lactic acid oligomer: a novel nanofiller to improve the performance of biodegradable poly(L-lactide). Polymer 48(6):1688–1694

Zheng X, Zhou S, Xiao Y, Yu X, Li X, Wu P (2009) Shape memory effect of poly(D, L-lactide)/ Fe_3O_4 nanocomposites by inductive heating of magnetite particles. Colloids Surf B 71(1): 67–72

Zhu A, Diao H, Rong Q, Cai A (2010) Preparation and properties of polylactide–silica nanocomposites. J Appl Polym Sci 116(5):2866–2873

Zhuang W, Liu J, Zhang JH, Hu BX, Shen J (2009) Preparation, characterization, and properties of TiO_2/PLA nanocomposites by in situ polymerization. Polym Compos 30(8):1074–1080

Biopolymers-Based Nanocomposites: Properties and Applications

Naruemon Sumrith, Sanjay Mavinkere Rangappa, Rapeephun Dangtungee, Suchart Siengchin, Mohammad Jawaid and Catalin Iulian Pruncu

Abstract The actual environmental policy, together with the strong interest to develop novel materials with better performances, makes the new eco-friendly and biodegradable composites a very attractive research field. Additionally, in this branch of composite materials, substantial attention is focused on the biopolymers made from "nanocomposites" particles that were considered both at academic level and industrial plant. The biopolymer-based nanocomposites provide an opportunity to replace the conventional non-biodegradable polymers, due to their user-friendliness, great biodegradability, and relatively good mechanical properties. This chapter contains a robust overview of a new class of biopolymers nanocomposites.

Keywords Biopolymers · Nanocomposites · Blends · Properties · Application

N. Sumrith · S. M. Rangappa (✉) · R. Dangtungee · S. Siengchin
Department of Mechanical and Process Engineering, The Sirindhorn International Thai-German Graduate School of Engineering (TGGS), King Mongkut's University of Technology North Bangkok, Bangkok, Thailand
e-mail: mcemrs@gmail.com

M. Jawaid
Department of Biocomposite Technology, Institute of Tropical Forestry and Forest Products (INTROP), Universiti Putra Malaysia (UPM), 43400 Serdang, Selangor, Malaysia

C. I. Pruncu
Department of Mechanical Engineering, Imperial College London, Exhibition Rd, London SW7 2AZ, UK

C. I. Pruncu
Department of Mechanical Engineering, School of Engineering, University of Birmingham, Birmingham B15 2TT, UK

M. L. Sanyang and M. Jawaid (eds.), *Bio-based Polymers and Nanocomposites*,
https://doi.org/10.1007/978-3-030-05825-8_12

1 Introduction

In the present scenario, the research dedicated to the polymer nanocomposites materials plays an important role for engineering applications such as electronic, food packaging, automotive, and aerospace (Essabir et al. 2018). It is well known that the composite materials have better properties (e.g., in terms of mechanical properties, thermal stability, permeability) generated from a combination of different materials that bring distinctive characteristics. Normally, composite material consists of two matrix components and fillers. To produce composite materials from carbon fibers, Kevlar and natural fibers are being used with different types of fillers including micro- and nanoparticles that improve the novel matrix properties in terms of mechanical properties, thermal stability, conductivity, isolation, elasticity, and flame retardant. The petroleum-based polymers are extensively used in various applications depending on requirements (Bordes et al. 2009; Lau et al. 2018).

The polymers compounds can be extremely advantageous in many aspects of engineering technology. However, they can also significantly damage the environment. Three main drawbacks were noted in the literature (Lau et al. 2018):

(i) the recycling of composite materials is very complicated due to toxic release during the decomposing process,
(ii) carbon fiber reinforced polymer composites are too strong, therefore difficult to decay,
(iii) the cost of using advanced composites is expensive for domestic products.

Moreover, the process of plastic degradation can occur in natural manner as the effect of sunlight, seawater, moisture, and temperature. Therefore, the contamination by plastic waste is regularly very detrimental to the natural environment. Hence, many researchers and the general public are highly concerned about this problem. They investigated potentially solutions to find the right recycling process and to develop an eco-friendly material (Li et al. 2016).

To overcome this problem, an alternative was proposed. There, the biopolymers were investigated and manufactured to replace synthetic polymers. The degradation mechanism of the biopolymer should respond to at least one step from natural process; for example, they may need to degrade as effect of sunlight, temperature, oxygen availability, and moisture. In addition, their decomposition needs to avoid the release of toxic compounds to the environment. Biopolymers are used as matrix for biocomposite materials, for instance, poly(L-lactide) (PLA), poly(vinyl alcohol) (PVA) poly(3-hydroxybutyrate-co-3-hydroxyvalerate) (PHBV), etc. Nevertheless, the variety of their applications is quite limited due to the lower mechanical properties, poor thermal resistance, and cost (Rhim et al. 2013). Various types of natural fibers are used as a substitute for synthetic fibers such as jute, hemp, sisal, and coir (Thakur and Thakur 2014).

Nowadays, in order to improve the material properties, the biocomposites are tremendously explored. Different nanoparticle fillers are proposed in order to help in enhancing the mechanical and thermal properties of biocomposites. It is obvious that some nanoparticle brings new properties. For example, introducing blend silica nanoparticles in polymer matrix nanocomposites (PMNCs) is generating an increase in the creep resistance, because of homogeneous distribution of nanoparticle (Hassanzadeh-Aghdam et al. 2018). Further, nanoparticle ZnO–N is prepared and mixed with sage starch matrix. The result indicated that the samples' properties are improved in terms of water vapor permeability, UV protection, and antimicrobial and make the novel biocomposite suitable for medical application, food packaging, and pharmaceutical industries (Nafchi et al. 2012).

2 Biopolymers or Green Polymers

Renewable materials are necessary for a sustainable society where people can live a decent life. Biomass materials can be used as sustainable raw materials for many industrial applications to replace petroleum products that are the most harmful in generating carbon dioxide causing global warming (Kargarzadeh et al. 2017).

Currently, technical biopolymers or green polymers play an important role in all engineering branches due to environmental benefits. Different eco-friendly materials and methods are proposed as replacement for toxic synthetic materials (Fatehi et al. 2018). Biopolymer is produced from biological sources. In fact, biopolymers are generally used in everyday life, for example, cellulose and starch. Polylactic acid (PLA) and polyhydroxyalkanoate (PHA) were authorized to be manufactured in 1990 in large quantities for the industry as novel bio-based polymers. They have been considered by many businesses mainly for solving water and soil pollution problems (Meraldo et al. 2016). These are being decomposed using microorganisms, and the effect is less environmental damage compared to the decomposition of petroleum-based materials (Thomas et al. 2015).

However, the biopolymer presents some limitation such as brittleness, low strength, and low stiffness. Therefore, the properties of the biopolymer have to be modified by adding in the matrix some compounds to form the biocomposite materials. The most advanced composites are the carbon fiber reinforced polymer composite (CFRP) produced from syntactic fiber (glass and carbon fiber). It contains a reinforcement and petroleum-based polymer that have been used for engineering applications such as automotive, aerospace, and construction. The carbon fiber can be used in conjunction with the biopolymer nonetheless the material cannot be used anymore as biodegradable (Thomas et al. 2015). It is noted that the functioning of biopolymer is limited. On the other hand, the design and life cycle of biopolymers help in some critical applications such as medical industry, where degradation has to be very fast. In this way, the biopolymer becomes meaningful for these functions (John and Thomas 2008).

3 Classification, Source, Properties, and Blends of Biopolymers

The biopolymer can be classified into four categories depending on its origin. The first group is biodegradable polymers achieved from biomass products such as polysaccharides, proteins, and lipids. The second group is the green polymer obtained by a variety of microorganisms, for instance, polyhydroxybutyrate (PHB), polyhydroxybutyrate-valerate (PHBV). The polymers fabricated from monomers of biomass such as polylactic acid (PLA) are the third group of biopolymer. The last group is the polymer obtained by synthesis from the petrochemical process, for example, polycaprolactone (PCL), polyesteramides (PEA), and aliphatic co-polyesters (John and Thomas 2008).

Currently, the biopolymers are the alternative solution of engineering materials because of limited fossil resources. Some concerns are related to their poor mechanical properties and difficulty in the manufacturing process. Accordingly, various additives are blended into biopolymer to enhance their properties. Zhou et al. investigated the preparation and characterization of polylactic acid (PLA) carbon nanotube nanocomposite by mixing circa 0.5% wt carbon nanotube in the PLA nanocomposites to improve their maximum tensile strength (Zhou et al. 2018).

Morelli et al. studied the antibacterial properties of natural oil from *copaifera multijuga* blended with bio-based materials for packaging application. The two bio-based materials, paper and polylactic acid (PLA), were mixed with the copaiba oil. The result showed that for the same concentration of 20% wt of the copaiba oil in both specimens can work against *B. subtilis.* In conclusion, these materials are possible to be used as biodegradable packaging with antibacterial and also allow to improve the shelf life of food products (Morelli et al. 2015). The properties of the biopolymer can be improved by blending with some additive into biopolymer. Some additives are used to enhance mechanical properties depending on the types of biopolymer and the substance. In addition, the researchers have to further investigate the influence of adding various alternative natural substances in the matrix to create novel biopolymers. A summary of classification of biopolymers is plotted in Fig. 1 (John and Thomas 2008).

4 Sustainable and Bio-Based Packaging

Eighty percentage of the plastic contamination in the marine environment is caused by littering, plastic bags usage, industrial activities, and solid waste disposal. Remaining 20% of marine plastic debris is formed from commercial fishing; for example, annually 640,000 t of waste are represented by discarded fishing gear (Li et al. 2016).

The biggest quantity of world production dedicated to plastic materials (EU28 +NO/CH) is packaging sector covering around 40% of its product. The issue

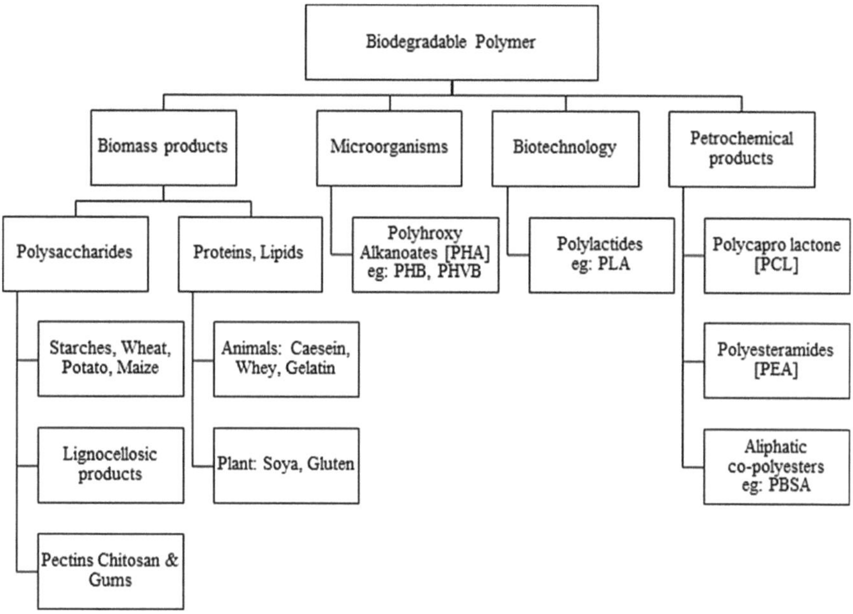

Fig. 1 Classtification of biodegradable polymer (John and Thomas 2008)

derives from short-term usability of the packaging products that is normally one-time usage. Thus, it is expected that the plastic packaging to be one of the highest potential of natural environment contamination (Bordes et al. 2009).

The duty of packaging products is not only to enhance the shelf life and quality of food products but also to have the ability to attract the attention of customers by its printed color. Moreover, increasing efficiency in their moisture, gas barriers, shrinking and sealing characteristics are also a variety of factors to consider (Youssef and El-Sayed 2018). However, reducing plastic packaging waste is the main desire for a friendly environment. It is well known that biopolymers play an important role in reducing pollution in both water and land. Biodegradability and non-toxicity are the most important properties to decrease the contamination in environment. Different types of biopolymers can be used as raw materials for packaging such as polysaccharides, lipids, proteins, polyhydroxybutyrates, polylactic acid, polycaprolactone, polyvinyl alcohol, and polybutylene succinate. The polyglycolic acid (PGA) is the biopolymer which has good properties for packaging, for example, great barrier properties and production of glycolic acid. Surely, the following properties of biopolymer such as the relatively high molecular weights and viscosity, hydrophilicity, crystallization behavior, brittleness, or melt uncertainties represent the main issues for the manufacturing process. Thus, the combination with additional biopolymers, plasticizer, and compatibilizers is necessary in order to improve their features (John and Thomas 2008).

Many researchers have investigated different potential solutions to enhance the properties of biopolymer for packaging. Thin wheat gluten protein film was studied as eco-friendly material for food packing. The result showed that at high humidity conditions, the wheat gluten film has good sensitive making it ideal for foreseen applications of packed food products. In addition, a cost-effective low energy consuming and reliable solution appeared to be the wheat gluten coupled with ultra-high frequency radio (Bibi et al. 2017). For other properties such as antimicrobial function, there are many compounds used such as metal ions (silver, copper, gold, platinum), metal oxide (TiO_2, ZnO, MgO), organically modified nanoclay, natural biopolymers (chitosan), and enzymes. However, there are also still many other substances, which should be studied to improve the abilities of biopolymer (Rhim et al. 2013).

5 Versatile Applications of Biopolymers

Figure 2 shows some desired functionalities for a bio-based product. Biopolymer can be used as raw material to produce the end product (Chaunier et al. 2017). Their different properties of each type of biopolymers affect the design of the product applications. For example, gas transmission and mechanical properties are necessary for biofilm product, whereas the design of packaging requires low fracture toughness, high fragmentation, high stiffness and strength for transportations and long shelf life. The bio-based product requirements are very strict in the area of tissue engineering, especially when they are used for antimicrobial cell growth (Chaunier et al. 2017).

These experiments are designed to bring improvements in material properties in order to achieve the final objective of product functioning. To produce complex objects, 3D printing plays an important role. It was noted that 3D manufacturing is growing very fast due to its diversity in design and complex production. Therefore, improving the biopolymer is a novel challenge that needs to be tackled, while using extrusion based on 3D print biopolymers. Thermomechanical control is a key factor in the ability of material printing. Some type of compounds may help as plasticizers to improve the thermal stability of biopolymer such as protein (zein from maize) (Miller et al. 2018).

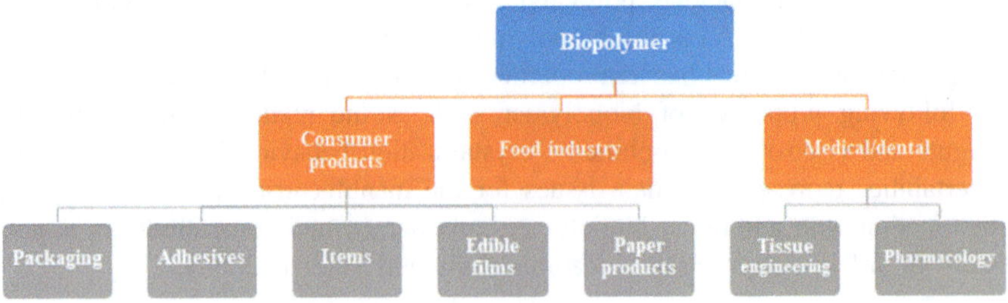

Fig. 2 Diagram presenting desired functionalities for a bio-based product (Chaunier et al. 2017)

Other applications such as post-consumer packaging use conventional expanded polystyrene made of petroleum-based polymer and have direct practicability for the protection of goods during transportations. After usage, the EPS wastes are difficult to decompose and to recycle because of the complex process and also the cost-effectiveness. A biodegradable and compostable packaging is a solution that can attract much attention for alternative raw material due to their higher recycling rate and parallel decrease in landfilling. However, up to now, the most development in the biopolymer was made only at the lab scale. Further optimization is required to upscale the manufacturing process for commercial level. There are couples of suggestions to be considered for expanding biopolymer as following (Razza et al. 2015):

- Reduction in the use of raw material by minimizing the density of material packaging,
- Optimization of energy efficiency by developing the expansion process,
- Applying renewable energy resources in industrial electricity system,
- Proposing renewable materials to reduce the damage to natural environment.

The composite made of biopolymer such as poly(L-lactic acid) (PLLA), polyhydroxybutyrate-co-hydroxyvalerate (PHBV), poly-ε-capolactone (PCL), and starch-poly-ε-caprolactone (SPCL) was investigated and blended with cork. The result showed that the cork–polymer biocomposite is promising in order to develop a sustainable composite material. They can bring as well some improvements in the mechanical properties, stiffness, and thermal properties (Fernandes et al. 2015).

The biopolymers can be used as remediation of heavy hydrocarbon impacted soils. To improve the soil washing efficiency, various mobilizing is blended with plastic sorbents as absorbent of heavy hydrocarbon from soil. Plant-based biopolymer is made from corn, hemp, and recycled material, which are efficient, sustainable non-toxic, environment-friendly, and reusable (Wilton et al. 2018).

The biopolymer composite can also be used as supercapacitor because of its good properties. These allow improving the potential of electrode material due to the capacitance efficiency. Lignin is the most used biopolymer, mainly used as dopant and surfactant to double increase the specific capacitance. Furthermore, carbon materials are necessary for supercapacitors design such as activated carbon, activated carbon fiber, carbon nanotubes, and graphene. The range of biopolymer composites application is spread on supercapacitors for military, automotive, and electronics applications (Okonkwo et al. 2017).

6 Biopolymers for Nanocomposite

In recent years, many types of nanoparticles were blended for improving the effective properties. The dimension of nanoparticle should be very small, in the nanoscale range (<100 nm). Advanced material used in automotive, aerospace, medicine, and another engineering applications are generally produced by blending with nanoparticles. Adding the nanoparticles allows improvements of mechanical properties, enhancement of temperature resistance and generates some antimicrobial properties (Hassanzadeh-Aghdam et al. 2018).

Researchers have investigated the biopolymer blended with nanoparticle to obtain solutions for industrial applications. PLA is one type of biopolymer, which can play an important role due to its mechanical properties, renewability, biodegradability, and low cost. The effect of mixing the PLA with carbon nanotube indicated that the T_g of the composite material is higher than near PLA and is possible to obtain stronger mechanical properties with higher strength and fracture toughness (Zhou et al. 2018).

Another research considered the *Lallemantia iberica* mucilage as a new source of polysaccharide that is blended with titanium dioxide (TiO_2) nanoparticle by ultrasonic-assisted treatment and casting method. The result showed that the mechanical, thermal, and barrier properties were improved and the bionanocomposite can be stable at different levels of humidity. The bionanocomposite materials produced are considered for the biodegradable packaging industries (Sadeghi-Varkani et al. 2018).

Clay nanoparticle is made of nanosized compounds, which are used in food packaging to enhance thermal resistance in production process, transportation, and storage. The surface-to-volume ratio of the nanoparticle may influence the quality of material. The preparing process of the composite should be meticulous to prevent aggregation and increase the surface area between filler and matrix (Rhim et al. 2013). Table 1 presents typical biodegradable polymers and inorganic or antimicrobial materials investigated for the preparation of bionanocomposites (Rhim et al. 2013).

For these materials, the most important property is the biodegradability, thus fragmentation, loss of mechanical properties, and degradation through microorganisms

Table 1 Typical biodegradable polymers and antimicrobial materials (Bordes et al. 2009)

Biodegradable polymers	Antimicrobial materials
Starch or thermoplastic starch (TPS)	Clay [e.g., montmorillonite (MMT)]
Chemically modified cellulose [e.g., cellulose acetate(CA) and cellulose acetate butyrate (CAB)]	Organically modified nanoclay (quaternary ammonium modified MMT, Ag-zeolite)
Polylactic acid (PLA)	Metal ions (e.g., silver, copper, gold, platinum)
Polycaprolactone (PCL)	Metal oxide (e.g., TiO_2, ZnO, MgO)
Polyhydroxyalkanoate (PHA)	Natural biopolymers (e.g., chitosan)
Polyhydroxybutyrate (PHB)	Natural antimicrobial agents (e.g., nisin, thymol, carvacrol, isothiocyanate, antibiotic)
Poly(butylene succinate) (PBS)	Enzymes (peroxidase, lysozyme) Synthetic antimicrobial agents quaternary ammonium salts, EDTA, propionic acid, benzoic acid, sorbic acid

such as bacteria, fungi, and algae. The test on the nanocomposite based on PCL with clay nanoparticle showed that the biodegradability improves when is compared with near PCL material. The results were confirmed by other studies. The PLA-based nanocomposite was blended with organoclay. When comparing the biodegradability of PLA and PLA nanocomposite, it is found that the PLA nanocomposite can decrease weight faster than near PLA, and within two months, the PLA nanocomposite is completely degraded by compost (Rhim et al. 2013).

Antimicrobial function is a critical one in the field of industries such as packaging products for medicine. There are many kinds of nano-substance for antimicrobials bionanocomposite, for example, silver, copper, gold, platinum, TiO_2, ZnO, MgO, modified nanoclay, chitosan, nisin, thymol, carvacrol, isothiocyanate, peroxidase, lysozyme and quaternary ammonium salts, EDTA, propionic, benzoic, sorbic acid. Silver ions and metallic silver nanoparticles are mainly used as antimicrobial filler. Ag-zeolite-incorporated chitosan film has the ability of anti-both Gram-positive and Gram-negative bacteria (Rhim et al. 2013).

7 Processing Techniques of Biopolymers-Based Nanocomposites

So far, the environmental pollution causes the phenomenon of the greenhouse effect. Governments of many countries are concerned about the impact of environment problems. The eco-materials are the alternative materials for the synthetic compounds. However, the environment-friendly materials are costly and some properties cannot be compared with petroleum-based materials. These are the reasons, why eco-friendly products are not widely used. The solutions are to improve the properties and to reduce the production cost.

Processing techniques are key factors affecting the qualities and the product cost. The filler is added to increase material features. Some processing such as isolation of the nanofibers results in improving mechanical properties and thermal stability of polyvinyl alcohol biocomposites. Casting is a conventional process for producing film biocomposites by mixing the starch, water, glycerol, and filers with a stirrer then heating and stirring until become homogeneous. This procedure may cause air bubbles, consolidation of the nanocellulose and coagulates that cannot merge together. Hence, ultrasonication is applied during mixing process because of the ability to destroy the agglomerations of nanocellulose and to distribute nanoparticle in the biopolymeric matrix. The result showed that the ultrasonic process enhances the properties of the bionanocomposite film (Abral et al. 2018). A variety of synthetic transformations can occur during the manufacturing of a number of thermoplastic and thermosetting oil-based biomaterial. Natural vegetables play an important role in renewable resources because of their low chemical potential, while the price is quite low. The vegetable oil derived from renewable thermoset is prepared by Thiol-Michael and adding of pentaerythritol tetrakis (3-mercaptopropionate)

(PE3MP) to high oleic sunflower oil, and then, the nano-biopolymer is mixing with cellulose nanocrystals (CNC), which later is modified with Beycostat A B09 as surfactant. The process shows that the nanoparticle can disperse with a low polarity reactive in the matrix. Mechanical properties of bionanocomposite as Young's Modulus increases by modifying the CNC (Moreno et al. 2016).

The quality of bionanocomposites is closely related to the nanoscale interface strength. In the biomedical engineering, the composite hydroxyapatite (HA) nanoparticles and polylactic acid (PLA) matrix are the materials which have the ability to repair bones and can be used as materials replacement. Nevertheless, blending between HA nanoparticles and PLA matrix is limited because of physical absorption. Besides, were noted that these composites have poor mechanical properties for "low load" of bending applications. To improve such properties, the interface between both matrix and filler must be modified. The plasma surface technology is the process that permits to enhance the interfacial strength. After using this technique, the result showed that the elasticity and damping behavior of the biocomposites is increased (Petisco-Ferrero et al. 2018).

In the engineering field that covers biosensors, tissue implants, and drug delivery system are used the superabsorbent polymers and/or hydrogel. The hydrogel is a macromolecular polymeric gel which was synthesized by crosslink of polymer chains by physical, ionic, or covalent bond. Due to their great abilities such as large amount of water absorption on H-bonding, hydrogels are used as drug transfer. It permits to release drugs or other agents in some situations controlling the temperature and pH value The hydrogel made of nanocomposite can be defined as three-dimensional network polymers which cross-linked chemically or physically with each other or with nanoparticles/ nanostructure. The chemical cross-linker can release the toxic agent with potential effect on the drugs. Therefore, the physical cross-linker is more suitable to use as drug delivery. The physical crosslinking technique is used for the preparation of the biodegradable carboxymethyl cellulose/ graphene oxide nanoparticle (CMC/GO) nanocomposite hydrogel beads using $FeCl_3 \cdot 6H_2O$. The investigation indicates that the swelling ratio of nanocomposite hydrogels is increased by incorporating the graphene oxide nanoparticles. The drug loading depends on the amount of the nanoparticles. However, the drug that is releasing from pure hydrogel is more than a nanocomposite hydrogel because of the strong interaction of nanoparticle in the nanocomposite (Rasoulzadeh and Namazi 2017).

8 Modifications of Biopolymers-Based Nanocomposites

In the past, the use of petroleum-based polymer was very famous for many products. However, the studies demonstrated that this causes plastic waste contamination in water source and soil. Thus, many people raised concerns about this problem. Therefore, the production and the use of biodegradable materials have significantly increased. Poor mechanical and barrier properties of biopolymer are the main factors which challenged the researchers to further study and to find some methods to improve and enhance their properties for industrial applications (Tabatabaei et al. 2018).

Different types of fillers were used as reinforcement and improvement of the biopolymer properties. Nanosized fillers are quite promising because they generate large interface area permitting an increase in the mechanical, thermal, and barrier properties when the distribution of nanoparticle is consistent. Biopolymer from agriculture, such as starch, is mainly used as biodegradable and renewable polymer. The plasticized starch was called "thermoplastic starch" (TPS) which has really poor mechanical properties and high hygroscopicity. To reduce this disadvantage, blending the composite is the main method to enhance the biocomposite materials features. Polybutylene adipate-co-terephthalate (PBAT) is an aromatic-aliphatic biodegradable copolymer with good properties and work well in conjunction with TPS blend. Nanoparticles of sepiolite were added in this blended biopolymer. The result confirms that the dynamic mechanical behavior of nano-biocomposites increased, while the water absorption rate and capacity decreased significantly (Olivato et al. 2017).

The combination of the nanoparticles, for example, clay, titanium dioxide (TiO_2) with biopolymers allows to improve not only their mechanical, thermal, and barrier properties but also to obtain other abilities and applications such as biosensor, antimicrobial, oxygen scavenger, and biodegradable features (Majid et al. 2018). K-carrageen is one of the biodegradable and renewable polymers extracted from red seaweeds. It belongs to polysaccharides family and owns high gel-forming capacity. Gelatin is an animal protein and is obtained by hydrolysis of the fibrous insoluble collagen from skins, bone of animal and fish. To obtain the effect of nano-SiO_2 of blended polymer that contains nanoparticle of SiO_2, it is necessary a mixing in the forming process. The biocomposite material produced from this combination indicated significant increases in the tensile strength and Young's Modulus. However, it is observed that the water vapor permeability and oxygen permeability decreases. Moreover, the rate of ultraviolent light absorbance of the bionanoparticles was much better than the biocomposite, up to four times (Singh et al. 2018).

Poly(L-lactide) (PLA) is the most common synthetic biopolymer which is widely used as raw material matrix. It is because of its excellent mechanical properties such as tensile strength that is about 44–59 MPa. Degradation time of PLA is about 18–24 months. It is noted that not only by adding of nanoparticle can improve the abilities of the material but also the solid-state drawing is one effective factor, which has possibility to enhance the properties of the nanocomposite. Thus, it was investigated the orientation effect of the PLA following drawing under different conditions. Chitin nanocrystals (ChNC) were mixed with PLA. The results indicated that the mechanical and thermal properties of the PLA nanocomposites increased because of the effective solid-state drawing in the polymer. The orientation of the macromolecular chains follows a "shish kebab" pattern. The amount of ChNC in the nanocomposite induces an increase of the degree of crystallinity and of the glass transition temperature after drawing (Singh et al. 2018).

9 Properties of Biopolymers-Based Nanocomposite

Table 2 shows an overview of the thermal and mechanical properties of various described biopolymers which are widely used as matrix for bionanocomposite (Park et al. 2017). In the Table 2, polysaccharides group consists of hyaluronic acid, chitosan, alginate, and starch. Hyaluronic acid (HA) is extracted from animal such as synovial fluid, vitreous humor of eye, and umbilical tissue. To produce the biofilm for healing wounds, hyaluronic acid isolated from chicken comb is mixed with the bacterial cellulose (BC). The effect of the compound in membrane synthesis, structure, morphology, and thermal stability was investigated. The results showed that the properties of membrane increased the thermal properties and roughness but decreased crystallinity. In conclusion, this bionanocomposite can be used as biomaterials for tissue engineering (de Oliveira et al. 2017).

Chitosan is wildly used as a component of scaffolding material because of its biodegradability, water sorptivity, oxygen permeability, blood coagulation, and cytokine induction. Chitosan may dissolve very well at a pH below 6. Chitosan-based material is mixed with nanoclay type. The result indicated that the mechanical properties of chitosan-based nano-biocomposite are higher than neat chitosan because of the strong affection between the nanoclay and the chitosan-based material and excellent dispersion. This material is suitable for biodegradable packaging applications because of 100% biodegradability within 160 days (Xie et al. 2013).

Table 2 Thermal and mechanical properties of biopolymers (Park et al. 2017)

Polymer	Glass transition temperature T_g (°C)	Melting temperature T_m (°C)	Tensile strength (MPa)	Elongation at break (%)
Hyaluronic acid	–	–	102–120	37–43
Chitosan	156–170	99–131	82–166	5–8
Alginate	3–113	>300	31–37	11–17
Starch	52–60	110–115	35–80	580–820
Collagen	40	70	202–224	37–39
Gelatin	50–70	25–85	0.66	62.5
Silk	178	192–203	16–20	5.29–5.79
Poly(lactic acid)	40–70	130–180	44–59	20–240
Poly-(γ-glutamic acid)	35–40	225–230	890	30
Poly(3-hydroxybutyrate)	−5 to 15	140–182	25–40	5–8
Poly(3 hydroxybutyrate-co-3-hydroxyvalerate)	0–30	100–190	25–30	7–15
Poly(3-hydroxyoctanoate)	−35	60	10	300
Bacterial cellulose	−14 to 48.82	–	55–363	5–55
Dextran	197	–	3.30	–

Alginates are extracted from brown seaweed and belong to polysaccharides family, which consist of copolymer containing 1,4-glycosidically linked β-D-mannuronic acid (M-blocks) and α-L-guluronic acid (G-blocks) monomers. Cellulose nanoparticles (CN) were mixed with alginate biopolymers using solution casting method. The result showed that the novel produced bio-based nanocomposites reduced the water solubility and water vapor permeability. The tensile strength of these materials is enhanced from 18.03 to 22.4 MPa (Abdollahi et al. 2013).

Starch is biodegradable and renewable material that is very attractive in academic and industrial sectors. Thermoplastic starch (TPS) has poor mechanical properties, thermal stability, and high water sensibility. The strengthening by additive is required to improve their properties. Cellulose nanofibers (CNFs) were mixed with thermoplastic starch. The result indicated that the mechanical and thermal properties of TPS with the addition of CNFs were improved because of CNGs interaction with the TPS and excellent dispersion of CNFs in the TPS matrix. In summary, we highlight some advantages by reinforcing TPS with CNFs because it is natural, cheap, and abundant green nanocomposite (Ghanbari et al. 2018).

Collagen and gelatin are biodegradable polymers, which are wildly used in tissue engineering and bioactive molecule delivery because of their great abilities; for example, good biocompatibility, low antigenicity, biodegradability, high mechanical properties, and water-uptake properties. Silica nanoparticles were added in collagen. The result showed that the surface contraction in short term (1 week) is lower compared to long term (3 weeks). It was detected that the bionanocomposite matrix is suitable for fibroblast adhesion (Desimone et al. 2010).

Silk is a natural protein fiber that can be synthesized by cocoon of silkworm (*Bombyx mori*). The structure of silk consists of glycine, alanine, and serine. To improve the properties of biopolymer such as PLA, silk was prepared to nanoparticle size [silk nanocrystals (SNCs)] and added into PLA. The result indicated the melt viscosity and thermal properties of PLA are improved by addition of silk nanocrystals (Tesfaye et al. 2017).

Polylactic acid (PLA) is a biopolymer which is produced from renewable resource and can be used for various applications. To decrease the photooxidative degradation, halloysite nanotubes (HNT) is added in PLA matrix to produce PLA nanocomposites. The results prove a good distribution of nanofiller, high tensile strength, and good mechanical rigidity. HNT generate an impressive pro-degradant effect on the PLA photooxidation (Therias et al. 2017).

10 Applications of Biopolymers-Based Nanocomposites in Energy, Environmental, Biomedical, and Food Packaging

The application of the bionanocomposite materials is large; it depends on the purpose and objective of area of applications such as paper, board, automotive, building and construction, food, textile, biomedical, energy, sensors, paints, cosmetics, hygiene, absorbent products, foams and aerogels, water purification, electronic and optoelectronic devices, and surface coating (Bajpai 1999).

Polylactic acid (PLA) is the widely used biopolymer in food packaging. Many products such as trays, tableware, cups, and cutleries are produced for customers using PLA in order to protect the environment. However, for food packaging application, the near PLA needs radical improvements. ZnO particles were added in PLA biopolymer. The result indicated good dispersion of ZnO particles in PLA matrix. The produced biocomposite films with ZnO may improve the mechanical properties and has the potential of fighting against *Escherichia coli*; this stops the disorder in great percentage (around 99.99%) only after 24 h (Marra et al. 2016). The medical products such as implants, stents, and bone support splints are produced because of their strong abilities (of PLA), for instance, safety and degradation by hydrolysis that is released in the body's normal metabolism. Interface modified is an efficient method for improving many properties of the material. In this case, nanohydroxyapatite (nHA) was mixed with polylactic acid/starch/poly ε-caprolactone (PLASCL20). The result indicated that the overall properties such as hydrolytic degradation, hydrophilicity, antibacterial activity, and the drug release are higher than in the absence of nHA nanoparticle. This bionanocomposite is used for many medical applications (Davachi et al. 2017).

The automotive industry is the sector influenced by the advanced materials development, where the nanotechnology and nanocomposites played an important role in the improvement of the efficiency of automobiles. The composite properties feature such as lightweights, good mechanical properties, and thermal stabilities allow reduction of oil fuels and permit increasing of the engine performance of automotive. By adding cellulose nanocrystals into PLA helps to improve the mechanical, barrier properties, and also abrasion resistance (Bajpai 1999).

Lightweight composites play an important role in the field of engineering. The properties of materials such as biodegradability, lightweight, low cost, good mechanical properties, and thermal stability have gained a lot of interest in the market. To create a material that protects traditional toxic lead from X-ray, gadolinium (III) oxide (Gd_2O_3) nanoparticles is added in epoxy composite. The research demonstrated that 16 mm thickness of epoxy nanocomposite can protect up to 99% of transmitted X-ray intensity. The weight of this material is 36–48% lighter than other material such as wood, concrete, and glass. In summary, this is an alternative product to produce an environmentally safe, lightweight composite with prominent X-ray attenuation abilities (La et al. 2018).

Reinforced polymeric hydrogel is produced from many types of nanoparticles such as inorganic/ceramic, polymeric, metal/metal oxide nanoparticles, and carbon-based nanomaterials. The effect of adding nanoparticle into hydrogel demonstrated increases in the mechanical properties and the biodegradability, non-toxic characters, and drug delivery features (Garland et al. 2017).

11 Challenges and Opportunities of Biopolymers-Based Nanocomposites in Food Packaging

So far, the nanotechnology played an important role in the advanced material improvement. In 2008, the value of the food packaging application was 4.3 billion US dollar worldwide. In the future, the number of products that will include nanotechnology can increase considerably because of great advantage of novel technology and due to consumer's requirements. In the past, the vacuum packaging technology was widely used for packaging fresh meat. A proper functioning of food packaging is effective preserving the quality of food products, as well as it enhances the shelf life. Besides, the raw material properties dedicated food packaging needs to be attractive in terms of mechanical, thermal, and gas barrier properties. However, the properties of biopolymer are normally poor in terms of mechanical properties, low water resistance, and low thermal stability. Some material improvement has been considered to obtain reliable characteristics (Youssef and El-Sayed 2018). There, two main types of cellulose nanostructures were investigated. They are nanocrystals (CNCs) and nanofibrils (CNFs). The size of CNCs studied is around 4–25 nm in diameter and 100–1000 nm in length, while the dimension of CNFs is around 10–100 nm in diameter. Cellulose nanofibers were investigated as new function of food packaging. Coating enables to generate a layer of coat on the paper that has the protection role in the food packaging. It generates a product that is 100% biodegradable (Lavoine et al. 2016). The coating films of wheat gluten bionanocomposite consisting of 7.5% CNC and 0.6% TiO_2 were coated on unbleached kraft paper sheets. The result showed the mechanical properties increased by 50% and the materials itself permits to reduce 98.5% of bacteria (*Saccharomyces cerevisiae, E. coli, and S. aureus*) (El-Wakil et al. 2015).

High performance, lightweight, and eco-friendly materials are the main advantages of bionanocomposite materials, which can be used as raw materials production to replace the petroleum-based composite materials. The biopolymers are the alternative raw materials to resolve the plastic waste issue and reduce pollution from water sources and soil because of their biodegradableability and non-toxicity. However, their overall disadvantages are poor mechanical properties, low thermal stability, water resistance, and reduced gas resistance. Therefore, nanoparticles were added to improve such properties without hindering their biodegradableability and non-toxic characters. For packaging applications, there are many nanoparticles, which were added such as montmorillonite (MMT), kaolinite, zinc oxide (ZnO-NPs), titanium dioxide (TiO_2-NPs), and silver nanoparticles (Ag-NPs). The effects of such substance are to improve mechanical properties, barrier properties, antibacterial ability and to reduce oxygen, carbon dioxide, flavor compounds diffusion. The dispersion of nanoparticle in the biopolymer matrix is the challenge for bionanocomposite preparation because of the excellent distribution of nanoparticles and the interaction of surface effect that allows improving the overall properties. Although there are many studies to enhance the properties of bionanopolymers for food application, there is yet open process for creating better bionanocomposites material properties and environmentally-friendly solution (Youssef and El-Sayed 2018).

12 Concluding Remarks

The new policy regarding environmental concerns and advanced technologies has growing interest to reduce the use of non-biodegradable and non-renewable polymer materials and replace these with bio-based polymers for nanocomposites in various applications. This review is focused on the biopolymers-based nanocomposites and their applications. It was demonstrated that the biopolymer-based nanocomposites features are mainly dependent on various factors such as compatibility between biopolymer matrix and type of reinforcement and nanocomposite preparation methods. To improve the biodegradable polymers properties, is necessary the addition of nanofillers with strong potential for designing eco-friendly materials for several applications.

References

Abdollahi M, Alboofetileh M, Behrooz R, Rezaei M, Miraki R (2013) Reducing water sensitivity of alginate bio-nanocomposite film using cellulose nanoparticles. Int J Biol Macromol 54:166–173

Abral H, Anugrah AS, Hafizulhaq F, Handayani D, Sugiarti E, Muslimin AN (2018) Effect of nanofibers fraction on properties of the starch based biocomposite prepared in various ultrasonic powers. Int J Biol Macromol 116:1214–1221

Bajpai P (1999) Application of enzymes in the pulp and paper industry. Biotechnol Prog 15 (2):147–157

Bibi F, Guillaume C, Gontard N, Sorli B (2017) Wheat gluten, a bio-polymer to monitor carbon dioxide in food packaging: electric and dielectric characterization. Sens Actuators B Chem 250:76–84

Bordes P, Pollet E, Avérous L (2009) Nano-biocomposites: biodegradable polyester/nanoclay systems. Prog Polym Sci 34(2):125–155

Chaunier L, Guessasma S, Belhabib S, Valle GD, Lourdin D, Leroy E (2017) Material extrusion of plant biopolymers: opportunities & challenges for 3D printing. Addit Manufact 21:220–233

Davachi SM, Heidari BS, Hejazi I, Seyfi J, Oliaei E, Farzaneh A, Rashedi H (2017) Interface modified polylactic acid/starch/poly ε-caprolactone antibacterial nanocomposite blends for medical applications. Carbohyd Polym 155:336–344

de Oliveira SA, da Silva BC, Riegel-Vidotti IC, Urbano A, de Sousa Faria-Tischer PC, Tischer CA (2017) Production and characterization of bacterial cellulose membranes with hyaluronic acid from chicken comb. Int J Biol Macromol 97:642–653

Desimone MF, Hélary C, Rietveld IB, Bataille I, Mosser G, Giraud-Guille MM, Coradin T (2010) Silica–collagen bionanocomposites as three-dimensional scaffolds for fibroblast immobilization. Acta Biomater 6(10):3998–4004

El-Wakil NA, Hassan EA, Abou-Zeid RE, Dufresne A (2015) Development of wheat gluten/nanocellulose/titanium dioxide nanocomposites for active food packaging. Carbohyd Polym 124:337–346

Essabir H, Raji M, Laaziz SA, Rodrique D, Bouhfid R, el kacem Qaiss A (2018) Thermo-mechanical performances of polypropylene biocomposites based on untreated, treated and compatibilized spent coffee grounds. Compos Part B Eng 149:1–11

Fatehi H, Abtahi SM, Hashemolhosseini H, Hejazi SM (2018) A novel study on using protein based biopolymers in soil strengthening. Constr Build Mater 167:813–821

Fernandes EM, Correlo VM, Mano JF, Reis RL (2015) Cork–polymer biocomposites: mechanical, structural and thermal properties. Mater Des 82:282–289

Garland NT, McLamore ES, Gomes C, Marrow EA, Daniele MA, Walper S, Claussen JC (2017) Synthesis and applications of cellulose nanohybrid materials. In: Hybrid polymer composite materials, vol 4. pp 289–320

Ghanbari A, Tabarsa T, Ashori A, Shakeri A, Mashkour M (2018) Preparation and characterization of thermoplastic starch and cellulose nanofibers as green nanocomposites: extrusion processing. Int J Biol Macromol 112:442–447

Hassanzadeh-Aghdam MK, Ansari R, Mahmoodi MJ, Darvizeh A (2018) Effect of nanoparticle aggregation on the creep behavior of polymer nanocomposites. Compos Sci Technol 162:93–100

John MJ, Thomas S (2008) Biofibres and biocomposites. Carbohydr Polym 71(3):343–364

Kargarzadeh H, Mariano M, Huang J, Lin N, Ahmad I, Dufresne A, Thomas S (2017) Recent developments on nanocellulose reinforced polymer nanocomposites: a review. Polymer 132 (2017):368–393

La LB, Leatherday C, Leong YK, Watts HP, Zhang LC (2018) Green lightweight lead-free Gd2O3/epoxy nanocomposites with outstanding X-ray attenuation performance. Compos Sci Technol 163:89–95

Lau KT, Hung PY, Zhu MH, Hui D (2018) Properties of natural fibre composites for structural engineering applications. Compos B Eng 136:222–233

Lavoine N, Guillard V, Desloges I, Gontard N, Bras J (2016) Active bio-based food-packaging: diffusion and release of active substances through and from cellulose nanofiber coating toward food-packaging design. Carbohyd Polym 149:40–50

Li WC, Tse HF, Fok L (2016) Plastic waste in the marine environment: a review of sources, occurrence and effects. Sci Total Environ 566:333–349

Majid I, Nayik GA, Dar SM, Nanda V (2018) Novel food packaging technologies: innovations and future prospective. J Saudi Soc Agric Sci 17(4):454–462

Marra A, Silvestre C, Duraccio D, Cimmino S (2016) Polylactic acid/zinc oxide biocomposite films for food packaging application. Int J Biol Macromol 88:254–262

Meraldo A (2016) Introduction to bio-based polymers. In: Multilayer flexible packaging, 2nd edn. pp 47–52

Miller DB, Glisson WB, Yampolskiy M, Choo KKR (2018) Identifying 3D printer residual data via open-source documentation. Comput Secur 75:10–23

Morelli CL, Mahrous M, Belgacem MN, Branciforti MC, Bretas RES, Bras J (2015) Natural copaiba oil as antibacterial agent for bio-based active packaging. Ind Crops Prod 70:134–141

Moreno M, Armentano I, Fortunati E, Mattioli S, Torre L, Lligadas G, Cádiz V (2016) Cellulose nano-biocomposites from high oleic sunflower oil-derived thermosets. Eur Polymer J 79:109–120

Nafchi AM, Alias AK, Mahmud S, Robal M (2012) Antimicrobial, rheological, and physicochemical properties of sago starch films filled with nanorod-rich zinc oxide. J Food Eng 113(4):511–519

Okonkwo PC, Collins E, Okonkwo E (2017) Application of biopolymer composites in super capacitor. In: Biopolymer composites in electronics. pp 487–503

Olivato JB, Marini J, Yamashita F, Pollet E, Grossmann MVE, Avérous L (2017) Sepiolite as a promising nanoclay for nano-biocomposites based on starch and biodegradable polyester. Mater Sci Eng C 70:296–302

Park SB, Lih E, Park KS, Joung YK, Han DK (2017) Biopolymer-based functional composites for medical applications. Prog Polym Sci 68:77–105

Petisco-Ferrero S, Álvarez LP, Ruiz-Rubio L, Vilela JV, Sarasua JR (2018) Plasma poly (acrylic acid) compatibilized hydroxyapatite-polylactide biocomposites for their use as body-absorbable osteosynthesis devices. Compos Sci Technol 161:66–73

Rasoulzadeh M, Namazi H (2017) Carboxymethyl cellulose/graphene oxide bio-nanocomposite hydrogel beads as anticancer drug carrier agent. Carbohyd Polym 168:320–326

Razza F, Degli Innocenti F, Dobon A, Aliaga C, Sanchez C, Hortal M (2015) Environmental profile of a bio-based and biodegradable foamed packaging prototype in comparison with the current benchmark. J Clean Prod 102:493–500

Rhim JW, Park HM, Ha CS (2013) Bio-nanocomposites for food packaging applications. Prog Polym Sci 38(10–11):1629–1652

Sadeghi-Varkani A, Emam-Djomeh Z, Askari G (2018) Morphology and physicochemical properties of a novel *Lallemantia iberica* mucilage/titanium dioxide bio-nanocomposite. Polym Testing 67:12–21

Singh AA, Wei J, Herrera N, Geng S, Oksman K (2018) Synergistic effect of chitin nanocrystals and orientations induced by solid-state drawing on PLA-based nanocomposite tapes. Compos Sci Technol 162:140–145

Tabatabaei RH, Jafari SM, Mirzaei H, Nafchi AM, Dehnad D (2018) Preparation and characterization of nano-SiO_2 reinforced gelatin-k-carrageenan biocomposites. Int J Biol Macromol 111:1091–1099

Tesfaye M, Patwa R, Gupta A, Kashyap MJ, Katiyar V (2017) Recycling of poly (lactic acid)/silk based bionanocomposites films and its influence on thermal stability, crystallization kinetics, solution and melt rheology. Int J Biol Macromol 101:580–594

Thakur VK, Thakur MK (2014) Processing and characterization of natural cellulose fibers/ thermoset polymer composites. Carbohyd Polym 109:102–117

Therias S, Murariu M, Dubois P (2017) Bionanocomposites based on PLA and halloysite nanotubes: from key properties to photooxidative degradation. Polym Degrad Stab 145:60–69

Thomas MG, Abraham E, Jyotishkumar P, Maria HJ, Pothen LA, Thomas S (2015) Nanocelluloses from jute fibers and their nanocomposites with natural rubber: preparation and characterization. Int J Biol Macromol 81:768–777

Wilton N, Lyon-Marion BA, Kamath R, McVey K, Pennell KD, Robbat A Jr (2018) Remediation of heavy hydrocarbon impacted soil using biopolymer and polystyrene foam beads. J Hazard Mater 349:153–159

Xie DF, Martino VP, Sangwan P, Way C, Cash GA, Pollet E, Avérous L (2013) Elaboration and properties of plasticised chitosan-based exfoliated nano-biocomposites. Polymer 54(14):3654–3662

Youssef AM, El-Sayed SM (2018) Bionanocomposites materials for food packaging applications: concepts and future outlook. Carbohydr Polym 193:19–27

Zhou Y, Lei L, Yang B, Li J, Ren J (2018) Preparation and characterization of polylactic acid (PLA) carbon nanotube nanocomposites. Polym Testing 68:34–38

MIX
Papier aus verantwortungsvollen Quellen
Paper from responsible sources
FSC® C105338

If you have any concerns about our products,
you can contact us on
ProductSafety@springernature.com

In case Publisher is established outside the EU,
the EU authorized representative is:
Springer Nature Customer Service Center GmbH
Europaplatz 3, 69115 Heidelberg, Germany

Printed by Libri Plureos GmbH
in Hamburg, Germany